Movement Ecology of Neotropical Forest Mammals

Rafael Reyna-Hurtado • Colin A. Chapman
Editors

Movement Ecology of Neotropical Forest Mammals

Focus on Social Animals

Springer

Editors
Rafael Reyna-Hurtado
El Colegio de la Frontera Sur (ECOSUR)
Department of Biodiversity Conservation
Lerma, Campeche, Mexico

The Wildlife Conservation Society (WCS)
Bronx, NY, USA

Colin A. Chapman
Department of Anthropology
McGill University
Montreal, QC, Canada

School of Life Sciences
University of KwaZulu-Natal
Scottsville, Pietermaritzburg, South Africa

Shaanxi Key Laboratory for Animal Conservation
Northwest University
Xi'an, China

ISBN 978-3-030-03462-7 ISBN 978-3-030-03463-4 (eBook)
https://doi.org/10.1007/978-3-030-03463-4

Library of Congress Control Number: 2018965602

This Springer imprint is published by the registered company Springer Nature Switzerland AG
The registered company address is: Gewerbestrasse 11, 6330 Cham, Switzerland

Foreword

Félix Samuel Rodríguez de la Fuente (1928–1980) was a Spanish naturalist who made an outstanding impact on the global public through his radio and TV programs on wildlife and conservation. In the early 1970s, he led a group of Spanish biologists in an incredible project, the *Encyclopedia Salvat of the World Fauna* (later collated as "World of Wildlife" or "Fauna Mundial"), publishing a 24-page illustrated volume every week for 3 successive years. As an Israeli teenager eager to learn about wildlife and nature, I waited impatiently every Thursday morning for the next Hebrew-translated Wildlife volume sold in only one newsstand in the center of Eilat, my hometown. Every Wildlife volume brought wonderful stories and beautiful pictures to Eilat about all sorts of incredible creatures and landscapes from far away. As I am sure was the case for many other nature-lovers at my age, the Wildlife volumes dedicated to tropical regions captured my imagination more than anything else. Volumes 95–107 focused on the neotropics, describing and explaining their diverse faunas and citing Marston Bates as labeling neotropical animals as the most bizarre creatures on Earth. Volumes 102 and 103 vividly described the primates and carnivores of the neotropics, highlighting their locomotion skills, social grouping, hunting tactics, and territorial behavior. The curiosity that the Wildlife neotropic volumes instilled in a teenaged nature-lover in the desert town of Eilat is perhaps best understood when considering the so-called naïve drawings of jungle scenes by Henri Rousseau, famous for fiercely delivering the fantasy of the wild tropics despite never leaving France.

More than four decades have passed since I first read about the life of neotropical animals on the move in the Wildlife encyclopedia. Unlike Rousseau, I have been fortunate enough to witness neotropical animals moving throughout their natural habitats in Panama, Peru, and Brazil. My childhood fascination for wildlife on the move has meanwhile matured into a profession called *movement ecology*. Movement of organisms, the subject of this book, characterizes all species, affects individual fitness, determines evolutionary pathways, and shapes ecological processes, including the most stressful global environmental problems and conservation concerns we face today. The movement ecology framework offers a conceptual platform for guiding and uniting research on movement of organisms by delineating the basic mechanistic components common to all types of movement and all kinds of organisms. Beyond the theoretical integration, movement ecology has experienced rapid progress over the last decade thanks to technological advancements allowing accurate tracking of wildlife by GPS and other tracking technologies, as well as advanced data analysis tools. In this book, Rafael Reyna-Hurtado, Colin A. Chapman, and their colleagues aimed to illustrate the importance of movement ecology research on neotropical mammals to achieve a better understanding of key ecological processes for advancing effective conservation efforts in these hyperdiverse yet fragile and endangered tropical ecosystems.

This book includes 3 useful introductory/review/concluding chapters and 12 research chapters describing studies of 9 different mammalian species from 4 different orders and 6 families. Common to all research chapters is the focus on the effects of various external factors – including hurricanes, hunters, landscape structure, food resources, seasonality, and topography – on the movement paths of individual animals quantified by direct observational tracking, VHF collars, and mostly GPS devices. External factors and movement paths are two of the five basic components of the movement ecology framework; the three other components – internal state (determining why to move), motion capacity (determining how to move), and navigation capacity (determining when and where to move) – describe the key features of the focal individual that likely interact with the external factors and with each other to jointly shape the resulting movement path. Reported associations between external factors and movement path, with some insights about the internal state, include, for example, the longer travel distances (movement path) of fawning female (internal state) white-tailed deer during the hot dry season (external factor) compared to the flooding season (Chap. 8). Additionally, fruit availability (external factor) did not affect group cohesion (external effect on the internal state) and daily travel distance (movement path) among four groups of woolly monkeys, but mating opportunities (external factor) significantly increased daily travel distance, suggesting that spatial dynamics in these groups are governed by competition over mates rather than food (Chap. 10). Furthermore, male (internal state) jaguars covered twice as much area compared to females, and one female (internal state) jaguar favored a relatively small patch of rugged forested terrain (external factor), presumably to provide a refuge while rearing cubs (Chap. 13).

Insights into the other two basic components of movement ecology that could further elucidate the mechanisms underlying variation in movement paths were usually not estimated directly due to data limitations (further discussed below) but were demonstrated or suggested in some studies. For example, larger groups of howler monkeys tended to move over extended areas and more frequently moved toward isolated trees in a fragmented landscape by walking on the ground; such "terrestrial locomotion," a rather unusual mode of motion for this canopy-dwelling species, occurred mostly in the rainy season and was initiated mostly by the oldest male in one group and an adult female carrying an infant in another (external and internal effects on motion capacity) (Chap. 9). Furthermore, the tendency toward more restricted movements of two GPS-tracked Baird's tapirs after hurricane Otto was suggested to reflect either higher food availability (external effects on the internal state), difficulty in moving in a complex terrain further induced by fallen trees (external effects on motion capacity), or lower visibility due to enhanced growth in canopy gaps (external effects on navigation capacity) (Chap. 2).

The immense use of GPS devices by humans has greatly facilitated the development of GPS devices for wildlife tracking, arguably constituting the most important technological development thus far in movement ecology research. GPS devices have been applied mostly to track mammals and birds of relatively large (>100 g) body mass, including many mammalian species studied in this collection. As illustrated for Colombian woolly monkeys (Chap. 11), GPS tracking could be advantageous over direct observations in estimating movement and home range due to difficulties in directly tracking monkey groups in areas of steep topography and across inaccessible landscape features. Direct tracking, though, can be advantageous by providing behavioral and ecological data not easily available by GPS. Other limitations of GPS tracking reported in various chapters occur due to problems in data download via satellites in closed canopy habitats and complex terrain, failure of tag drop-off mechanisms, and nearly inaccessible areas necessitating tremendous efforts to recover deployed data loggers. Overall, I congratulate the authors for all of these incredible efforts and their determination to pursue their goals, proven compulsory for making this book possible. In addition, many chapters in this book have applied novel data analysis tools such as the semivariance approach capable of accommodating irregularly sampled data (Chap. 6), the biased random bridge, and a step selection function (Chap. 13). It is worth stressing that problems resulting from data limitations were not neglected in these studies; on the contrary, the authors have carefully interpreted their results, highlighted small sample size and/or other data-related uncertainties, and repeatedly emphasized that further insights about their systems and research questions should be pursued by obtaining richer data sets through more effective tracking devices and data download procedures. Auxiliary data on energy expenditure, behavior, and environmental factors might also be obtained in future studies through other bio-logging technologies including accelerometers and various other sensors, to further enrich our ability to quantify the internal state, motion and navigation constraints, and behavioral response to the variable external environment the animal encounters en route.

In summary, I warmly congratulate this excellent team of authors for covering a wide range of topics, questions, species, and ecosystems and for proving that insightful movement ecology research is feasible even in such challenging environments and under very difficult working conditions. I am confident that the important research projects presented in this book will prove critical for further advancement of the study of animal movement ecology in the neotropics and in other challenging environments to ensure that threatened wildlife will prevail.

Jerusalem Ran Nathan

Acknowledgments

Rafael Reyna wants to thank his wife Edith Rojas-Flores and his two kids, Aranza and Emiliano, for their always invaluable support toward his academic career. They have been the source of inspiration to overcome long hours in the field or behind a desk. Thanks go to Nicolas Arias for the invaluable field help and to many students that have shared the passion for studying tropical mammals' movement ecology. El Colegio de la Frontera Sur (ECOSUR) has been a very open institution that allowed Rafael's academic growth and provided resources, time, and tools to facilitate research with tropical mammals. Thank you to Consejo Nacional de Ciencia y Tecnologia (CONACYT, National Council of Science and Technology of Mexico) for providing multiple sources of funding for research for completing this book (mainly, project number 182386 of "Ciencia Basica 2012" and a grant of National System of Researchers, SNI 47455, to Rafael); thank you to National Geographic for two grants (9189-12 and 9839-16) that were essential to conduct research that contributed to this book.

Colin Chapman would like to thank all the great collaborators/friends that he has worked with over the years, but there is not enough room to thank them all. He is particularly grateful to Lauren Chapman, Patrick Omeja, Dennis Twinomugisha, and all his great field assistants who he has worked with in the field over the years. Thanks is also given to the funding from the IDRC grant "Climate change and increasing human-wildlife conflict: How to conserve wildlife in the face of increasing conflicts with landowners," the Canada Research Chairs Program, the Natural Sciences and Engineering Research Council of Canada, Fonds Québécois de la Recherché sur la Nature et les Technologies, and the National Geographic Society.

Contents

Chapter 1
Why Movement Ecology Matters

Colin A. Chapman and Rafael Reyna-Hurtado

The scientific discipline of "Movement Ecology" (Nathan et al. 2008) has played an important role in advancing our understanding of almost every ecological and evolutionary process, from nutrient cycling, to habitat selection, to population dynamics and community ecology. Interestingly, it has been almost a quarter of a century ago since Rodgers and Anson (1994) stated that GPS-based animal-location systems would become the standard for habitat selection studies. They were right! The data made available from GPS telemetry (i.e., sequence of GPS locations) quickly boosted the field of "Movement Ecology" (Nathan et al. 2008), and this field was also greatly advanced when the Max Planck Institute of Ornithology developed a free online database, Movebank (movebank.org), that allowed movement data from many, many species to be freely accessed and analysed (millions and millions of travel routes). Further advancements became possible with the development and use of new analytical tools to understand the rules used by the study animals to move (Ropert-Coudert and Wilson 2005; Sengupta et al. 2018).

In 2008 a Special Feature of the Proceedings of the National Academy of Science was published that was based on an international project held at the Institute for Advanced Studies in Israel. The Special Feature aimed to generate a conceptual

C. A. Chapman (✉)
Department of Anthropology, McGill University, Montreal, QC H3A 2A7, Canada

School of Life Sciences, University of KwaZulu-Natal,
Scottsville, Pietermaritzburg, South Africa

Shaanxi Key Laboratory for Animal Conservation, Northwest University, Xi'an, China
e-mail: colin.chapman@mcgill.ca

R. Reyna-Hurtado
El Colegio de la Frontera Sur (ECOSUR), Department of Biodiversity Conservation,
Lerma, Campeche, Mexico

The Wildlife Conservation Society (WCS), Bronx, NY, USA
e-mail: rreyna@ecosur.mx

R. Reyna-Hurtado, C. A. Chapman (eds.), *Movement Ecology of Neotropical Forest Mammals*, https://doi.org/10.1007/978-3-030-03463-4_1

framework of movement and of ways of generating and analysing movement paths (Nathan 2008; Nathan et al. 2008). In addition, the Special Feature illustrated the application of the framework to different types of questions and illustrated the scope of the Movement Ecology field, both from theoretical and taxonomic perspectives (Fryxell et al. 2008; Holyoak et al. 2008).

The study of animal movement progressively has become more important as the world has become more and more aware of how human actions were endangering natural systems. Today, the loss of tropical forest is causing the extinction and endangerment of many species (Estrada et al. 2017; Pimm et al. 2014). Globally, it is estimated that biodiversity is being lost at an accelerating rate, with current extinction rates approximately 1000 times higher than background rates (Pimm et al. 2014). Recent estimates suggest that 11,000–58,000 species are lost each year and that surviving vertebrate species have declined in abundance by 25% since 1970 (Dirzo et al. 2014). Humans are clearly responsible for this accelerating loss of biodiversity, particularly in the neotropics. Between 2000 and 2012, 2.3 million km^2 of forest was lost globally, and in the tropics forest, loss increased each year (Hansen et al. 2013). To put this in perspective, this area is approximately the size of Mexico. Global estimates of the extent of wildlife over-exploitation are very poor. However, Bennett et al. (2000) estimated that six million mammals were hunted annually in Malaysian Borneo. With respect to climate change, temperatures are predicted to increase by 1.5 °C by the end of the twenty-first century (IPCC 2014), and using moderate greenhouse gas emission estimates, it is projected that by 2100 75% of all tropical forests present in 2000 will experience temperatures that are higher than the temperatures presently supporting closed canopy forests (Peres et al. 2016; Wright et al. 2009).

This volume represents the culmination of a discussion that stated at our field site 4 years ago. We were both adamant that a greater understanding of animal movement would advance tropical conservation efforts, and we were determined to illustrate this. As a result, we gathered together an amazing group of scholars who worked on animal movement and had them contribute papers to this book. We sincerely hope that when readers finish examining the contributions we have gathered together, they will be convinced of the importance of "Movement Ecology" advancing a myriad of academic questions and addressing many of the most important conservation/management questions. Most importantly, we hope that the chapters in this volume inspire the next generation to devote the huge amounts of time to collect and analyse animal movement data to conserve the amazing mammals that we find in the neotropics.

References

Bennett EL, Nyaoi A, Sompud J (2000) Saving Borneo's bacon: the sustainability of hunting in Sarawak and Sabah. In: Hunting for sustainability in tropical forests. Columbia University Press, New York, pp 305–324

Dirzo R, Young HS, Galetti M, Ceballos G, Isaac NJB, Collen B (2014) Defaunation in the anthropocene. Science 345:401–406

Estrada A, Garber PA, Rylands AB, Roos C, Fernandez-Duque E, Di Fiore A, Nekaris KA-I, Nijman V, Heymann EW, Lambert JE (2017) Impending extinction crisis of the world's primates: why primates matter. Sci Adv 3:e1600946

Fryxell JM, Hazell M, Börger L, Dalziel BD, Haydon DT, Morales JM, McIntosh T, Rosatte RC (2008) Multiple movement modes by large herbivores at multiple spatiotemporal scales. Proc Natl Acad Sci 105:19114–19119

Hansen MC, Potapov PV, Moore R, Hancher M, Turubanova SA, Tyukavina A, Thau D, Stehman SV, Goetz SJ, Loveland TR, Kommareddy A, Egorov A, Chini L, Justice CO, Townshend JRG (2013) Highresolution global maps of 21st-century forest cover change. Science 342:850–853

Holyoak M, Casagrandi R, Nathan R, Revilla E, Spiegel O (2008) Trends and missing parts in the study of movement ecology. Proc Natl Acad Sci 105:19060–19065

IPCC (2014) Climate change 2014: synthesis report. Contribution of working groups I, II and III to the fifth assessment report of the intergovernmental panel on climate change [Core Writing Team, R.K. Pachauri and L.A. Meyer (eds.)]. IPPC, Geneva, 151p

Nathan R (2008) An emerging movement ecology paradigm. Proc Natl Acad Sci 105:19050–19051

Nathan R, Getz WM, Revilla E, Holyoak M, Kadmon R, Saltz D, Smouse PE (2008) A movement ecology paradigm for unifying organismal movement research. Proc Acad Natl Acad Sci 105:19052–19059

Peres CA, Emilio T, Schietti J, Desmoulière SJ, Levi T (2016) Dispersal limitation induces long-term biomass collapse in overhunted Amazonian forests. Proc Natl Acad Sci 113:892–897

Pimm SL, Jenkins CN, Abell R, Brooks TM, Gittleman JL, Joppa LN, Raven PH, Roberts CM, Sexton JO (2014) The biodiversity of species and their rates of extinction, distribution, and protection. Science 344:1246752

Rodgers A, Anson P (1994) Animal-borne GPS: tracking the habitat. GPS World 5:20–32

Ropert-Coudert Y, Wilson RP (2005) Trends and perspectives in animal-attached remote sensing. Front Ecol Environ 3:437–444

Sengupta R, Chapman CA, Sarkar D, Bortolamiol S (2018) Automated extraction of movement rationales for building agent-based models: example of a red Colobus monkey group. In: Perez L, Kim EK, Sengupta R (eds) Agent-based models and complexity science in the age of geospatial big data. Advances in geographic information science. Springer, Cham

Wright SJ, Muller-Landau HC, Schipper J (2009) The future of tropical species on a warmer planet. Conserv Biol 23:1418–1426

Chapter 2
The Impact of Hurricane Otto on Baird's Tapir Movement in Nicaragua's Indio Maíz Biological Reserve

Christopher A. Jordan, Brendan Hoover, Armando J. Dans, Cody Schank, and Jennifer A. Miller

2.1 Introduction

Hurricanes have played an important role in the successional pathways of Nicaragua's Caribbean Coast forests throughout history. Records show that between 1892 and 1996, Nicaragua was hit by 40 tropical cyclones, 18 of which made landfall as hurricanes (Centro Humboldt and Fundación del Río 2017). The Category 5 Hurricane Joan that battered Nicaragua in 1989 is perhaps the country's best known hurricane due to the damage it brought on south Caribbean Coast communities and cities and to the extensive research carried out after the hurricane to document the regeneration of the region's lowland tropical forests (Yih et al. 1991). While post-hurricane forest regeneration has been studied comprehensively, the impact of hurricanes on wildlife is comparatively unknown. The few studies that exist address how rapidly certain species assemblages recover over time (Will 1991). Yet the drastic changes in habitat structure and resource availability caused by hurricanes affect not only the species richness of the forest but also the behavior and movement ecology of species, which could have implications for species' carrying capacities, survival rates, and reproduction in sites damaged by hurricanes. For example, in some forests affected by Hurricane Joan, only 27% of trees remained standing and

C. A. Jordan (✉)
Global Wildlife Conservation, Austin, TX, USA

Panthera, New York, NY, USA
e-mail: cjordan@globalwildlife.org

B. Hoover · C. Schank · J. A. Miller
Department of Geography and The Environment, The University of Texas at Austin, Austin, TX, USA
e-mail: brendanhoover@utexas.edu; jennifer.miller@austin.utexas.edu

A. J. Dans
Global Wildlife Conservation, Austin, TX, USA

R. Reyna-Hurtado, C. A. Chapman (eds.), *Movement Ecology of Neotropical Forest Mammals*, https://doi.org/10.1007/978-3-030-03463-4_2

only 18% of trees had leaves in the immediate aftermath of the storm. For primates, this almost certainly shifted their general foraging patterns and habitat use. For large mammals such as jaguars (*Panthera onca*) and Baird's tapirs (*Tapirus bairdii*), the 73% of trees that fell during Hurricane Joan must have turned the forest floor into a messy labyrinth of fallen trees and limbs that affected their ability to move efficiently and thus forced them to adjust their hunting and browsing strategies, respectively (Yih et al. 1991). While it is difficult to collect data on animal movements before and after hurricanes due to the unpredictable nature of the storms, these data are critical to understanding how hurricanes affect animal movement and ultimately their capacity to thrive in hurricane-damaged forests.

Understanding the full impact of hurricanes on neotropical forest wildlife is even more critical considering recent studies that indicate rising sea surface temperatures in the Western Caribbean are expected to produce larger cyclones with higher intensity throughout the region (Baldini et al. 2016; Overland et al. 2016). In effect, this means that in the remainder of this century, the Western Caribbean is likely to see more intense storms capable of damaging more extensive areas of forest.

In this chapter, we report on a unique dataset collected with GPS collars that were installed on two Baird's tapirs before the Category 2–3 Hurricane Otto in Nicaragua's Indio Maíz Biological Reserve during 2016. These data gave us the opportunity to examine how hurricane damage to these tapirs' home ranges affected their movement, which in turn gives us insight into how hurricanes and similar extreme weather events may affect the survival of this globally endangered large mammal in the current century. We hypothesized that tapir home ranges would decrease in size due to the increased energy required for individual tapirs to move long distances in hurricane-damaged forest, in addition to an increase in food density and availability on the forest floor due to post-hurricane forest recovery. We also expected the data to show that tapirs would avoid areas with the combination of large slopes and many downed trees due to the energy required to move around these areas. Likewise, given that rivers and streams post-hurricane are relatively open areas compared with the damaged forest, we hypothesized that tapirs might be using rivers and streams simply because they offer an efficient means of moving around their home ranges.

2.2 Study Site

The 2639 km^2 Indio Maíz Biological Reserve in southeastern Nicaragua constitutes important habitat for a host of regionally endangered and threatened species, such as Baird's tapirs, jaguars, and white-lipped peccaries (*Tayassu pecari*), great green macaws (*Ara ambiguus*), and wild almond trees (*Dipteryx panamensis*) (Fig. 2.1). The reserve receives more than 4000 mm of annual rainfall and is comprised of lowland tropical rainforest, *Raphia* palm swamps, and seasonally flooded forests, and the lower reaches of its rivers harbor freshwater grasses that sustain local manatee (*Trichechus manatus*) populations. The reserve has no road access; its heart is only accessible by boat or dugout canoe. Indio Maíz is a core area of the over 4000 km^2 terrestrial portion of the Rama-Kriol indigenous territory and is inhabited

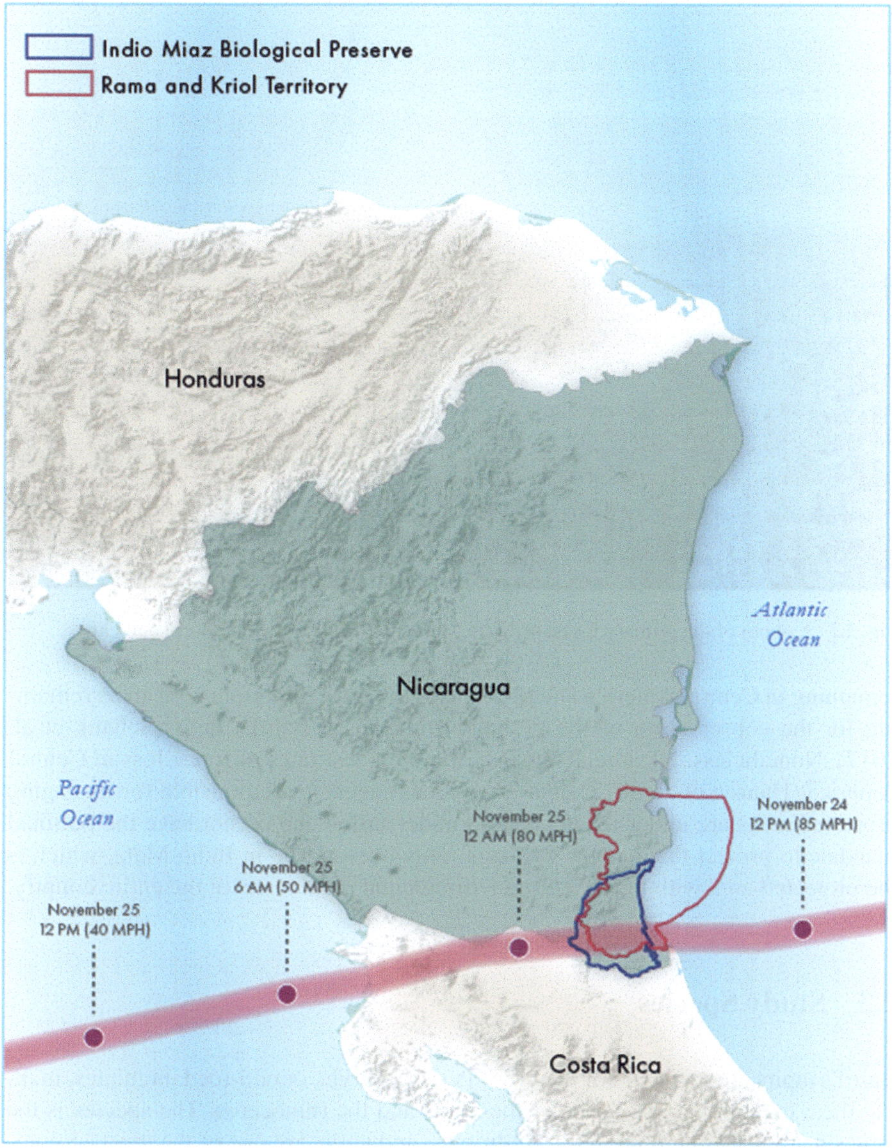

Fig. 2.1 Path of 2016's Hurricane Otto through the Indio Maíz Biological Reserve in the southeast corner of Nicaragua

by Rama indigenous peoples, afro-descendant Kriol, and, increasingly, illegal cattle ranchers but continues to have an extremely low human density.

Historically Indio Maíz has been Nicaragua's best-preserved rainforest and was comprised almost entirely of primary forest (Fig. 2.2). Before 2016 there was no known historical record of Indio Maíz having been affected by a hurricane (Centro Humboldt and Fundación del Río 2017). Indio Maíz is one of the five largest forests

Fig. 2.2 A photo of the primary forest of Indio Maíz in 2015, long before Hurricane Otto

remaining in Central America and also one of the most important core areas remaining for the conservation of the globally endangered Baird's tapir (Schank et al. 2017). Nonetheless, Nicaragua has one of the highest rates of forest loss in Central America (Hansen et al. 2013), and government agencies responsible for managing protected areas are under-resourced and understaffed and do not have the political mandate to protect the country's forests. This is even true in Indio Maíz, which is the protected area with the strictest environmental regulations in the entire country.

2.3 Study Species

Baird's tapirs are members of the order Perissodactyla, or odd-toed ungulates, making them evolutionary relatives of the horse and the rhinoceros. The species is the largest terrestrial mammal in the neotropics, and tapirs are one of the most ecologically important species in the forested landscapes they inhabit (Peres et al. 2016). Tapirs have large daily food requirements and are prodigious browsers, consuming leaves, seeds, and fruits from an estimated 100–200 different species of plants (Jordan, unpublished data). Through their selective browsing behavior and dispersal of seeds through their feces, they have a large impact on the successional pathways of neotropical forests, and for this reason we often refer to them as the "farmers of the forest" and "ecosystem engineers" (O'Farrill et al. 2013).

There have been few studies on Baird's tapir movement. The most frequently cited study is from Corcovado National Park (Costa Rica) where radio telemetry data collected over the course of approximately 1 year on five adults (three females and two males) were used to estimate a mean home range size of only 1.25 km^2 for all tapirs combined using 95% minimum convex polygon (MCP) (Foerster and Vaughan 2002). A more recent study provided much larger estimates ranging from 4.1 to 39.9 km^2, but the dataset differs markedly from our study in that mostly camera-trap data were used to estimate home range size rather than telemetry data (Reyna-Hurtado et al. 2016).

The Baird's tapir is classified as globally endangered by the IUCN (Garcìa et al. 2016). Experts estimate that the current global population could have as few as 4500 mature adults (Garcìa et al. 2016). The main threats to the survival of the species include extensive deforestation, unsustainable poaching, and climate change (Garcìa et al. 2016). Tapirs are solitary animals with large spatial requirements and a slow reproductive cycle given their gestation period of approximately 400 days (Brooks et al. 1997; Garcìa et al. 2016). Massive forest loss across most of Central America in the last 15 years combined with significant hunting has diminished the global Baird's tapir population by more than 50% in the past three generations and left the remaining tapirs in mostly isolated subpopulations, many of which may not be genetically viable (Garcìa et al. 2016; Hansen et al. 2013; Jordan et al. 2014). In the Selva Maya, the species' northernmost stronghold and one of its most important core habitats, there is concern that climate change may decrease water availability and decrease the suitability of the habitat for tapirs over the course of this century (O'Farrill et al. 2014). Without a long-term, large-scale effort to reverse recent trends in the species' populations, Baird's tapirs are vulnerable to becoming critically endangered (Garcìa et al. 2016). In this context, the potential for hurricanes to negatively affect Baird's tapirs' survival and carrying capacity in important core areas like the Indio Maíz Biological Reserve could further threaten the survival of the species.

2.4 Hurricane Otto

Hurricane Otto was an unusually late-forming hurricane during 2016's hurricane season. The storm first achieved hurricane status on November 23, 2016 approximately 240 km from the Nicaraguan/Costa Rican border (Brown 2017). Hurricane Otto made landfall as a Category 3 storm with sustained winds of approximately 175 km/h on November 24, 2016 near Greytown in the southeastern corner of Nicaragua and then began its path across the Indio Maíz Biological Reserve (Brown 2017).

Hurricane Otto was a relatively small storm, with hurricane force winds extending only 16–32 km from the center of the storm (Brown 2017). Although the storm lost strength fairly quickly as it moved across Indio Maíz, the eye of Otto remained intact across almost the entirety of the Central American isthmus and caused extensive damage to reserve's primary forest (Centro Humboldt and Fundación del Río

Fig. 2.3 A photo of Cucaracha Hill in the Indio Maíz Biological Reserve approximately 1 week after Hurricane Otto passed through Indio Maíz. Before the hurricane, this hill was covered with primary forest. (Photo courtesy of Camilo de Castro)

2017) (Fig. 2.3). For instance, in seven sites surveyed in the immediate aftermath of the hurricane, 72.7% of trees had fallen to the forest floor (Centro Humboldt and Fundación del Río 2017). Local researchers estimated that Hurricane Otto damaged 1667.9 km^2 of forest in Indio Maíz (Centro Humboldt and Fundación del Río 2017).

2.5 Data and Methodology

2.5.1 Tapir Captures

Our team captured and immobilized Baird's tapirs as a part of a GPS telemetry project during two separate capture expeditions in 2016, the first from March 29 to May 8 and the second from August 5 to 24. During both expeditions, our team used pitfall traps camouflaged with leaf litter and dirt to capture tapirs.

We built 18 traps during the first expedition then monitored them over a period of 35 days. Total effort was 385 trap nights during the first expedition during which we captured two adult male tapirs and one adult female tapir (0.0078 capture rate). The female's collar was not fitted properly and fell off after a few days.

We built 12 traps during the second expedition, but only monitored them for 9 days due to unexpected torrential rains that forced our team to end the expedition early. Total effort was 75 trap nights. During this expedition we captured one adult male Wes (0.013 capture rate). No nontarget species were captured in pitfalls during either expedition with the exception of a single Tome's spiny rat (*Proechimys semispinosus*) that we removed from the trap and released.

All captured tapirs were immobilized and then fitted with a Telonics Iridium collar programmed to attempt a GPS fix of the animal's location every hour and to make periodic satellite transmissions. All collars had automatic release mechanisms programmed to release approximately 1 year after collar installation; however the automatic release mechanisms malfunctioned and did not drop off. After multiple and ongoing expeditions to recapture collared individuals and recover the information stored on board, we managed to recapture one of the males collared in April 2016 (Almuk) and the male (Wes) captured in August 2016. Expeditions continue to capture the final individual.

Almuk was captured and collared on April 8, 2016. From this date the collar continued to attempt GPS fixes every hour until September 7, 2017. During this interval, Almuk's collar successfully recorded a total of 8830 locations, including 4217 before the hurricane and 4613 after the hurricane. Almuk's collar had an overall fix acquisition rate of 70.6%.

Wes was captured and collared on August 16, 2016 with hourly GPS fixes attempted until January 23, 2018. During this interval, a total of 8532 locations were recorded, including 1464 before the hurricane and 7068 after the hurricane. Wes' collar had an overall fix acquisition rate of 64.8%.

2.5.2 Data Analysis

We investigated the relationship of tapir movement with slope and distance to streams and rivers to test our hypothesis that tapirs would utilize areas of the forest that maximized the efficiency of their movements, such as along waterways and in areas with lower slopes. We digitized all visible streams and rivers from topographic maps (US National Imagery and Mapping Agency, Series E751) and calculated the distance from GPS fixes to these waterways (units = meters). We then compared these samples before and after the hurricane. We also downloaded a 30-m DEM (Tachikawa et al. 2011), calculated slope (units = degrees), and did another comparison of before and after values. This was done to determine how tapir movement changed in relation to these environmental variables.

To analyze tapir movement patterns in general, we compared the size of home ranges before and after the hurricane. We used a fixed kernel density estimation (KDE) (Worton 1989; Seaman and Powell 1996) to calculate the 95% (home range) and 50% (core area) utilization distributions for each tapir before and after the hurricane. To smooth the data, we used the default smoothing factor, which is generated for each of the trajectory datasets (Worton 1995). Smoothing is beneficial because it reduces variance in the low sample areas and reduces bias in areas with many observations (Worton 1989). Home range estimators like KDE potentially misrepresent home ranges because they treat movement data, which is inherently spatially and temporally autocorrelated, as a point pattern rather than as a movement process (Hemson et al. 2005). Autocorrelation can be reduced by filtering points so they are statistically independent (Welch et al. 2015); however, the spatiotemporal autocorrelation can also emerge as a function of an animal's preference for specific areas within their home range. Therefore, instead of filtering data, we used time local convex hull (T-LoCoH) which is a time-scaled home range estimator (Lyons et al. 2013), in addition to KDE. T-LoCoH converts the time difference between point pairs into a time-scaled distance metric used with spatial proximity to construct nearest neighbor convex hulls (Lyons et al. 2013). A scaling factor, s, determines the maximum amount of time from which a spatial neighbor is still considered correlated to the focal location and is therefore calculated as a nearest neighbor (Lyons et al. 2013). As the scaling factor, s, increases, the time factor is weighted more heavily; $s = 0$ means time is not considered at all. To determine our s factor, we plotted our data to see the natural frequencies in the data for the distance of each point to the centroid of the entire datasets over time (Lyons 2014). Using that technique, we set used $s = 0.015$. We then compared the areas calculated using both KDE and T-LoCoH before and after the hurricane to determine if change had occurred. Finally, we also investigated how movement velocity changed before and after the hurricane, calculated as the distance between subsequent GPS fixes, divided by the time elapsed between them.

2.6 Results

Welch's t-test results between samples of environmental variables (slope and distance to waterways) for GPS fixes before and after the hurricane show that (1) both individuals used areas with lower slopes after the hurricane and (2) Wes used habitat farther from rivers with a higher frequency after the hurricane, whereas Almuk used habitat closer from rivers with a higher frequency after the hurricane (Table 2.1, Fig. 2.4). During recent recapture expeditions for Wes, however, we noted that there were several small creeks in the areas far from the main river that he appeared to frequent. These small creeks were not included in our rivers and streams layer, so

Table 2.1 Comparison, using Welch's t-test, between samples of environmental variables for GPS fixes before and after the hurricane (negative values in the confidence intervals indicate higher values after the hurricane)

	Individual	Mean before	Mean after	95% Conf. intervals for difference of means (before–after)
Slope	Almuk	4.22	3.59	0.499–0.772
	Wes	6.81	5.86	0.713–1.187
Distance to waterways	Almuk	426	287	124–153
	Wes	134	278	−154 to −134

All differences were statistically significant at $p < 0.001$

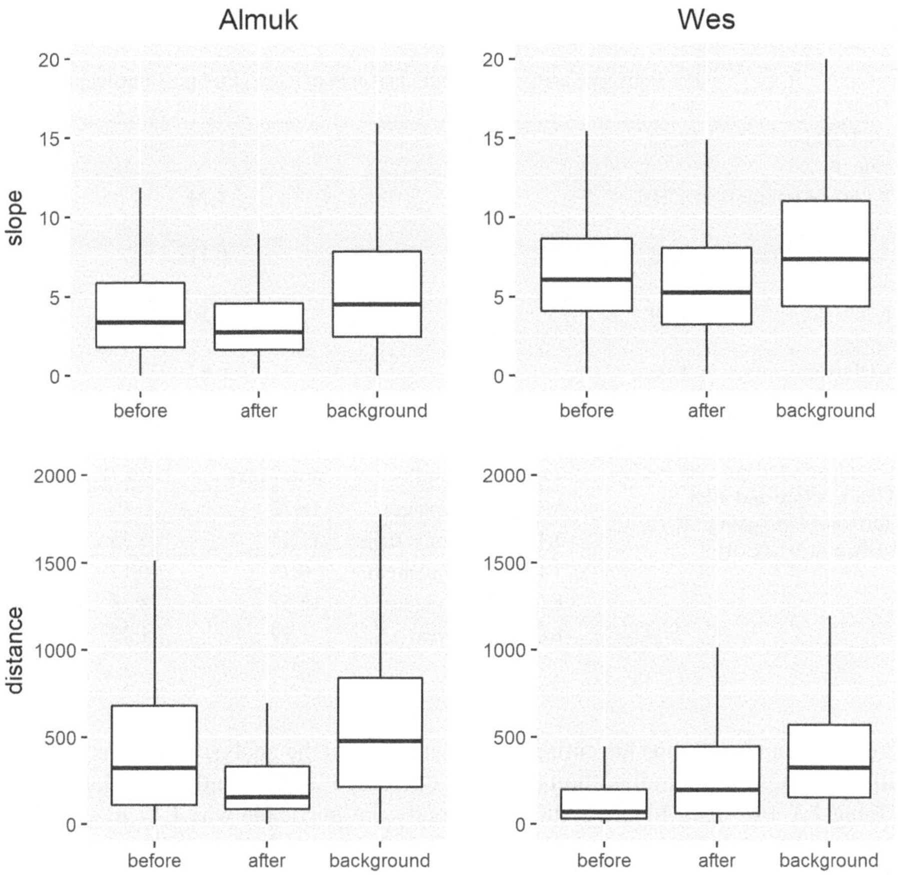

Fig. 2.4 Boxplots for the comparison between slope and distance to waterways for GPS fixes before and after the hurricane. "Background" represents all of the values in the study area – the extent that includes a buffer of 3 km from all GPS points for each individual. Outliers have been removed

there is the chance that our analyses were unable to capture the true relationship between Wes' movements and local waterways.

Given that our data for Wes only included approximately 3 months pre-hurricane and 14 months post-hurricane, we do not believe that his pre-hurricane data were enough to capture his entire home range. For this reason, we restricted our comparisons pre- and post-hurricane to the area of overlap between his pre-hurricane and post-hurricane home ranges. Therefore, results for Wes do not represent his entire home range but rather his home range use within this area of overlap. In all analyses, Wes' use of this area of home range decreased significantly after the hurricane (Table 2.2). However we also provide estimates for Wes' full post-hurricane dataset as this likely does represent his full home range after the storm (Table 2.2). In contrast to Wes, Almuk's dataset included a relatively equal number of points both pre- and

Table 2.2 Comparison of home range (95% isopleth) and core areas (50% isopleth) of Wes before–after Hurricane Otto using both KDE and T-LoCoH

Tapir2 (Wes)	Pre-hurricane area (km^2)	Post-hurricane area of overlap (km^2)	Full post-hurricane dataset (km^2)
MCP (Home range)	18.32	9.61	11.90
T-LoCoH (Home range)	4.21	0.87	1.24
T-LoCoH (Core area)	0.35	0.09	0.11
KDE (Home range)	14.99	2.85	4.50
KDE (Core area)	1.96	0.51	0.82

Table 2.3 Comparison of home range and core areas of Almuk before and after Hurricane Otto using MCP, KDE, and T-LoCoH

Tapir 1 (Almuk)	Pre-hurricane area (km^2)	Post-hurricane area (km^2)
MCP (Home range)	18.72	16.12
T-LoCoH (Home range)	11.25	3.35
T-LoCoH (Core area)	0.78	0.13
KDE (Home range)	13.57	8.73
KDE (Core area)	1.99	1.22

post-hurricane, and thus his entire dataset was used in the analyses. Similar to Wes, all estimates of Almuk's home range decreased significantly post-hurricane (Table 2.3, Fig. 2.5). Finally, Almuk's velocity pre-hurricane was 1.17 m/s, which decreased by 67–0.39 m/s post-hurricane, while Wes' velocity pre-hurricane was 0.42 m/s and decreased slightly to 0.40 m/s post-hurricane.

2.7 Discussion

The home range estimates for these two tapirs are interesting given that the estimates from the most commonly cited paper on Baird's tapir home ranges collected from animals in Corcovado National Park (Costa Rica) range from 0.968 to 1.25 km^2 using 95% minimum convex polygon (MCP) (Foerster and Vaughan 2002). According to our estimates at 95% isopleth, in the primary forest of Indio Maíz before the hurricane, the two tapirs reported on in this chapter had home ranges that measured an order of magnitude more than what was previously reported. For example, using KDE (95% isopleth) which is a traditional, non-time-bound method that can be used for historical comparisons, Almuk had a home range of 13.57 km^2 and Wes 14.99 km^2. Using MCP, pre-hurricane estimates for their home ranges are 18.72 km^2 and 18.32 km^2, respectively. Even in the hurricane-damaged forest, which we assume is similar to secondary forest, home ranges are still larger than those reported in Corcovado, although similar to those reported by Reyna-Hurtado et al. (2016).

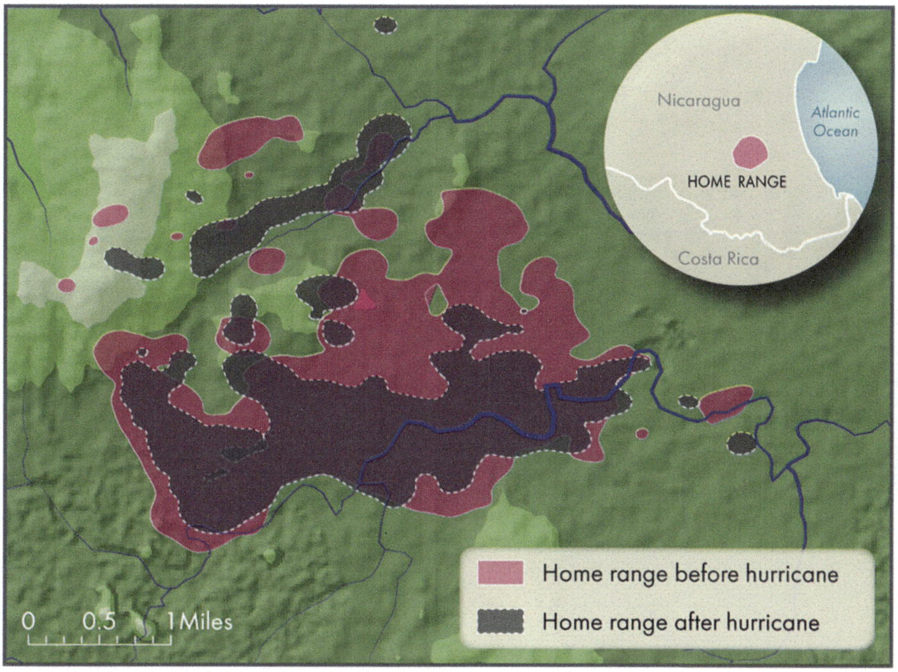

Fig. 2.5 KDE home range results pre- and post-hurricane for Almuk . The pre- and post-hurricane home range results for Wes are not included in this figure because we are unsure if the pre-hurricane data was sufficient for capturing his entire home range

Using MCP with post-hurricane data, our estimates for their home ranges were 16.12 km^2 and 11.90 km^2, respectively. In the case of Wes, these estimates are particularly interesting because we are not confident that his pre-hurricane home range represents his entire home range given that we only collected data on him for 3 months before the storm. Importantly, the pre-hurricane home ranges are consistent with other adult tapirs that we have captured and collared in Indio Maíz. Thus, these new estimates for Baird's tapir home range are significant in that they show that the most frequently referenced estimates for the species are not consistent across the species' entire distribution. This, along with the fact that home ranges for the same individuals in the same locations changed along with significant changes in the condition of the forest, suggests that a home range estimate from one location cannot be extrapolated and used to estimate population size or density for Baird's tapirs in other sites throughout the species' global distribution.

Previous research has shown that Baird's tapirs prefer secondary forest over primary forest due to the greater availability of vegetation close to the forest floor (Foerster and Vaughan 2002; Fragoso 1983). Thus, tapirs are likely to have smaller home ranges in secondary forest, as they have to move less to consume the same amount of food. We predicted smaller home ranges post-hurricane considering that tree fall caused by Hurricane Otto triggered a massive release of nutrients on the forest floor which resulted in extremely rapid regeneration and therefore a spike in

Fig. 2.6 Cucaracha Hill (the same hill pictured in Fig. 2.2) approximately 1 year after Hurricane Otto. (Photo courtesy of Camilo de Castro)

tapir food availability (Fig. 2.6). Furthermore, we optimistically predicted that in the short term, this might even increase the carrying capacity for Indio Maíz's tapirs and thus results in a larger population. Our results showed that there was a significant decrease in both of our collared tapirs' home ranges post-hurricane.

However, a look at the individual movement patterns in relation to slope and distance to streams and rivers puts our prediction of a subsequent increase in population into question. For instance, in the case of Almuk, movements post-hurricane occurred close to rivers and streams and in areas with more gentle slopes than pre-hurricane, most likely in an attempt to move more efficiently around his home range. Anecdotally, increased difficulty of moving in the study area was observed by our field team, who estimated that walking the same distance can take five to six times longer post-hurricane simply because one has to constantly climb over and under a mess of fallen trees and limbs. This has been exacerbated even further as the forest regenerates because new growth has filled the gaps in the fallen trees and reduced visibility to almost nothing (this also increased the difficulty of recapturing Almuk and Wes). Therefore following rivers and streams as much as possible because they tend to provide clearer paths allowed for more efficient and faster movement in general. While tapirs are certainly more adept at moving through the forest than our field team, it seems plausible that tapirs would do the same and look for the most efficient way to travel around their home ranges, and this is consistent with research on other species (Shepard et al. 2013; Wall et al. 2006). We hypothesized that optimizing efficiency would include concentrating movements: (a) along rivers and streams and/or (b) in lowland areas that do not cross hills of significant size covered in hurricane-damaged hardwood forest but rather stick to lowland areas (i.e., areas of lower slope) that likely have more palms and potentially less woody debris covering the forest floor. Our results support this hypothesis. Therefore, while tapirs do appear to have reduced home ranges in

Indio Maíz, we are unable to unequivocally conclude that this will lead to an increase in carrying capacity in the reserve. This is because our data suggest that Almuk's and Wes' home ranges do not appear to have reduced solely due to an increase in food availability, but also because tapirs are selecting habitat that allows them to move efficiently. It is unclear if tapirs can now use the entirety of the landscape of Indio Maíz, such as the forested hills away from rivers and streams post-hurricane, and thus more research is needed to determine if the reduction in home range size will result in an increase in carrying capacity. It is also unclear if the restrictions on movement post-hurricane influences the dispersal capabilities of juveniles and/or the genetic connectivity of tapirs within Indio Maíz, but given our findings this is possible.

In addition to results on home ranges and habitat use, velocity post-hurricane was lower than pre-hurricane for both tapirs, with a more substantial decrease in the case of Almuk. This is likely due to the combination of higher food availability on the forest floor resulting in less movement for foraging overall and the increased difficulty of moving around the forest due to fallen trees. We assume that there was a smaller decrease in Wes' velocity because his home range, while drastically altered by the hurricane, was visibly less damaged than Almuk's. Future research should include using state-space models to differentiate behavioral modes like foraging and general movement before and after the hurricane to assess the significance of this decrease in velocity (Patterson et al. 2008).

Although two individuals represent a very small sample size for more robust movement pattern analyses, the unique nature of our dataset have allowed us a novel look into the immediate impact of hurricanes on the movement of Baird's tapirs. In the coming years as climate change produces stronger hurricanes capable of causing more extensive damage to the lowland forests of the Western Caribbean, it will be important to continue to monitor these and other tapirs in Indio Maíz to observe how Baird's tapirs home ranges change along with the regeneration of Indio Maíz's forests in order to better understand the long-term impacts of these storms.

This chapter shows that research on tapir movement is crucial to understanding changes in tapir behavior and movement when faced with natural perturbations to their habitat such as the damage caused by hurricanes. It also suggests that data on movement ecology of endangered species in fragmented, degraded, or damaged habitat are essential to understanding how these species survive outside of core areas of primary forest (Welch et al. 2015; Whittington et al. 2005). Such data will allow us to develop more effective and appropriate conservation plans for endangered species that are subject to human and/or natural perturbations.

References

Baldini LM, Baldini JUL, McElwaine JN, Frappier AB, Asmerom Y, Liu K-B, Liu KB, Prufer KM, Ridley HE, Polyak V, Kennett DJ, Macpherson CG (2016) Persistent northward North Atlantic tropical cyclone track migration over the past five centuries. Sci Rep 6:37522

Brooks DM, Bodmer RE, Matola S (1997) Tapir action plan. IUCN/SSC Tapir Specialist Group

Brown DP (2017). National Hurricane Center Tropical Cyclone Report: Hurricane Otto, 20–26 November 2016. National Hurricane Center. Retrieved from https://www.nhc.noaa.gov/data/tcr/AL162016_Otto.pdf

Centro Humboldt and Fundación del Río (2017) Evaluación de daños socio-ambientales causados por el Huracán Otto en la zona sureste de Nicaragua. Centro Humboldt and Fundación del Río
Foerster CR, Vaughan C (2002) Home range, habitat use, and activity of Baird's Tapir in Costa Rica. Biotropica 34(3):423–437
Fragoso J (1983) The ecology and behavior of Baird's tapir in Belize. Documento del centro de Documentación de la Universidad de Heredia, Costa Rica
Garcìa M, Jordan CA, O'Farrill G, Poot C, Meyer N, Estrada N et al (2016) *Tapirus bairdii*. The IUCN red list of threatened species. https://doi.org/10.2305/IUCN.UK.2016-1.RLTS.T21471A45173340.en
Hansen MC, Potapov PV, Moore R, Hancher M, Turubanova SA, Tyukavina A et al (2013) High-resolution global maps of 21st-century forest cover change. Science 342(6160):850–853
Hemson G, Johnson P, South A, Kenward R, Ripley R, MacDonald D (2005) Are kernels the mustard? Data from global positioning system (GPS) collars suggests problems for kernel home-range analyses with least-squares cross-validation. Journal of Animal Ecology, 74(3), 455–463
Jordan CA, Galeano MR, Alonzo AS (2014) La Cacería Histórica de Tapires Centroamericanos (*Tapirus bairdii*) en la RAAS, Nicaragua. Estud Ambient 1(1):73–87
Lyons AJ, Turner WC, Getz WM (2013) Home range plus: a space-time characterization of movement over real landscapes. Movement Ecology 1 (1):2
Lyons AJ (2014) T-LoCoH for R: tutorial and users guide. Google Scholar, 2–53.
O'Farrill G, Galetti M, Campos-Arceiz A (2013) Frugivory and seed dispersal by tapirs: an insight on their ecological role. Integr Zool 8(1):4–17
O'Farrill G, Gauthier Schampaert K, Rayfield B, Bodin Ö, Calmé S, Sengupta R, Gonzalez A (2014) The potential connectivity of waterhole networks and the effectiveness of a protected area under various drought scenarios. PLoS One 9(5):e95049
Overland JE, Dethloff K, Francis JA, Hall RJ, Hanna E, Kim SJ et al (2016) Climate science special report: fourth National Climate Assessment. Clim Dyn 46(7):2115–2122
Patterson TA, Thomas L, Wilcox C, Ovaskainen O, Matthiopoulos J (2008) State–space models of individual animal movement. Trends Ecol Evol 23(2):87–94
Peres CA, Emilio T, Schietti J, Desmoulière SJM, Levi T (2016) Dispersal limitation induces long-term biomass collapse in overhunted Amazonian forests. Proc Natl Acad Sci 113(4):892–897
Reyna-Hurtado R, Sanvicente-López M, Pérez-Flores J, Carrillo-Reyna N, Calmé S (2016) Insights into the multiannual home range of a Baird's tapir (*Tapirus bairdii*) in the Maya Forest. THERYA 7(2):271–276
Schank CJ, Cove MV, Kelly MJ, Mendoza E, O'Farrill G, Reyna-Hurtado R et al (2017) Using a novel model approach to assess the distribution and conservation status of the endangered Baird's tapir. Divers Distrib 23(2):1–13
Seaman ED, Powell RA (1996) An Evaluation of the Accuracy of Kernel Density Estimators for Home Range Analysis. Ecology 77 (7):2075–2085
Shepard ELC, Wilson RP, Rees WG, Grundy E, Lambertucci SA, Vosper SB (2013) Energy landscapes shape animal movement ecology. Am Nat 182(3):298–312
Tachikawa T, Kaku M, Iwasaki A, Gesch DB, Oimoen MJ, Zhang Z, Danielson J, Krieger T, Curtis B, Haase J, Abrams M (2011) ASTER global digital elevation model version 2-summary of validation results. NASA. Retrieved from https://pubs.er.usgs.gov/publication/70005960
Wall J, Douglas-Hamilton I, Vollrath F (2006) Elephants avoid costly mountaineering. Curr Biol 16(14):R527–R529
Welch RJ, Tambling CJ, Bissett C, Gaylard A, Müller K, Slater K, Parker DM (2015) Brown hyena habitat selection varies among sites in a semi-arid region of southern Africa. J Mammal 97(2):473–482
Whittington J, St Clair CC, Mercer G (2005) Spatial responses of wolves to roads and trails in mountain valleys. Ecol Appl 15(2):543–553
Will T (1991) Birds of a severely hurricane-damaged Atlantic Coast rain forest in Nicaragua. Biotropica 23(4):497–507

Worton BJ (1989) Kernel Methods for Estimating the Utilization Distribution in Home-Range Studies. Ecology 70 (1):164–168

Worton BJ (1995) Using Monte Carlo Simulation to Evaluate Kernel-Based Home Range Estimators. The Journal of Wildlife Management 59 (4):794

Yih K, Boucher DH, Vandermeer JH, Zamora N (1991) Recovery of the rain forest of Southeastern Nicaragua after destruction by Hurricane Joan. Biotropica 23(2):106–113

Chapter 3
White-Lipped Peccary Home-Range Size in the Maya Forest of Guatemala and México

José Fernando Moreira-Ramírez, Rafael Reyna-Hurtado, Mircea Hidalgo-Mihart, Eduardo J. Naranjo, Milton C. Ribeiro, Rony García-Anleu, Roan McNab, Jeremy Radachowsky, Melvin Mérida, Marcos Briceño-Méndez, and Gabriela Ponce-Santizo

3.1 Introduction

Understanding the spatial requirements of animals is crucial for addressing effective conservation planning, especially for endangered species that have important socio-economic values for rural communities and play key ecological roles, such as seed

J. F. Moreira-Ramírez (✉)
El Colegio de la Frontera Sur, Ciudad Industrial, Lerma,
Ciudad de Campeche, Campeche, México

Wildlife Conservation Society, Programa para Guatemala, Flores. Petén, Guatemala
e-mail: jfmoreira@ecosur.edu.mx

R. Reyna-Hurtado
El Colegio de la Frontera Sur (ECOSUR), Department of Biodiversity Conservation,
Lerma, Campeche, México

The Wildlife Conservation Society (WCS), Bronx, NY, USA

M. Hidalgo-Mihart
División Académica de Ciencias Biológicas, Universidad Juárez Autónoma de Tabasco,
Villahermosa, Tabasco, Mexico

E. J. Naranjo · M. Briceño-Méndez
El Colegio de la Frontera Sur, Carretera Panamericana y Periférico Sur s/n,
San Cristóbal de Las Casas, Chiapas, Mexico

M. C. Ribeiro
Laboratório de Ecologia Espacial e Conservação, Departamento de Ecologia, Universidade
Estadual Paulista, Rio Claro, São Paulo, Brazil

R. García-Anleu · R. McNab · M. Mérida · G. Ponce-Santizo
Wildlife Conservation Society, Programa para Guatemala, Flores. Petén, Guatemala

J. Radachowsky
Wildlife Conservation Society, Bronx, New York, USA

R. Reyna-Hurtado, C. A. Chapman (eds.), *Movement Ecology of Neotropical Forest Mammals*, https://doi.org/10.1007/978-3-030-03463-4_3

dispersers and seed predators (Beck 2006; Keuroghlian et al. 2009a, b). Habitat loss and overhunting are two of the main activities affecting wildlife in protected and communal areas (Bodmer et al. 1997; Fritz et al. 2009). Hunting in tropical forests can exert a strong effect on the behavior and abundance of ungulates (Peres 2000; Naranjo and Bodmer 2007; Reyna-Hurtado and Tanner 2007; Endo et al. 2010; Reyna-Hurtado et al. 2010). Previous studies documented that hunting can have a negative impact on vertebrate populations and combined with habitat destruction can decrease Neotropical prey species to the point of causing local population extinction (Alvard et al. 1997; Reyna-Hurtado and Tanner 2007; Andrade Melo et al. 2015). In contiguous forests, where the persistence of populations in the sink habitat depends on their migration from high-quality source habitats (Pulliam 1988), the establishment of reserves can function as wildlife refuges for prey populations (Naranjo and Bodmer 2007), which is vital for the survival of hunted species (Peres 2001). Additionally, primary forests distant from human settlements are effective in maintaining biodiversity, even if not officially protected (Peres et al. 2003; Peres and Palacios 2007). Hunters often seek out larger animals, such as white-lipped peccaries (*Tayassu pecari*, Link 1795), which are favorite prey among hunters in the Neotropics (Peres 1996; Reyna-Hurtado and Tanner 2007; Santos-fita et al. 2012). White-lipped peccaries are social ungulates that form large and cohesive groups between 10 and 300 individuals (Altrichter et al. 2012; Moreira-Ramírez et al. 2015; Reyna-Hurtado et al. 2016). White-lipped peccaries are confined to the Neotropical Region, from southeastern Mexico to northern Argentina and Rio Grande do Sul in southern Brazil, moving across large areas of up to 200 km^2 (Fragoso 1998; Altrichter et al. 2012), and they are an important seed predator with a unique role as an ecosystem engineer (Beck 2006; Keuroghlian et al. 2009b). Local populations have declined at alarming rates throughout Mesoamerica in the last 20 years as a result of hunting pressure, habitat loss, and forest fragmentation (Peres 1996; Escamilla et al. 2000; Endo et al. 2010; Reyna-Hurtado et al. 2017), and the species is now confined to few large tropical forest reserves (Altrichter et al. 2012; Reyna-Hurtado et al. 2017). Currently, it is listed as vulnerable throughout all its distributional range (Keuroghlian et al. 2013).

In the Maya Forest of Mexico, hunting of white-lipped peccaries occurs mainly in the dry season (Reyna-Hurtado et al. 2010), resulting in lower abundances and smaller group sizes than in protected areas (Reyna-Hurtado and Tanner 2007; Reyna-Hurtado 2009; Briceño-Méndez et al. 2016). Previous studies have analyzed the home-range sizes of the white-lipped peccaries in fragmented and protected areas (Fragoso 1998; Carrillo et al. 2002; Keuroghlian et al. 2004, 2015; Reyna-Hurtado et al. 2009; Jacomo et al. 2013); however the effect of hunting on the groups' movement pattern is unknown. The only study carried out in the Calakmul Biosphere Reserve focused on estimating the home-range size and habitat preferences of white-lipped peccary (Reyna-Hurtado et al. 2009). This study showed that peccaries will undertake long-range movements to locate preferred types of habitats and water ponds on a landscape scale. To understand how the landscape dynamics of this region influence space use by white-lipped peccaries, we investigated, using GPS satellite collars, whether group home-range size is influenced by season and hunting. The present study reports white-lipped peccary's home-range sizes for four groups at three sites in the Maya Forest, and for the first time, information on the temporal and spatial movements of white-lipped peccaries in a hunting site is described. We explored the following question: Are there differences in home-range size between seasons in hunted and non-hunted sites? We hypothesized that home ranges would be larger in the rainy season compared to the dry season and that home ranges would be larger in protected areas than the hunted site.

3.2 Materials and Methods

3.2.1 Study Site

The Maya Forest is the largest continuous remaining tropical rainforest in Mesoamerica. The area includes a ~31,000 km² protected forest that stretches across Belize, northern Guatemala, and throughout Mexico's Yucatan Peninsula. The forest is extensive and diverse, supporting large populations of many rare and endangered species such as white-lipped peccaries, jaguars (*Panthera onca*, Linnaeus, 1758), and tapirs (*Tapirus bairdii*, Gill 1975). Guatemala's 21,100 km² Maya Biosphere Reserve is adjacent to Mexico's 7238 km² Calakmul Biosphere Reserve to the north.

The study was carried out in three areas of the Maya Forest: Laguna del Tigre National Park (LTNP) in Guatemala and Calakmul Biosphere Reserve (CBR) and "*ejido*" Nuevo Becal (NB) in Mexico. An *ejido* is defined as a piece of land farmed communally under a system supported by the Mexican state. The LTNP and CBR are protected areas without the presence of hunting, and NB is a communal forest with human presence and where hunting is common (Fig. 3.1). Laguna del Tigre National Park is located to the west of the Maya Biosphere Reserve (MBR) in the vicinity of latitude 17°10′30″ N and longitude −90°25′22″ W. It has an extension of 3370 km², being the largest Core Zone of the MBR and the largest National Park in Guatemala. When the MBR was established in 1990, LTNP was declared a national park and became a Ramsar site (Moreira-Ramírez et al. 2016). The mean temperature in the dry season and rainy season is 30 °C and 25 °C, respectively. The accumulated rainfall for the year 2015 was 1483 mm (Estación Meteorológica Mactún, San Andrés, Petén, Instituto de Sismología, Vulcanología, Meteorología e Hidrología). The altitude varies between 40 and 200 m. There are two climatic seasons: the dry season is from December to May and the rainy season from June to November (Moreira-Ramírez et al. 2016). The LTNP faces serious threats in the center and west side of the park, as forest fires in the dry season are caused primarily by unplanned human settlements, land speculation for agriculture and ranching, and oil activities without mitigation (Consejo Nacional de Áreas Protegidas and Wildlife Conservation Society 2015).

The CBR, which was established in 1989, covers an area of 7238 km² and conserves the largest tropical forest in Mexico. It is located in the southeastern corner of the state of Campeche in the vicinity of latitude 18°07′21″ N and longitude 89°48′56″ W (Garcia-Gil 2003). In this region, subhumid warm climate prevails; the average annual temperature is 24.6 °C. The rainy season is between June and November with

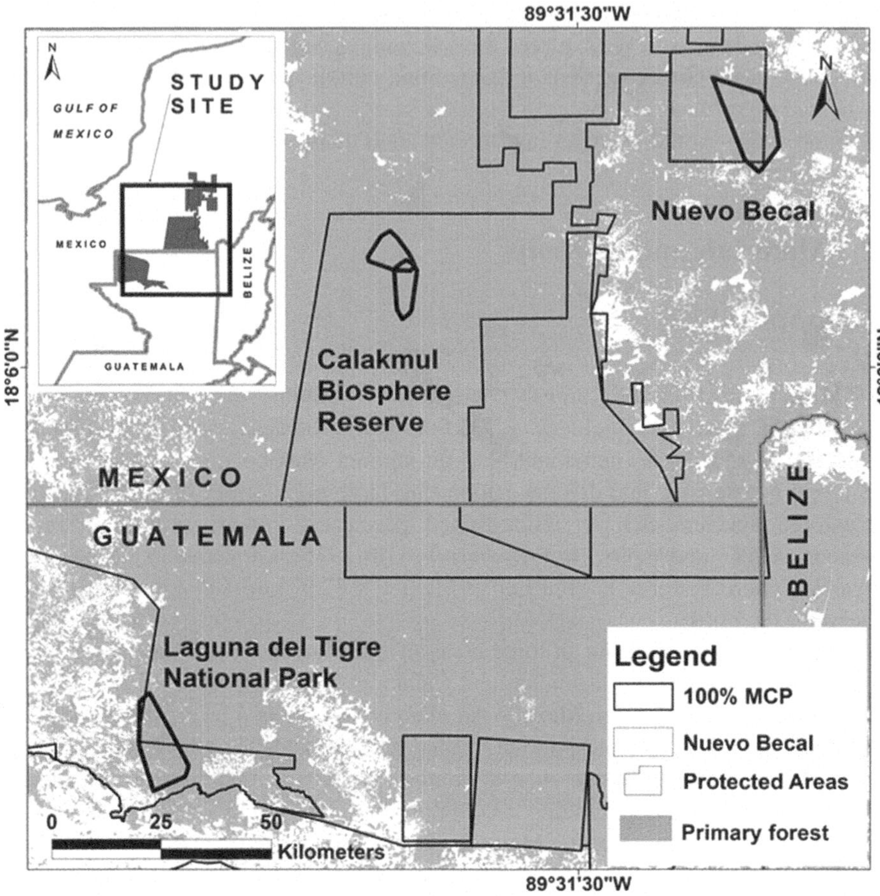

Fig. 3.1 Location of Laguna del Tigre National Park, Guatemala, and Calakmul Biosphere Reserve and "ejido" Nuevo Becal, Mexico, with the minimum convex polygon (100% MCP) home ranges of four white-lipped peccary groups

an annual rainfall average of 1076 mm, varying from 900 mm in the northern part to more than 1400 mm in the southern part (Martínez and Galindo-Leal 2002). The accumulated rainfall in 2015 was 633 mm (Conhuás Weather Station, Calakmul, Comisión Nacional del Agua). The topography in the area is very flat (approximately 250 m above mean sea level) with the highest hills reaching 340 m above sea level. Since 1989, hunting has not been allowed within the CBR, and human activities inside the reserve are limited to tourism and research (Reyna-Hurtado and Tanner 2007).

The *ejido* NB is located in the vicinity of 18°40′07″ N, 89°12′34″ W, adjacent to the CBR. Nuevo Becal has an extension of 520 km^2, of which half is under forest management. The altitude varies between 100 and 380 m. The predominant climate is warm subhumid with summer rains and less than 60 mm of precipitation in the driest month. The annual average temperature is 25 °C, and the average annual rainfall ranges from 1200 to 1500 mm in the center and 1500–2000 mm in the south (García-Gil 2003). The accumulated rainfall for 2015 was 762.7 mm (Zoh Laguna Weather Station, Calakmul, Comisión Nacional del Agua). Approximately 80% of the land is still forested (Reyna-Hurtado 2009). The main productive activities include farming, beekeeping, livestock production, and coal production. However, hunting is practiced year-round (Escamilla et al. 2000; Reyna-Hurtado and Tanner 2007; Santos-fita et al. 2012). Previous studies have shown that white-lipped peccary hunting occurs frequently in this community, which has resulted in smaller group sizes and relative abundance of this species in NB in comparison to CBR (Reyna-Hurtado and Tanner 2007; Briceño-Méndez et al. 2016).

3.2.2 White-Lipped Peccary Monitoring

In the dry season, white-lipped peccaries visit ponds frequently in search of water, mud, and food (Moreira-Ramírez et al. 2016). This behavior allowed us to capture peccaries in ponds near field camping sites (5–15 km) in the three study areas. The capture was done by chemical immobilization using ketamine (7.7 mg/kg) and xylazine hydrochloride (4.3 mg/kg) delivered by a dart rifle (Telinject Inc., Santa Clarita, California, and Pneu-Dart Inc., Williamsport, Pennsylvania) from a tree nearby a pond (Reyna-Hurtado et al. 2009). We weighed and recorded the age and sex of each animal, as well as standard morphometric measurements. All animals were captured and handled following the guidelines of the American Society of Mammalogists (Sikes et al. 2011) and from El Colegio de la Frontera Sur. We used satellite global positioning system (GPS) collars. We captured and radio-collared four and two white-lipped peccaries in the dry season of 2015 and 2016, respectively. The radio-collared animals belonged to 4 different groups corresponding to 1 in LTNP (minimum group size 46 individuals), 2 in CBR (each minimum group size: 30 individuals; CBR1, CBR2), and 1 in NB (minimum group size 25 individuals). We fitted 2 individuals with Telonics collars model TGW-4470-4 (1 in CBR and 1 in LTNP; Telonics Inc., Mesa, Arizona) and 4 with Vectronic collars model GPS PLUS Vertex Survey (1 in CBR, 2 in LTNP and 1 in NB; Vectronic Aerospace GmbH, Berlin, Germany). Animals recovered from immobilizations within 1–1.3 h, after which they usually rejoined their original group 1–3 days after capture. GPS collars were programmed to acquire a location every 2 h (12 locations/day, Telonics collars) and 12 h (2 locations/day, Vectronic collars) and to send data through the IRIDIUM and Globalstar system every 2 and 4 days for Telonics and Vectronic collars, respectively. Telonics collars included a programmable release mechanism (model CR-2A, Telonics Inc.), and we scheduled the drop-off mechanism to release 11–12 months after capture. The GPS collars were recovered when possible using the mortality locations obtained through the IRIDIUM and Globalstar system and by searching the VHF pulse of the collars using a receiver (Daninject Dart Guns, Austin, Texas). We were unable to recover the GPS collar of the peccary captured in NB, and therefore we were unable to download additional fixes stored onboard. Due to this we obtained a variation in the fixes for this group.

3.2.3 Home-Range Estimation and Analysis

Many problems and limitations in home-range estimation were overcome by the recently developed autocorrelated kernel density estimates (AKDE) that can handle large autocorrelated movement datasets without the need of thinning the data or excluding inherent information from the data structure (Fleming et al. 2015). Autocorrelated kernel density estimate (AKDE) home ranges were calculated using R 3.3.2, package ctmm 0.3.6 (R Development Core Team 2012; Fleming et al. 2015; Calabrese et al. 2016; Moßbrucker et al. 2016), following the procedure of Fleming et al. (2015). After visualizing the autocorrelation structure to obtain starting values for the variance and autocorrelation timescales, we fitted three different continuous-time movement models to each white-lipped peccary dataset: (i) independent and identically distributed (IID), (ii) Ornstein–Uhlenbeck (OU, including autocorrelation in location; Dunn and Gipson 1977), and (iii) Ornstein–Uhlenbeck-F (OUF, including autocorrelation in both location and velocity; Fleming et al. 2014). From these three models, we selected the one with the best fit by comparing second-order Akaike information criterion values and then proceeded to calculate AKDE home ranges and confidence limits (Calabrese et al. 2016). These data were managed and stored in Movebank (http://www.movebank.org Wikelski and Kays 2017). We also used the fixed kernel estimator (KDE) (Kernohan et al. 2001), as this estimator is the recommended for investigations focusing on home-range boundaries (Millspaugh and Marzluff 2001) and has previously used to estimate the white-lipped peccaries (WLP): home range in Mexico (Reyna-Hurtado et al. 2009) and Brazil (Jacomo et al. 2013). KDE home ranges were estimated with the bandwidth (*h*) selected using the least square cross-validation method (LSCV; Seaman and Powell 1996; Seaman et al. 1999). For AKDE and KDE analyses, we considered 50% and 95% of the locations, the former to represent the core area of a peccary group's home range, the latter to represent its full range (Harris et al. 1990). Because of this differential spatial sampling effort and to facilitate comparisons with other studies, we further estimated home ranges for groups of white-lipped peccary by the minimum convex polygon (MCP) using the full dataset (100% of fixes). Both KDE and MCP were calculated using Geospatial Modelling Environment (Beyer 2012). We defined the dry season as December–May and the rainy season from June to November. Seasonal 95% and 50% KDE home ranges were estimated for both the dry and rainy seasons in 2015 and 2016 only for LTNP group. For groups CBR1 and CBR2, we estimated the home range only for the dry seasons of 2015 and 2016, respectively. For NB we estimated the home range for rainy season of 2015 and the dry season 2016. We compared home-range size during the dry and rainy seasons with a Mann-Whitney-Wilcoxon Test and evaluated if the home-range sizes were influenced by the number of fixes obtained per individual using Spearman's correlation analysis. All spatial data were prepared and compiled using ArcMap 10 (ESRI 2011) and R software (R Development Core Team 2017) performed to the 0.05 confidence level.

3.3 Results

The monitoring of four groups of white-lipped peccary over periods of 2–16 months resulted in a total of 1322 GPS localizations available for analysis (Table 3.1). Mean error was 2.5 ± 1.6 m for localizations of Telonics collars and 2 ± 0.6 m for Vectronic collars. After recovering the collars and downloading the data stored, there we found that the fix success of the two Telonics collars varied between 85.1% and 96.3%. For the four Vectronic collars, the fix success was much lower and varied between 3.7% and 56.7% (mean = 21.7%). The home-range sizes of the LTNP and NB white-lipped peccaries groups when plotted against the number of tracking months (LTNP: 17 months; NB: 13 months) indicated that our sampling effort was adequate to describe the annual 100% MCP home-range sizes for these groups (Fig. 3.2). The home ranges for CBR1 and CBR2 were likely underestimated

Table 3.1 Satellite telemetry period, continuous-time movement model selection for autocorrelated kernel density estimation (AKDE) using the difference for finite sample size-corrected Akaike information criterion (ΔAICC) among the three models, independent and identically distributed (IID), Ornstein–Uhlenbeck (OU), and Ornstein–Uhlenbeck-F (OUF), and home-range estimates in Km² for four groups of white-lipped peccaries using 95% AKDE, including the 50% core area and confidence limits Cllow and Clup, the 95% kernel density estimation (KDE; including the 50% core area), and the 100% minimum convex polygon (MCP) estimates

Group	Telemetry period	ΔAICC IID	ΔAICC OU	ΔAICC OUF	95% AKDE {50% core area} (Cllow, Clup) [km²]	95% KDE {50% core area} [km²]	100% MCP [km²]
LTNP	March 2, 2015–July 4, 2016	2548.97	0	2.05	156 {31} (96, 231) *n* = 508	99 {17}	138
CBR1	March 12, 2015–June 10, 2015	138.09	0	2.88	142 {28} (92, 203) *n* = 102	77 {15}	47
CBR2	March 15, 2016–May 13, 2016	ND	36.73	0	62 {15} (39, 89) *n* = 682	40 {7}	54
NB	May 10, 2015–June 19, 2016	34.88	0	3.1	312 {80} (178, 483) *n* = 30	140 {38}	154

ND No Data
n number of relocations available for analysis

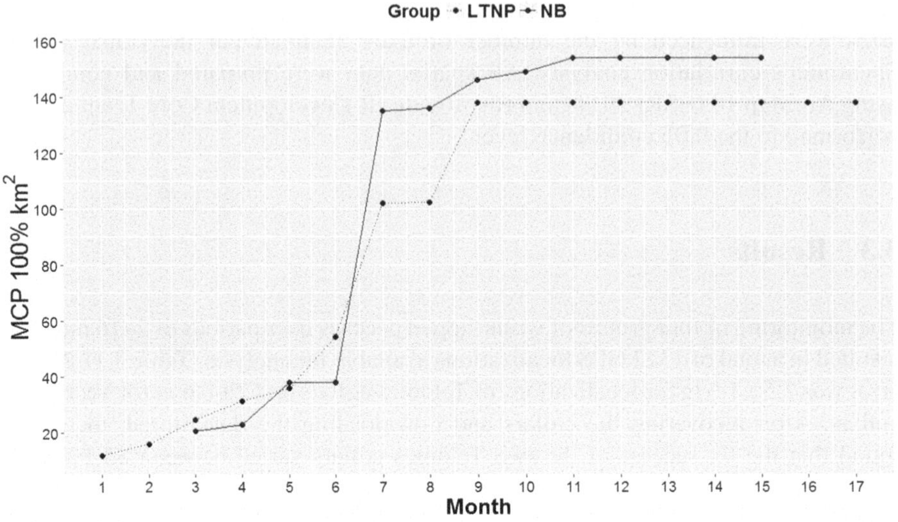

Fig. 3.2 The 100% minimum convex polygon home ranges of Laguna del Tigre National Park (LTNP, dotted line) and Nuevo Becal (NB, solid line) groups that were tracked in the Maya Forest obtained by cumulative sequential samples and then plotted versus the number of months tracked

because they were only monitored for 4 and 2 months, respectively. Of three groups, we were able to recover the collars to download the stored localizations. The collar from the NB group was not retrieved in the field because it did not generate a mortality signal possibly due to the battery running out. This collar only sent sporadic locations, on average three locations per month. This could be due to several factors, such as canopy cover, mud in the collar antenna, or simply failure to connect with the satellites. In spite of this situation, we considered that the number of locations obtained for this group and the time span tracked allowed for a reliable home-range estimate of the NB group.

AKDE were obtained for all four white-lipped peccary groups (Fig. 3.3, Table 3.1), with the model OU performing best for all but one group (CBR2, Table 3.1), for which the OUF model reached a smaller AKDE value, most likely owing to the more continuous movements observed for this particular group and for the shorter time interval that it was followed. The home-range estimation using AKDE, KDE, and MCP was lower for the three groups of the non-hunted sites than that of the group at the hunted site. The 95% AKDE area estimates for non-hunted sites ranged from 62 km^2 (Cllow = 39, Clup = 89) for CBR2 to 156 km^2 (Cllow = 96, Clup = 231) for LTNP. For the hunted site (NB), AKDE estimate was 312 km^2 (Cllow = 178, Clup = 483). Core area estimates using the 50% AKDE in non-hunted sites ranged from 15 km^2 for CBR2 to 31 km^2 for LTNP and were smaller than that of the hunted site in NB (80 km^2). The 95% KDE home-range area estimates for non-hunted sites ranged from 40 km^2 (CBR2) to 99 km^2 (LTNP). Considering the estimate from LTNP as more reliable due to the fact that this group was monitored

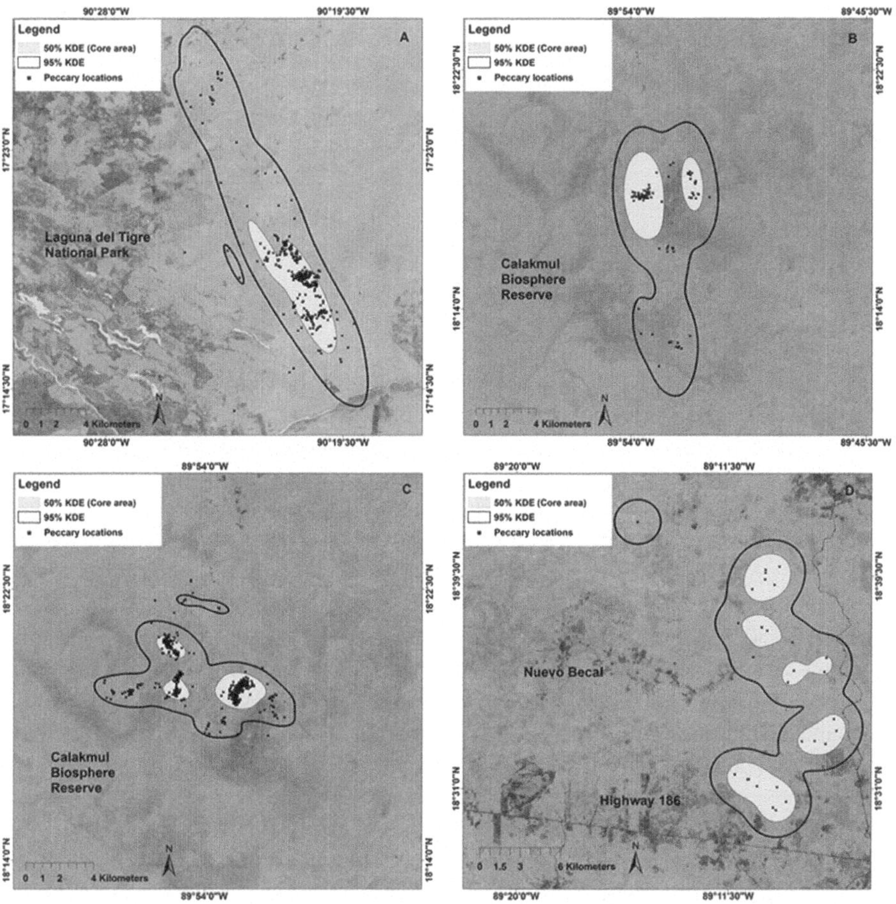

Fig. 3.3 Kernel density estimation (95% KDE) home ranges and 50% core area of four white-lipped peccary groups. (**a**) Laguna del Tigre National Park. (**b**) Calakmul Biosphere Reserve 1. (**c**) Calakmul Biosphere Reserve 2. (**d**) Nuevo Becal

for more than a year, we still found this estimate substantially smaller than the hunted site with 140 km^2. Core area estimates using the 50% KDE in non-hunted sites ranged from 7 km^2 for CBR2 to 17 km^2 for LTNP and were smaller than that of the hunted site in NB with 38 km^2. Similarly, the 100% MCP area estimates for non-hunted sites varied between 47 km^2 (CBR1) and 138 km^2 (LTNP), which were substantially smaller than the MCP area estimates at the hunted site with 154 km^2 (Table 3.1). When we compared the two sites where movement data exceeded 10 months, we observed that in the NB area, where the level of hunting pressure was higher, the 95% AKDE, 95% KDE, and 100% MCP home-range estimates were 156 km^2 (100%), 41 km^2 (41%), and 16 km^2 (12%) larger, respectively, than the non-hunted site in LTNP. In addition, when we compared the 95% KDE home-

Table 3.2 White-lipped peccary home-range estimates for four groups in the Maya Forest of Guatemala and Mexico in dry and rainy seasons using 95% kernel density estimation (KDE)

	95% KDE {50% core area} [km^2]			
Group	Dry season 2015 {50% core area}	Rainy season 2015 {50% core area}	Dry season 2016 {50% core area}	Rainy season 2016 {50% core area}
LTNP	36 {8}	106 {42}	4 {1}	14 {3}
CBR1	57 {11}	ND	ND	ND
CBR2	ND	ND	40 {7}	ND
NB	ND	103 {15}	132 {29}	ND

ND No data

range size between the LTNP and NB groups only for the rainy season of 2015 and between the LTNP, CBR2, and NB groups only for the dry season of 2016, the hunted site still presents a larger home-range size compared with the two non-hunted sites (Table 3.2).

There was no correlation between the home-range size and the number of fixes obtained per group for 95% AKDE ($rho = -0.8$, $S = 18$, $P = 0.33$), 95% KDE ($rho = -0.8$, $S = 18$, $P = 0.33$), and 100% MCP ($rho = -0.4$, $S = 14$, $P = 0.75$). There were no significant differences in home-range size between the AKDE and KDE estimators (50%: Mann-Whitney-Wilcoxon Test $W = 11.5$, $P = 0.38$; 95%: Mann-Whitney-Wilcoxon Test $W = 3$, $P = 0.2$). For the LTNP and NB groups, we obtained an asymptote in the 100% MCP size for 7 and 5 months, respectively, because we were able to obtain information on the movements of the two groups for more than 12 months. The LTNP group showed an annual pattern of seasonal movements by visiting the capture pond at the beginning of February 2016, 1 year after the first white-lipped peccary was captured. Later they moved 18 km toward the northwest in August. The group in LTNP, a non-hunted site, increased their home-range movements substantially during the rainy season (Table 3.2). We estimated a home range of 36 km^2 and 106 km^2 for the 2015 dry season and 2015 rainy season, respectively, for the LTNP group using 95% KDE. This same pattern was observed in 2016 in the same place where we estimated a home range of 4 km^2 and 14 km^2 for the dry season and rainy season, respectively. In contrast, the NB group (hunted site) presented a larger home range in the dry season (132 km^2) compared to the rainy season (103 km^2). After capture, the NB group made a displacement of 12 km southeast of the *ejido*, remaining in that area from May to August. In September, the group traveled further south, staying just 3 km away from federal highway 186, which may have served as a barrier. At the end of March, the group returned to the central area of their home range, remaining there until June, when contact was lost with the group. However, we did not find significant differences in home-range sizes between the two seasons for non-hunted sites (Mann-Whitney-Wilcoxon Test; $W = 3$, $P = 0.8$). At non-hunted sites, we recorded white-lipped peccary group sizes as 46 individuals in LTNP and 30 individuals in CBR, both slightly larger than the estimated group size of 25 individuals at the hunted site NB.

3.4 Discussion

Home-range size of white-lipped peccary groups in non-hunted sites (CBR and LTNP) was smaller than in the group of the hunted site (NB), contrary to our hypothesis. This movement pattern suggests that the hunting has some influence on the behavior of this group in NB. A similar pattern was found in Mato Grosso do Sul, Brazil, where the impact of forest deforestation and fragmentation on mammals was compared by analyzing home ranges of two white-lipped peccary groups between a pristine and a disturbed area. In the disturbed habitat, the home-range area was 51% larger than that of the pristine area because animals resorted to changing their routes and using larger areas to find enough resources (Keuroghlian et al. 2015).

In the Maya Forest of México, previous studies showed that the status of white-lipped peccary populations outside the protected areas is of special concern due to several factors such as overhunting (Reyna-Hurtado and Tanner 2007; Reyna-Hurtado 2009; Reyna-Hurtado et al. 2010). Estimates of white-lipped peccary group sizes in protected areas were bigger than in areas heavily used by humans, where there is hunting pressure by subsistence and sport hunters (Briceño-Méndez et al. 2016). The hunting season occurs mainly in the dry season when this species is highly vulnerable because groups congregate in the remaining water ponds that did not dry up. Subsistence hunting harvest rates are probably larger than is sustainable for this population (Reyna-Hurtado 2009; Briceño-Méndez et al. 2016).

In order to compare our results with other studies, we relied on the KDE values, as this method was better adjusted to the white-lipped peccary biology in our study area and showed a more precise contour of the areas where the groups move. Our home-range estimates in non-hunted sites (95% KDE: 40–99 km^2) were similar to those previously determined at CBR, between 39 and 98 km^2 using 95% fixed kernel and 23–122 km^2 using 100% MCP (Reyna-Hurtado et al. 2009), and at a relatively pristine area of the southern Pantanal, Mato Grosso do Sul, Brazil, where home ranges were estimated between 46 and 55 km^2 using 95% kernel method and the 95% harmonic mean (Keuroghlian et al. 2015). Home ranges observed in our study were larger than estimates from most other studies. At the Bladen Nature Reserve, Belize, home range was estimated at 55 km^2 for a single group using 95% AKDE (Hofman et al. 2016). In a fragmented area of the Atlantic Forest of Brazil, average home range was estimated to be 18.7 km^2 for three to four groups using a 90% harmonic mean estimator (Keuroghlian et al. 2004). In Corcovado National Park, Costa Rica, annual MCP home ranges were reported between 32 and 38 km^2 for one group (Carrillo et al. 2002). In Maracá Island, Roraima, Brazil, white-lipped peccaries' (WLP) home range was 21.8 km^2 (Fragoso 1998). In Panama in the Darien National Park, Meyer et al. (2018; this book) estimated between 42 and 76 km^2: studies that reported similar or larger home ranges compared to our study have also had larger group sizes, ranging from 56 to 130 individuals. Examples include the Cerrado ecosystem in Goias State, Brazil (277, 147, and 91 km^2 for groups of 65, 70, and 110 individuals, respectively—Jacomo et al. 2013); the southern Pantanal, Mato Grosso do Sul, Brazil, where the level of disturbance due to deforestation was higher (80 km^2 using 95% MCP for a group of 56 individuals—Keuroghlian et al. 2015); and Maracá Island, Roraima, Brazil with a home range of 110 and 200 km^2 using 100% MCP for groups of 130 and >200 individuals, respectively (Fragoso 1998, 2004).

Our results suggest that in the Maya Forest, groups are smaller in size (25–46 individuals) and move in larger areas compared to other sites in the Neotropics. This pattern is similar to that reported previously by Reyna-Hurtado et al. (2009). In hunted areas, group sizes tend to decrease considerably, while the group occupies larger areas in comparison to most preserved, non-hunted regions, possibly because of the need to evade hunters, especially during the dry season. Water ponds in the Maya Forest are seasonal and are likely the dominant environmental resources contributing to nomadic movements of white-lipped peccary in this area (Reyna-Hurtado et al. 2009). Two groups of white-lipped peccaries in Roraima, Brazil (Fragoso 1998), and four groups in CBR, México (Reyna-Hurtado et al. 2009), visited the same pond on a regular basis over one or two seasons, respectively. This pattern was also observed for the group in the protected site (LTNP). In the 2015 and 2016 dry seasons, the LTNP group visited the same pond where we captured and collared the animals, as this is a year-round pond, providing surface water in this critical season of the year.

The white-lipped peccary group in the LTNP increased their home range in the rainy season, suggesting that water availability is the most important factor that determines the movements of this species and consistent with our hypothesis. This pattern is similar to that reported previously in CBR, where four groups increased their home ranges at the beginning of the rainy season (Reyna-Hurtado et al. 2009), suggesting that water availability in non-hunted sites in the Maya Forest is the key factor that determines the movement of this species (Reyna-Hurtado et al. 2012). In contrast, in NB, where hunting of ungulates is common, especially in the dry season (Reyna-Hurtado et al. 2010; Briceño-Méndez et al. 2016), the white-lipped peccary group modified their movement patterns during this time, resulting in slightly larger home ranges in the dry season compared to the rainy season (Table 3.2). This change in the movement pattern of the NB group in the dry season suggests that hunting pressure (Reyna-Hurtado et al. 2012) may influence the movement of this social species. In other studies, food resources appear to be the main driver for the movements of white-lipped peccaries. In Corcovado National Park, the home ranges of white-lipped peccaries were smaller from June to September, which encompassed the majority of the rainy season and produced an increased abundance of fruits (Carrillo et al. 2002). Likewise, in the Atlantic Forest, Brazil, it was found that the seasonal movements of white-lipped peccaries were apparently driven by the supply of key fruits, rare habitats, and riparian zones (Keuroghlian et al. 2004, 2009b; Keuroghlian and Eaton 2008). In the Cerrado ecosystem in Goias State, Brazil, it seems that a combination of low food resources and reduction in the availability of surface water forced white-lipped peccary groups to occupy larger areas when foraging during the rainy season and to remain closer to the few water sources in the dry season (Jacomo et al. 2013).

The effect of hunting is demonstrated in a large-scale study across white-lipped peccary geographic distribution: hunting pressure, which was highly correlated with proximity to the nearest settlement, has a detrimental effect on group size and density for larger groups living in areas farther from human settlements (Reyna-Hurtado et al. 2016). In addition, it is important to address the potential risk of disease transmission by domestic animals to white-lipped peccaries, since they form large and cohesive groups which may speed up an epidemic outbreak with a reinfection cycle (Fragoso 2004). In the Maya Forest, white-lipped peccary skin problems have been reported through photographic records obtained with camera traps (Reyna-Hurtado et al. 2014). Assessment on the health and diseases in populations of white-lipped peccaries and domestic animals in close proximity to wildlife areas is an important subject to be addressed in the Maya Forest and Mesoamerica.

Our results indicate that although drier than NB, which has bigger water ponds that allowed people to colonize this area, CBR still maintains a greater population size than that of NB. While similar in water availability, the estimate for LTNP annual home range was smaller than and the group size was larger than the respective estimates for NB, perhaps due to the lack of hunting in LTNP. These results suggest that hunting in areas with human settlements puts greater pressure on the ecology of white-lipped peccary, resulting in smaller group sizes and use of larger home range compared with non-hunted sites.

Although we only monitored one group in NB and obtained relatively few GPS locations for several reasons, such as percent canopy cover, the results showed that the group has a large annual home range, increasing in the dry season when hunting is more frequent. In contrast, in non-hunted areas, the home ranges are smaller, and during the dry season, the groups remained close to water ponds, expanding their home range when the rainy season began. We recognize that our results could be biased by the small number of groups we monitored and that this sample size may not be representative of the population of white-lipped peccaries of our study area. Therefore, we consider that it is important to carry out further studies in areas with hunting pressure to estimate the movements of more white-lipped peccary groups to obtain higher fixes and compare them with our results and to promote conservation actions in protected and communities' areas. Our data suggest that white-lipped peccary groups required large areas to meet their spatial requirements in the Maya Forest. This has important implications for the management and conservation plans for this species at the landscape scale primarily because habitat destruction and fragmentation would have severe effects on white-lipped peccary population. Although the white-lipped peccary can persist in highly altered landscapes in South America (Jacomo et al. 2013), it prefers to use forested areas rather than sites close to human settlements and secondary forests associated with high human activities (Keuroghlian et al. 2015; Reyna-Hurtado et al. 2016).

White-lipped peccaries are experiencing population declines throughout their range (Altrichter et al. 2012; Reyna-Hurtado et al. 2017), specifically in Guatemala and México; its current distribution makes up only 10% and 16%, respectively, of its historical distribution (Altrichter et al. 2012; Moreira-Ramírez 2017). Due to the eminent threats to white-lipped peccaries and their habitat, we propose that efforts should be taken to improve monitoring and surveillance in protected and communities' areas to decrease excessive hunting. Furthermore, joint conservation and monitoring strategies should be developed with government institutions and civil society in the three countries that are part of the Maya Forest. The Maya Forest in Guatemala, México, and Belize embraces the largest contiguous area of tropical forest habitat available for white-lipped peccary in Mesoamerica (Sanderson et al. 2002; Altrichter et al. 2012). Effective conservation at tri-national level of protected areas and integrating local communities will allow for the maintenance of viable populations of white-lipped peccaries in the future. Consistent communication with communities adjacent to protected areas is essential. Dialogue with community members about the current status of the peccaries, managing subsistence hunting, and declaring areas with strict protection for wildlife within their communal lands are initiatives that will contribute to the conservation of the species. Low-impact ecotourism should be promoted in water ponds of national parks and communal areas to observe large mammals, such as peccaries, to generate additional economic income. Our data suggest that the following conservation steps should be taken: protection against hunting, restrictions on road construction, reduction of large-scale agriculture and forest conversion for grazing, and landscape conservation of large, continuous, and ecologically diverse areas containing a mosaic of habitat types that guarantee the survival of this species. Finally, this study shed light in differences in home-range size and movement patterns of a species that is under hunting pressure. Evaluating changes in movement are a research priority for this highly mobile species as these changes can indicate the impact of human activities in wildlife.

Acknowledgments To the National Council of Science and Technology of Mexico for the grant offered to the first author to carry out his doctoral dissertation. We would like to give thanks to the American Society of Mammalogists, the Wildlife Conservation Society Research Fellowship Program, El Colegio de la Frontera Sur, Unidad Campeche, the Council of Science and Technology of Mexico (CONACYT) for the support of the Mexican research part through the project number 182386 to RR-H, to McGill University for providing cameras, and to the Rufford Foundation and Idea Wild for financial support. We would also like to thank the National Council of Protected Areas of Guatemala, the National Commission of Protected Natural Areas of Mexico, the *ejidatarios* of Nuevo Becal, the National Institute of Anthropology and History and Las Guacamayas Biological Station for the support, permits, and facilities provided. Lastly, we would like to thank N. Arias, G. Castillo, A. Hettena, W. Martínez, K. Sánchez, K. Tut, P. Pérez, C. Umaña, Y. Polanco, A. Xol, R. Chatá, and anonymous reviewers for their help. MCR is funded by FAPESP (process 2013/50421-2) and receives the research grant from CNPq (process 312045/2013-1).

References

Altrichter M et al (2012) Range-wide declines of a key Neotropical ecosystem architect, the Near Threatened white-lipped peccary *Tayassu pecari*. Oryx 46:87–98

Alvard MS, Robinson JG, Redford KH, Kaplan H (1997) The sustainability of subsistence hunting in the Neotropics. Conserv Biol 11:977–982

Andrade Melo ÉR, Gadelha JR, da Silva M d ND, da Silva AP, Mendes AR (2015) Diversity, abundance and the impact of hunting on large mammals in two contrasting forest sites in northern amazon. Wildl Biol 21:234–245

Beck H (2006) A review of peccary–palm interactions and their ecological ramifications across the Neotropics. J Mammal 87:519–530

Beyer HL (2012) Geospatial modelling environment (version 0.7.3.0). www.spatialecology.com/gme

Bodmer RE, Eisenberg JF, Redford KH (1997) Hunting and the likelihood of extinction of Amazonian mammals. Conserv Biol 11:460–466

Briceño-Méndez M, Naranjo EJ, Mandujano S, Altricher M, Reyna-Hurtado R (2016) Responses of two sympatric species of peccaries (Tayassu pecari and *Pecari tajacu*) to hunting in Calakmul, Mexico. Trop Conserv Sci 9:1–11

Calabrese JM, Fleming CH, Gurarie E (2016) ctmm: an R package for analyzing animal relocation data as a continuous-time stochastic process. Methods Ecol Evol 7:1124–1132

Carrillo E, Saenz JC, Fuller TK (2002) Movements and activities of white-lipped peccaries in Corcovado. Biol Conserv 108:317–324

Consejo Nacional de Áreas Protegidas, Wildlife Conservation Society (2015) Monitoreo de la gobernabilidad en la Reserva de Biosfera Maya. Actualización a 2014 de la versión de septiembre de 2013. San Benito

Dunn JE, Gipson PS (1977) Analysis of radio telemetry data in studies of home range. Biometrics 33:85–101

Endo W et al (2010) Game vertebrate densities in hunted and nonhunted forest sites in Manu National Park. Biotropica 42:251–261

Escamilla A, Sanvicente M, Sosa M, Galindo-Leal C (2000) Habitat mosaic, wildlife availability, and hunting in the tropical forest of Calakmul, Mexico. Conserv Biol 14:1592–1601

ESRI (2011) ArcView GIS. Ver. 10.1. Environmental System Research Institute, Redlands

Fleming CH, Calabrese JM, Mueller T, Olson KA, Leimgruber P, Fagan WF (2014) From fine-scale foraging to home ranges: a semivariance approach to identifying movement modes across spatiotemporal scales. Am Nat 183:E154–E167

Fleming CH, Fagan WF, Mueller T, Olson KA, Leimgruber P, Calabrese JM (2015) A new autocorrelated kernel density estimator reports. Ecology 96:1182–1188

Fragoso JMV (1998) Home range and movement patterns of white-lipped peccary (*Tayassu pecari*) herds in the Northern Brazilian Amazon. Biotropica 30:458–469

Fragoso JMV (2004) A long-term study of white-lipped peccary (*Tayassu pecari*) population fluctuation in northern Amazonia. In: Silvius K, Bodmer RE, Fragoso JMV (eds) People in nature, wildlife conservation in South and Central America. Columbia University Press, New York, pp 286–296

Fritz SA, Bininda-Emonds ORP, Purvis A (2009) Geographical variation in predictors of mammalian extinction risk: big is bad, but only in the tropics. Ecol Lett 12:538–549

Garcia-Gil G (2003) Colonización humana reciente y formación del paisaje agrario en la Reserva de la Biosfera de Calakmul, Campeche, México. Ph.D. Dissertation, Universidad Nacional Autónoma de México. México, D. F

Harris S, Cresswell WJ, Forde PG, Trewhella WJ, Woollard T, Wray S (1990) Home range analysis using radio tracking data: a review of problems and techniques particularly as applied to the study of mammals. Mammal Rev 20:97–123

Hofman MPG, Signer J, Hayward MW, Balkenhol N (2016) Spatial ecology of a herd of white-lipped peccaries (*Tayassu pecari*) in Belize using GPS telemetry: challenges and preliminary results. Therya 7:21–37

Jacomo AT et al (2013) White-lipped peccary home-range size in a protected area and farmland in the central Brazilian grasslands. J Mammal 94:137–145

Kernohan BJ, Gitzen RA, Millspaugh JJ (2001) Analysis of animal space use and movement. In: Millspaugh JJ, Marzluff JM (eds) Radio tracking and animal populations. Academic, San Diego, pp 125–166

Keuroghlian A, Eaton DP (2008) Importance of rare habitats and riparian zones in a tropical forest fragment: preferential use by *Tayassu pecari*, a wide-ranging frugivore. J Zool 275:283–293

Keuroghlian A, Eaton DP, Longland WS (2004) Area use by white-lipped and collared peccaries (Tayassu pecari and *Tayassu tajacu*) in a tropical forest fragment. Biol Conserv 120:411–425

Keuroghlian A, Eaton DP, Desbiez ALJ (2009a) The response of a landscape species, white-lipped peccaries, to seasonal resource fluctuations in a tropical wetland, the Brazilian pantanal. Int J Biodiver Conserv 1:87–97
Keuroghlian A, Eaton DP, Desbiez AL (2009b) The response of a landscape species, white-lipped peccaries, to seasonal resource fluctuations in a tropical wetland, the Brazilian pantanal. Biodivers Conserv 1:87–97
Keuroghlian A et al (2013) *Tayassu pecari*. In: IUCN 2017. The IUCN Red List of Threatened Species. Version 2017.1. www.iucnredlist.org. Acceded 1 Aug 2017
Keuroghlian A, Andrade Santos MDC, Eaton DP (2015) The effects of deforestation on white-lipped peccary (*Tayassu pecari*) home range in the southern Pantanal. Mammalia 79:491–497
Martínez E, Galindo-Leal C (2002) La vegetación de Calakmul, Campeche, México: clasificación, descripción y distribución. Bol Soc Bot Méx 71:7–32
Meyer N, Moreno R, Martínez-Morales MA, Reyna-Hurtado R (2018). Spatial ecology of a large and endangered tropical mammal: the white-lipped peccary in Darién, Panama. In: Reyna-Hurtado R, Chapman C (eds) Movement Ecology of Neotropical Forest Mammals. Springer Nature
Millspaugh JJ, Marzluff JM (2001) Radio tracking and animal populations. Academic, San Diego
Moreira-Ramírez JF (2017) Movimientos del pecarí de labios blancos en relación con la disponibilidad de agua y cacería en la Selva Maya de Guatemala y México. Ph.D. Dissertation, El Colegio de la Frontera Sur. Campeche, México
Moreira-Ramírez JF, Lopez JE, García-Anleu R, Córdova F, Dubón T (2015) Tamaño, composición y patrones diarios de actividad de grupos de pecarí de labios blancos (*Tayassu pecari*) en el Parque Nacional Mirador-Río Azul, Guatemala. Therya 6:469–482
Moreira-Ramírez JF et al (2016) Importance of waterholes for white-lipped peccary (*Tayassu pecari*) in the Selva Maya, Guatemala. Therya 7:51–64
Moßbrucker MA, Fleming CH, Ali Imron M, Satyawan P, Sumardi (2016) AKDE C home range size and habitat selection of Sumatran elephants. Wildl Res 43:566–575
Naranjo EJ, Bodmer RE (2007) Source-sink systems and conservation of hunted ungulates in the Lacandon Forest, Mexico. Biol Conserv 138:412–420
Peres CA (1996) Population status of white-lipped Tayassu pecari and collared peccaries *T. tajacu* in hunted and unhunted Amazonian forests. Biol Conserv 77:115–123
Peres CA (2000) Effects of subsistence hunting on vertebrate community structure in Amazonian forests. Conserv Biol 14:240–253
Peres CA (2001) Synergistic effects on Amazonian forest vertebrates fragmentation. Conserv Biol 15:1490–1505
Peres CA, Palacios E (2007) Basin wide effects of game harvest on vertebrate population densities in Amazonian forests: implications for animal mediated seed dispersal. Biotropica 39:304–315
Peres CA, Barlow J, Haugaasen T (2003) Vertebrate responses to surface wildfires in a central Amazonian forest. Oryx 37:97–109
Pulliam DW (1988) Sources, sinks, and population regulation. Am Soc Nat 132:652–661
R Development Core Team (2012) R: a language and environment for statistical computing. R Foundation for Statistical Computing, Vienna www.R-project.org/:2009
R Development Core Team (2017) R: a language and environment for statistical computing. R Foundation for Statistical Computing, Vienna, Austria. www.R-project.org/, 2009
Reyna-Hurtado R (2009) Conservation status of the white-lipped peccary (*Tayassu pecari*) outside the Calakmul Biosphere Reserve in Campeche, Mexico: a synthesis. Trop Conserv Sci 2:159–172
Reyna-Hurtado R, Tanner GW (2007) Ungulate relative abundance in hunted and non-hunted sites in Calakmul Forest (Southern Mexico). Biodivers Conserv 16:743–756
Reyna-Hurtado R, Rojas-Flores E, Tanner GW (2009) Home range and habitat preferences of white-lipped peccaries (*Tayassu pecari*) in Calakmul, Campeche, Mexico. J Mammal 90:1199–1209
Reyna-Hurtado R, Naranjo E, Chapman C a, Tanner GW (2010) Hunting and the conservation of a social ungulate: the white-lipped peccary *Tayassu pecari* in Calakmul, Mexico. Oryx 44:89

Reyna-Hurtado R, Chapman CA, Calme S, Pedersen EJ (2012) Searching in heterogeneous and limiting environments: foraging strategies of white-lipped peccaries (*Tayassu pecari*). J Mammal 93:124–133

Reyna-Hurtado R et al (2014) White-lipped peccaries with skin problems in the Maya Forest. Suiform Soundings 13:29–31

Reyna-Hurtado R et al (2016) What ecological and anthropogenic factors affect group size in white-lipped peccaries (*Tayassu pecari*)? Biotropica 48:246–254

Reyna-Hurtado R et al (2017) White-lipped peccary in Mesoamerica: status, threats and conservation actions. Suiform Soundings 15:31–35

Sanderson EW, Jaiteh M, Levy M a, Redford KH, Wannebo AV, Woolmer G (2002) The human footprint and the last of the wild. Bioscience 52:891–904

Santos-fita D, Naranjo EJ, Rangel-salazar JL (2012) Wildlife uses and hunting patterns in rural communities of the Yucatan Peninsula, Mexico. J Ethnobiol Ethnomed 8:38

Seaman E, Powell RA (1996) An evaluation of the accuracy of kernel density. Ecology 77:2075–2085

Seaman DE, Millspaugh JJ, Kernohan BJ, Brundige GC, Raedeke KJ, Gitzen RA (1999) Effects of sample size on kernel home range estimates. J Wildl Manag 63:739–747

Sikes RS, Gannon WL, Care A, Committee U, Journal S (2011) Guidelines of the American Society of Mammalogists for the use of wild mammals in research guidelines of the American Society of Mammalogists for the use of wild mammals in research. J Mammal 92:235–253

Wikelski M, Kays R (2017) Movebank: archive, analysis and sharing of animal movement data. www.movebank.org

Chapter 4
White-Lipped Peccary Movement and Range in Agricultural Lands of Central Brazil

Maria Luisa S. P. Jorge, Alexine Keuroghlian, Jennifer Bradham, Júlia Emi F. Oshima, and Milton Cezar Ribeiro

4.1 Introduction

White-lipped peccaries (*Tayassu pecari*, Tayassuidae, Cetartiodactyla, hereafter "WLPs"), large herd-forming ungulates, can live in groups with more than 100 individuals (Sowls 1997) and often comprise the largest terrestrial mammalian biomass in Neotropical forests (Kiltie and Terborgh 1983). They are mainly frugivorous, with fruits and seeds encompassing 60–80% of their diet (Desbiez et al. 2009; Keuroghlian and Eaton 2008a, 2009), and have a preference for large-seeded species (Beck 2006). As ecosystem engineers, WLPs have strong and lasting effects on forest dynamics and structure at the local level, through selective seed and seedling predation, trampling, and soil plowing (Altrichter et al. 2001; Keuroghlian and Eaton 2008a, b, 2009; Beck et al. 2010). As a result of seasonal travel through large areas (2000–20000 ha – Fragoso 1998; Carrillo et al. 2002; Keuroghlian et al. 2004, 2015; Reyna-Hurtado et al. 2009; Jacomo et al. 2013), WLPs act as mobile links of ecological processes (such as movement of seeds and nutrients) and generate unique spatiotemporal patterns of landscape modification and disturbance.

M. L. S. P. Jorge (✉)
Department of Earth and Environmental Sciences, Vanderbilt University, Nashville, TN, USA

Department of Biological Sciences, Vanderbilt University, Nashville, TN, USA
e-mail: malu.jorge@vanderbilt.edu

A. Keuroghlian
Peccary Project/IUCN/SSC Peccary Specialist Group, Campo Grande, Brazil

J. Bradham
Department of Earth and Environmental Sciences, Vanderbilt University, Nashville, TN, USA

J. E. F. Oshima · M. C. Ribeiro
Programa de Pós Graduação em Zoologia, Instituto de Biociências, Departamento de Ecologia, Laboratório de Ecologia Espacial e Conservação (LEEC), Universidade Estadual Paulista (UNESP), Rio Claro, Brazil

R. Reyna-Hurtado, C. A. Chapman (eds.), *Movement Ecology of Neotropical Forest Mammals*, https://doi.org/10.1007/978-3-030-03463-4_4

Tropical ecosystems boast some of the highest levels of biodiversity in the world, yet they are increasingly threatened by agriculture intensification (Achard et al. 2002; Butchart et al. 2010; Hansen et al. 2013; Austin et al. 2017). In the Cerrado biome of Central Brazil, cattle ranching drives deforestation and has resulted in extensive replacement of native flora with exotic grasses (Klink and Machado 2005). Mechanized agriculture is also widespread in the Cerrado highlands, where only minor flooding occurs and soils are able to support extensive monocultures of soy, corn, cotton, and sugarcane (Leite et al. 2012). Since the 1960s, changes in land use and natural vegetation cover associated with the expansion of cattle ranching and agriculture have removed approximately half of the Cerrado's native vegetation (Françoso et al. 2015). Long-term conservation of a wide-ranging species, such as the WLPs, in a biome with increasing fragmentation and deforestation is dependent on understanding how the species, and consequently their ecosystem functions, respond to landscape changes within agroecosystem-dominated landscapes.

Studies have shown that WLPs are highly vulnerable to native habitat removal, fragmentation, and illegal hunting, which make them uniquely suited to guide conservation efforts throughout their biogeographic range (Cullen et al. 2001; Peres 2000; Jorge et al. 2013; Keuroghlian et al. 2017), and that removing WLPs from an ecosystem causes a cascade of negative impacts including reduction of jaguar populations (Paviolo et al. 2008), changes in fruit resources and vegetation structure (Painter 1998; Silman et al. 2003; Keuroghlian and Eaton 2009; Galetti et al. 2015b), increase in pest rodents and associated diseases (Galetti et al. 2015a), and an overall impoverishment of forest plant and wildlife communities (Terborgh et al. 2008; Beck et al. 2010, 2013; Azevedo and Conforti 2008; Galetti et al. 2015a, b).

Fine-scale information on WLP's movement and use of space as well as how they are affected by changes in landscape structure will help with management efforts of WLP populations in agricultural landscapes. Although some information on movement patterns of WLPs is available from VHF tracking studies (Fragoso 1998; Carrillo et al. 2002; Keuroghlian et al. 2004, 2015; Reyna-Hurtado et al. 2012; Jacomo et al. 2013), more systematic, continuous, and fine-scale movement data is still lacking. Here, we analyze WLP movement of seven GPS-tracked herds in two agricultural regions of Central Brazil. We expect WLP movement and range to be negatively affected by the removal of native forests resulting from the expansion of agricultural lands because WLPs are a wide-ranging species and depend on native fruiting trees for food. We hypothesize that herds will be able to continue moving through the landscape to find their food yet will move more and occupy larger ranges in areas with less native forest in order to support essential resource needs. Complementarily, we investigate the relationship of movement and range with seasons and the monthly variation of fruit diversity (as proxies of temporal variation of environmental conditions and availability of food resources), as we expect that seasonality and, more specifically, temporal variation of food availability may exacerbate the negative effects of forest removal.

4.2 Methods

4.2.1 Study Areas

The study region is located in the upper Paraguay River basin of Central-Western Brazil, between 23°22 and 23°23 of latitude south and 45°33 and 45°32 of longitude west. The region encompasses two municipalities that are part of the Cerrado highlands and borders the Pantanal floodplain to the east and south (Fig. 4.1). Savanna forests were the main original vegetation of the region, with headwater streams that drain into the Pantanal basin. The 3000 km^2 eastern highlands (hereafter called *Corguinho*) occur in the Maracaju mountain ridge and encompass the headwater stream basins of the Negro, Taboco, and Aquidauana rivers, which supply water to the southern Pantanal floodplain (Fig. 4.1).

Native vegetation in these regions consists of several types of seasonal savanna and semi-deciduous forest formations with high levels of plant species diversity and endemism, including unique habitats like *Mauritia flexuosa* swamps (locally called *buritizais*) that are characteristic of headwater springs and small streams. Rapid agricultural expansion characterized by extensive conversion of natural vegetation to planted pasture (exotic grass monocultures) has relegated the original vegetation

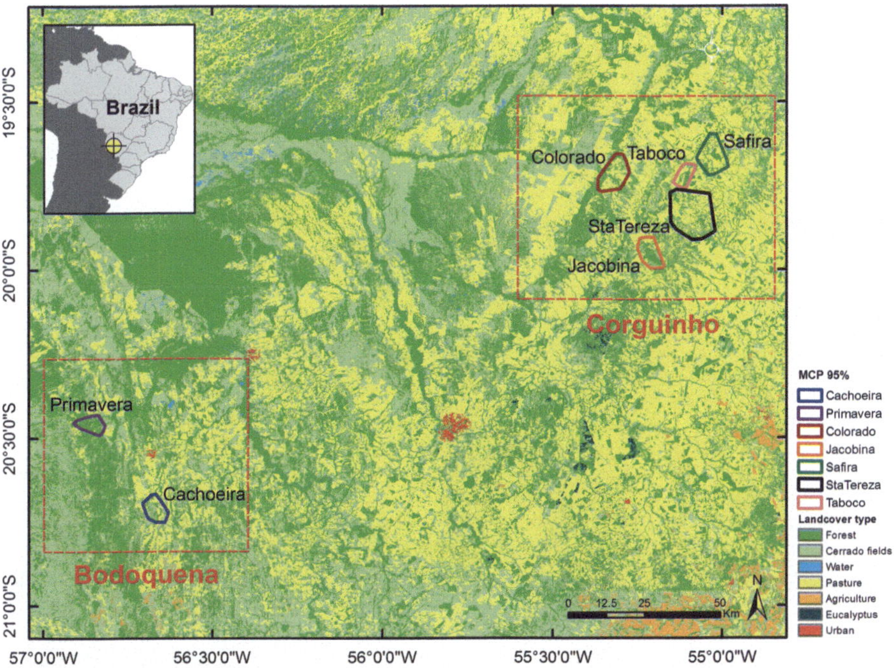

Fig. 4.1 Map of the study area showing the Corguinho and Bodoquena highlands (red dashed boxes) in the Cerrado biome of Central Brazil. Polygons outline the 95% MCP overall home range of each peccary herd included in this study

cover to isolated forest fragments (Santana 2015). The 10,000 km^2 southern highlands (hereafter called *Bodoquena*) occur in the Bodoquena mountain ridge, encompass the headwater streams of the Miranda river basin, and border the southern edge of the Pantanal floodplain. Bodoquena vegetation includes both deciduous and semi-deciduous seasonal savanna forests, with plant composition influenced by the neighboring Atlantic Forest biome (Pott and Pott 2003). Similar to other Cerrado regions, forests of *Bodoquena* that are not federally protected are fragmented and under threat from the expansion of monocultures, such as soybean and corn plantations, as well as mining (Silva 2008).

The climate of both the *Bodoquena* and *Corguinho* regions is Aw, or tropical subwarm according to Köppen classification (Köppen 1884), with medium annual temperatures between 22 and 26 °C and maximum temperatures around 35–40 °C. The prevailing climate is humid to subhumid, with annual rainfall ranging from 1500 to 1750 mm, a wet season occurring between October and April and a dry season ranging from May to September (Fig. 4.2 – SEMAC 2011).

Fig. 4.2 Jay, the first tracked white-lipped peccary (*Tayassu pecari*, Tayassuidae, Cetartiodactyla) of our study

4.2.2 *Fruit Diversity*

To examine whether there was monthly variation in food resources, we performed fruit censuses for both *Corguinho* and *Bodoquena*. In each area, we chose three distinct sites, at least 10 km away from each other, and used one to three 500 m trails. Every 15–30 days, we randomly chose a 150 m section to sample in each trail. Slowly walking those parts of the trail, we identified and counted all ripe fruits on the forest floor that fell within 1 m of the trail center. We then compiled the number of fruiting tree species by month and area by incorporating the information from all trails. Although not part of our main hypothesis, we tested whether fruit diversity was statistically related to rainfall to complement the underlying idea that temporal variation in food availability is associated with environmental conditions and, more specifically, rainfall. For both regions (*Corguinho* and *Bodoquena*), there was a significant positive relationship between monthly rainfall and number of fruit species (fruit diversity for *Corguinho*: adjusted-$R^2 = 0.33$, $F = 6.36$, df = 1, 10, $p = 0.03$; *Bodoquena*: adjusted-$R^2 = 0.34$, $F = 6.79$, df = 1, 10, $p = 0.03$).

4.2.3 *Animal Capture and Handling*

All captures followed procedures outlined in Keuroghlian et al. (2004, 2015), and capture licenses were provided by the Brazilian National Environmental Agency (ICMBio – Instituto Chico Mendes de Conservação da Biodiversidade – SISBIO license n. 31088 and n. 46131). For captures, temporary baiting stations were set up in areas used by WLP, as indicated by footprints and trail marks. Box traps and wire mesh panel traps were placed at these stations and baited with salt, corn, manioc (locally called "mandioca" – *Manihot* sp.), and local fruits that are known to be consumed by WLPs. When captured, individuals were sedated in the trap with a TeleDart RD206 injection pistol and a dart 3 cc syringe with zolazepam-tiletamine (Zoletil, 0.9 ml/10 kg). We took morphometric measures (e.g., weight and length) and collected blood and hair samples for all captured individuals. PIT tags (Biomark®) were inserted subcutaneously for further individual identification in the event of a recapture. All measuring, sampling and tagging procedures followed guidelines described by Keuroghlian and Desbiez (2010). GPS collars were placed on one to two adults per herd, one male and one female, whenever possible. After sedation and data collection, animals were kept in the trap for at least 6 h before being released to ensure full recovery from the anesthesia. Through VHF and GPS monitoring, we confirmed that all the individuals rejoined their herds shortly after release.

4.2.4 Movement Monitoring

A total of 12 individual WLPs from 7 herds were GPS-tracked between July 2013 and March 2018 in the *Bodoquena* and *Corguinho* regions, accumulating 12,568 GPS animal locations (Table 4.1, Fig. 4.2). Individuals were monitored with GPS/VHF collars weighing approximately 650 g through satellite transmission (Iridium) from two GPS collar brands (Followit and Telonics). Collars were programmed to receive 4, 6, 8, or 12 locations/individual/day for 12–24 months (365–730 days/individual). Satellite data transmission of relocation data varied from daily (Followit) to once a week (Telonics). For Followit collars, a drop-off mechanism was preprogrammed to automatically open the collar buckle after 1 year of monitoring or was remotely triggered through an UHF antenna. However, no collars were automatically dropped and later retrieved because all batteries expired prior to 1 year. For Telonics collars, the drop-off mechanism was programmed to open the buckle after 2 years. For one animal, the collar stopped working after 13 months, and for another, the collar is still collecting data since its initiation in March 2017.

4.2.5 Mapping and Percent Forest Cover Calculation

Land cover mapping was based on a 30 m resolution supervised classification of Landsat images for Mato Grosso do Sul State, Brazil (Reynolds et al. 2016), created through ArcGIS 10.2.1 (ESRI®). The map had a scale of 1:50,000 and included the following classes: (a) native forests, (b) wetlands, (c) native open vegetation, (d) exotic open vegetation, (e) forestry (exotic plant forested vegetation), (f) urban, and (g) water. Specifically in our study areas, the vegetation classes (b), (c), (e), (f), and (g) were negligible or nonexistent. Therefore, maps were reclassified into two categories: (a) native forests or (b) exotic open vegetation (pasture and crops), using the Reclassify function of the Spatial Analyst Toolbox of ArcGIS 10.2 (ESRI®). We then defined the boundaries of each herd's overall range (using locations from the entire monitoring period for each herd) by calculating the home range with the minimum convex polygon method (MCP – Mohr 1947; Hayne 1949) at the 95% limit.

To define herds, we combined data from individuals that overlapped locations. Individuals that did not overlap locations were considered to be from different herds, except when previous VHF tracking data indicated otherwise (A. Keuroghlian, unpublished data). MCP calculation and mapping were performed with adehabitatHR package for R (Calenge 2006). We then extracted the vegetation classification from the vegetation map within each herd's MCP 95% home range boundaries using the Extract function of the Spatial Analyst Toolbox of ArcGIS 10.2 (ESRI®). Finally, for each home range map, we computed the number of pixels that were classified as forest and calculated the percent forest cover by dividing it by the total number of pixels within each herd's overall MCP 95% home range.

Table 4.1 Summary data of the individuals GPS-tracked in this study, from 2013 to 2018, at the *Corguinho* and *Bodoquena* municipalities, Mato Grosso do Sul, Brazil

Name	Gender	Region	Location	Start date	End date	*N*. days	*N*. months	*N*. locations	Interval between consecutive locations (hours)	~Locations/month
Jay	Male	Corguinho	Colorado	07/12/13	10/27/13	107	3.6	377	6	106
Lurdes	Female	Corguinho	Colorado	06/05/14	12/12/14	190	6.3	452	6	71
Tony	Male	Corguinho	Safira	06/08/14	10/21/14	135	4.5	395	6	88
Malu-I[a]	Female	Corguinho	Safira	06/08/14	11/24/14	169	5.6	504	6	89
Julia	Female	Corguinho	Sta. Teresa	07/27/14	11/15/14	111	3.7	349	6	94
Tainara	Female	Corguinho	Jacobina	03/10/15	12/14/15	279	9.3	1642	4[b]	177
Cleide	Female	Corguinho	Jacobina	03/10/15	10/18/15	222	7.4	1759	3	238
Valentina	Female	Corguinho	Sta. Teresa	08/09/15	01/05/16	149	5.0	1165	3	235
Malu-II[a]	Female	Corguinho	Safira	10/01/15	01/03/16	94	3.1	732	3	234
Cida	Female	Corguinho	Safira	12/18/15	05/30/16	164	5.5	1292	3	236
Vagner	Male	Corguinho	Taboco	02/18/16	07/14/16	147	4.9	1135	3	232
Cintia	Female	Bodoquena	Cachoeira	10/27/16	11/27/17	396	13.2	1446	6	110
Primavera	Female	Bodoquena	Primavera	02/24/17	03/31/18	400	13.3	1320	6	99
Total/average						2563	6.6	12,568		154

[a]The same individual was captured once in 2014 and after a GPS collar malfunction was recaptured again in 2015. The information is reported separately in the table, but it was analyzed together and results presented for the individual

[b]The collar was programmed to collect data every 3 h but changed the schedule on its own accord after 5 days of tracking to collect data every 4 h

4.2.6 Movement and Range Calculations

We used the linear distance between two consecutive locations to quantify WLP movement (Getz and Saltz 2008) and the minimum convex polygon method for home range estimation. To account for possible differences in environmental effects on shorter- versus longer-term movement, we calculated linear distances at six separate time intervals: 3 h (the minimum interval that our dataset provided), 6 h, 12 h, 24 h, 168 h (7 days), and 720 h (30 days). Linear distances were calculated with the adehabitatLT package for R (Calenge 2006). MCP ranges were calculated for the percentages of 30, 50, 70, and 90, through the package adehabitatHR for R (Calenge 2006). The MCP method was used because it emphasizes range, as opposed to differential use within range, and is less sensitive to the spatial distribution of locations within the range. MCP calculates the size of the smallest convex polygon encompassing the known locations of an individual (Powell 2000).

To be consistent between MCP calculations for distinct herds, we used 6-h intervals for all individuals (even though some individuals had locations with higher temporal resolution – Table 4.1). We calculated ranges for each individual within a month to assess temporal changes in range. We used only months with at least 20 days of sampling and a minimum of 80 locations, as that sample size proved to be the minimum to render asymptotic range estimates (the maximum possible number of locations being 4 locations X 31 days = 124 locations/individual/month).

4.2.7 Statistical Analyses

To investigate our main hypothesis that forest cover affects movement and range, we tested whether linear distances and monthly MCPs varied between thresholds of percent forest cover (<40%, 40–60%, >60%). Complimentarily, to investigate possible effects of temporal variation of environmental conditions and resources on movement and range, we tested whether linear distances or monthly MCPs varied between seasons (wet, dry, and transitional) and monthly diversity of fruiting trees. We also investigated whether effects of forest removal could be more exacerbated in the wet or the dry season. Wet and dry seasons were defined as the 4 months with the highest (wet) and lowest (dry) monthly rainfall, such that November, December, January, and February defined the wet season while June, July, August, and September constituted the dry season. To normalize the dependent variables, we square-root transformed the distances and log-transformed the MCP ranges prior to statistical analyses. For all tests, we used linear models (RStudio, Inc. Version 1.0.136 – © 2009–2016, built-in function lm()). In all circumstances, we defined the significance level as 5%, but we also investigated the magnitude of the effect using the adjusted-R^2, as some of our datasets (specifically, short-term distances: 3 h, 6 h, 12 h, and 24 h) had sample sizes greater than 2000, and generated a large proportion of significant p-values with very little variation explained by the factor. Therefore, we considered results to be ecologically meaningful only if they had a p-value <0.05 and an adjusted-R^2 >0.05.

4.3 Results

Average monthly range sizes were 250, 470, 752, and 1447 ha for MCPs 30%, 50%, 70%, and 90%, respectively (Table 4.2a). Overall, range sizes showed a negative relationship with percent of forest cover, but the relationship was significant only at MCP 70% ($N_{70} = 63$, adjusted $- R^2{}_{70} = 0.10$, $F_{70} = 3.25$, $\mathrm{df}_{70} = 2.62$, $P_{70} = 0.05$) and 90% ($N_{90} = 63$, adjusted $- R^2{}_{90} = 0.12$, $F_{90} = 5.09$, $\mathrm{df}_{90} = 2.60$, $P_{90} = 0.01$). There was no significant relationship between monthly range sizes and seasons, but there was a small yet significant positive relationship between range sizes and number of fruiting tree species at all percentages but 90% ($N_{\mathrm{all}} = 63$, $\mathrm{df}_{\mathrm{all}} = 1.61$: adjusted-$R^2{}_{30} = 0.073$, $F_{30} = 4.77$, $P_{30} = 0.003$; $R^2{}_{50} = 0.065$, $F_{50} = 5.34$, $P_{50} = 0.02$; $R^2{}_{70} = 0.061$, $F_{70} = 5.01$, $P_{70} = 0.03$ – Fig. 4.3). In the wet season (but not in the dry season), herds occupying areas with less than 40% of native forest cover showed monthly ranges twice as large as the herds occupying areas with more than 60% of forest cover, and the relationship was significant for all range limits ($N_{\mathrm{all}} = 15$, $\mathrm{df}_{\mathrm{all}} = 1.13$: adjusted-$R^2{}_{30} = 0.58$, $F_{30} = 20.68$, $P_{30} = 0.001$; adjusted-$R^2{}_{50} = 0.68$, $F_{50} = 30.27$, $P_{50} = 0.0001$; adjusted-$R^2{}_{70} = 0.57$, $F_{70} = 19.95$, $P_{70} = 0.001$; adjusted-$R^2{}_{90} = 0.49$, $F_{90} = 14.68$, $P_{90} = 0.002$ – Fig. 4.4).

Table 4.2 Summary statistics for (a) herd monthly range sizes (minimum convex polygon, in hectares) and (b) movement (linear distances between consecutive locations, in meters)

(a)

MCP limits (%)	N	Minimum	1st quartile	Median	Mean	3rd quartile	Maximum
30	63	10	81	155	250	352	1223
50	63	31	133	381	470	595	2402
70	63	61	322	646	752	969	3755
90	63	153	666	1217	1447	1706	9832

(b)

Time interval (hours)	N	Minimum	1st quartile	Median	Mean	3rd quartile	Maximum	Median speed (m/h)	Average speed (m/h)
3	6025	0	92	268	431	611	4806	89	144
6	7141	0	192	456	647	922	4683	76	108
12	4524	2	314	749	993	1424	8104	62	83
24	2478	3	489	1093	1375	2001	8334	46	57
168	367	13	804	1716	2167	3001	13890	10	13
720	84	207	1301	2537	2951	4263	9141	4	4

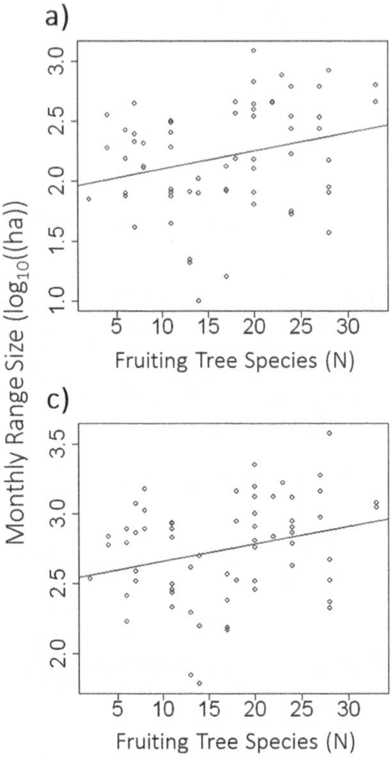

Fig. 4.3 Relationship between herds' monthly range sizes (MCP (**a**) 30%, (**b**) 50%, and (**c**) 70%) and the average number of fruiting tree species (N_{all} = 63, df_{all} = 1.61: adjusted-R^2_{30} = 0.073, F_{30} = 4.77, P_{30} = 0.003; R^2_{50} = 0.065, F_{50} = 5.34, P_{50} = 0.02; R^2_{70} = 0.061, F_{70} = 5.01, P_{70} = 0.03). Best-fit lines included

As expected, linear distances were larger, as time interval increased, yet smaller estimated speeds (distance/time interval – Table 4.2b), confirming that linear distances were less capable of taking into account the tortuosity of true trajectories at increased time intervals. Therefore, we considered that our best estimated speed was 90 m/h (median with 95% CI = 20–1393 m/h), calculated using linear distances between consecutive locations at 3 h intervals (N = 6025).

Long-term linear distances (720 h) were greater in areas with less percent cover (N_{720} = 84, adjusted-R^2_{720} = 0.11, F_{720} = 6.05 df_{720} = 2,81, P_{720} = 0.004 – Fig. 4.5a), with most of the effect occurring in the wet season, although in this case, probability value was not below the significance level, probably due to the low sample size (N_{720} = 19, adjusted-R^2_{720} = 0.12, F_{720} = 2.21, df_{720} = 2.16, P_{720} = 0.142 – Fig. 4.5b). Linear distances at 720 h interval were also positively related to the average of fruiting tree species (N_{720} = 84, R^2_{720} = 0.05, F_{720} = 5.65, df_{720} = 1,82, P_{720} = 0.02 – Fig. 4.5c). For some of the other comparisons, linear distances yield significant p-values yet negligibly adjusted R^2 (<0.05); therefore, they were not considered ecologically meaningful.

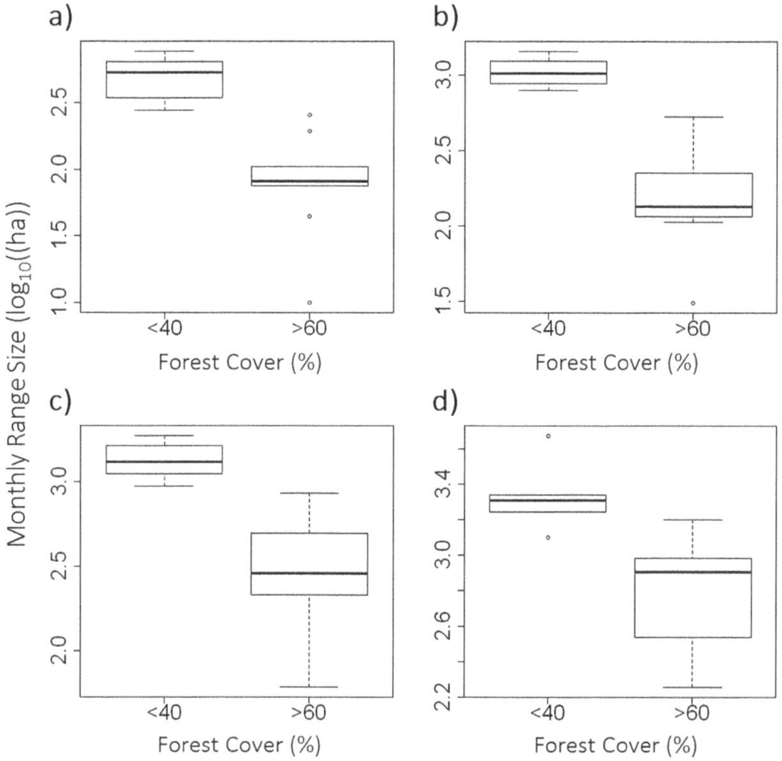

Fig. 4.4 Relationship between herds' monthly range sizes (MCP (**a**) 30%, (**b**) 50%, (**c**) 70%, and (**d**) 90%) and percent forest cover (<40% and >60%) at the core of the wet season (November, December, January, and February). Statistical results are $N_{all} = 15$, $df_{all} = 1.13$: adjusted-$R^2_{30} = 0.58$, $F_{30} = 20.68$, $P_{30} = 0.001$; adjusted-$R^2_{50} = 0.68$, $F_{50} = 30.27$, $P_{50} = 0.0001$; adjusted-$R^2_{70} = 0.57$, $F_{70} = 19.95$, $P_{70} = 0.001$; adjusted-$R^2_{90} = 0.49$, $F_{90} = 14.68$, $P_{90} = 0.002$. The single line within each box represents the median value, while the upper and lower boundaries of the box represent the first and third quartiles, respectively. Lines extending perpendicularly from the outer boundaries of the box encompass data falling outside of the first and third quartiles, with the end points of these lines not exceeding 1.5 * interquartile range (distance between lower and upper quartile)

4.4 Discussion

White-lipped peccaries are wide-ranging species with home ranges varying from 2000 to 20000 ha (Fragoso 1998, 2004; Carrillo et al. 2002; Keuroghlian et al. 2004, 2015; Reyna-Hurtado et al. 2009; Jacomo et al. 2013, Hofmann et al. 2016). Because of their large ranges and unique behavioral and ecological traits (e.g., frugivorous, gregarious, herd-forming, forest-associated ungulates that use a diversity of native habitats and resources - Keuroghlian et al. 2008a, b), it has always been expected and inferred that they are a "highly mobile" species. Nevertheless, due to the limitations of systematic data collection with VHF radiotelemetry collars, up to now, estimates of WLP displacement have been difficult to infer. Keuroghlian (2003)

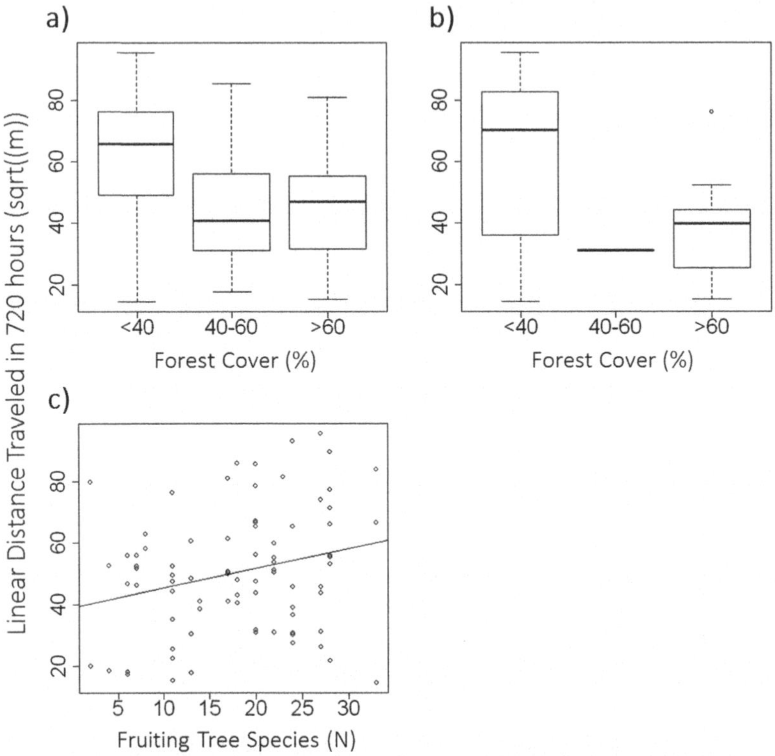

Fig. 4.5 Relationship between (**a**) linear distance traveled (m) in 1 month (720 h) and percent forest cover (<40%, 40–60%, and >60%; $N_{720} = 84$, adjusted-$R^2_{720} = 0.11$, $F_{720} = 6.05$, $\text{df}_{720} = 2.81$, $P_{720} = 0.004$); (**b**) linear distance traveled (m) in 1 month (720 h) and percent forest cover (<40%, 40–60%, and >60%) at the core of the wet season (November, December, January, and February; $N_{720} = 19$, adjusted-$R^2_{720} = 0.12$, $F_{720} = 2.21$, $\text{df}_{720} = 2.16$, $P_{720} = 0.142$); and (**c**) linear distance traveled (m) in 1 month (720 h) and average number of fruiting tree species ($N_{720} = 84$, $R^2_{720} = 0.05$, $F_{720} = 5.65$, $\text{df}_{720} = 1.82$, $P_{720} = 0.02$). For the box plots, the single line within each box represents the median value, while the upper and lower boundaries of the box represent the first and third quartiles, respectively. Dashed lines extending perpendicularly from the outer boundaries of the box encompass data falling outside of the first and third quartiles, with the end points of these lines not exceeding 1.5 * interquartile range (distance between lower and upper quartile). For the linear regression, best-fit line is included

estimated that daily distances traveled by VHF-monitored WLPs (calculated as the sum of the distances traveled between four or more fixes during a 24-h period) ranged from 950 to 6640 m (mean = 3892, *SD* = 1643, *N* = 31) in a 2178 ha fragment of semi-deciduous Atlantic Forest, Brazil. More recently Hofmann et al. (2016) presented preliminary data of one GPS-tracked individual traveling 1000 m/day (ranging from 270 to 3788 m/day). Our herds traveled an average of 3446 m/day (median = 2146 m/day, 95% CI = 166–11142 m/day, *N* = 6025), within the same order of magnitude of estimates from Keuroghlian (2003) and Hofmann et al. (2016). Nevertheless, we still believe these are underestimates, as our minimum time interval was 3 h and herds probably still move a lot within that time span. A comparison of our speed estimates and those from 57 other species of terrestrial mammal throughout the world (Tucker et al. 2018) confirms that white-lipped peccaries are among the "highly mobile" species of terrestrial mammals in the world.

Food (primarily fruits) and water sources are key environmental drivers in determining WLP movement and range (Altrichter et al. 2001; Keuroghlian and Eaton 2008a, b; Keuroghlian, Andrade Santos and Eaton 2015, Reyna-Hurtado et al. 2009, 2012). Because there are large monthly variations in the availability of these two main resources, we expected to see effects of month, season, and temporal fruiting patterns on movement and ranges. Reyna-Hurtado et al. (2009) in Calakmul Biosphere Reserve (Mexico) and Jacomo et al. (2013) in Emas National Park (Brazil) found that WLP herd range sizes were smaller in the dry season compared to the wet season and hypothesized the primary cause to be the reduction of water in their study sites during the dry season. On the other hand, Fragoso (1998) found an overall increase in home range size in the dry season in Maracá Island (Brazil). Carrillo et al. (2002) and Altrichter et al. (2001) found the same pattern in Corcovado National Park (Costa Rica), with smaller home ranges in the wet season, when there was also a peak of fruit availability. Furthermore, Altrichter et al. (2002) documented greater travel distances during lean fruit periods. Finally, Keuroghlian et al. (2004) found distinct seasonal shifts in home range use, but no significant difference in range size between wet and dry seasons in an Atlantic forest fragment, in São Paulo (Brazil). Our results indicate WLPs in Central Brazil are also responding to temporal variation in food resources and herds have greater monthly movement and range sizes when there are more species of trees fruiting. Nevertheless, this positive relationship is not reflected in a seasonal effect, probably because there is a great within-season variation in the number of tree species fruiting, and consequently in the monthly movement and range sizes.

Most relevant to conservation efforts, our results indicate that wide-ranging native species with high diversity of resource requirements, such as the white-lipped peccary, are negatively affected by deforestation in complex ways. We expected ranges and movement to be negatively affected by deforestation, since less forested landscapes should have a more sparse distribution of fruiting trees, thus forcing WLPs to travel further in order to find available food resources. Deforestation does not affect short-term distances, suggesting that short-term movement is not driven by external environmental factors. However, deforestation affects long-term WLP movement and monthly range size but mainly in the wet season, when ranges are twice as large in areas with less than 40% of forest cover than in areas with more than 60% of cover. During the wet season, WLPs use a greater variety of tree species. Our results suggest that the spatial distribution of those fruiting trees may become more sparsely distributed with deforestation, forcing WLPs to travel longer distances and expand their ranges to meet their resource needs. On the other hand, while the diversity of fruits is reduced in the dry season, fruiting tree species that are available are abundant and patchily distributed. For example, buriti (*Mauritia flexuosa*) and acuri (*Attalea phalerata*) are two palm fruits that form groves and are available for most of the year but comprise a greater portion of WLP diet during the dry season when there is less fruit diversity (Keuroghlian et al. 2009). These fruiting patterns likely explain the lack of range size difference between areas of different percent forest cover in the dry season.

Climate change models predict a low level of future climate stability for the Pantanal and a low to moderate level of climate stability for the Cerrado, meaning that as climate changes, existing species in both biomes will need to adjust their current distributions and habits to "track their climate niche" (Watson et al. 2013). Considering that the adaptive capacity, or level of vegetation intactness, is already considered "low" for the Cerrado, efforts to preserve remaining natural habitat cover and connectivity between habitat patches, as well as the Cerrado and Pantanal biomes, will prove critical for climate change adaptation and sustainability of native mammalian fauna (Watson et al. 2013). Our data clearly shows that low forest cover has spatial use implications for a wide-ranging species such as white-lipped peccaries. Continued unmanaged deforestation will most certainly cause the large-scale extinction of the species and possibly other wildlife.

Using high-resolution movement and range data, here we demonstrate that in agricultural landscapes with low amounts of native forest cover, white-lipped peccaries move more and occupy larger ranges. If changes in movement and range patterns have negative consequences for the local persistence of the species, it will be important for agricultural landscapes to have sufficiently high forest cover to maintain populations of white-lipped peccary and, consequently, the ecosystem services that the species provides.

Acknowledgments We thank Maria do Carmo Andrade Santos, Paulino Oliveira Ângelo, Renata Reinoso Rocha, and Marcello Schiavo Nardi (our project veterinarian) for their assistance in capturing, collaring, and monitoring white-lipped peccaries, sampling fruits, and plant identification. We thank Donald P. Eaton for previous edits. We thank all the volunteers that have participated on the project, especially teachers Jeff Wilford and Liz Kinzly, the Global Ecotours and Expeditions volunteers, and Chatham University. We acknowledge and appreciate the support from the local communities for access to land and key information regarding the local flora and fauna. This research was graciously funded by FAPESP (grant n. 2013/50421-2 and 2014/23132-2), CNPq (grant n. 312045/2013-1; 312292/2016-3 and scholarship: 161089/2014-3), Silicon Valley Community Foundation, the Overbrook Foundation, the International ReSource Award, WCS Brasil, and Vanderbilt University.

References

Altrichter M, Drews C, Sáenz JC, Carrilo E (2002) Presupuesto de tiempo del chancho cariblanco (*Tayassu pecari*) em un bosque humedo de Costa Rica. Biotropica 34:136–143

Altrichter M, Carrillo E, Sáenz J, Fuller TK (2001) White-lipped peccary (Tayassu pecari, Artiodactyla: Tayassuidae) diet and fruit availability in a Costa Rican rain forest. Rev de Biol Trop 49(3):1183–1192

Achard F, Eva HD, Stibig HJ, Mayaux P, Gallego J, Richards T, Malingreau JP (2002) Determination of deforestation rates of the world's humid tropical forests. Science 297(5583):999–1002. https://doi.org/10.1126/science.1070656

Austin KG, González-Roglich M, Schaffer-Smith D, Schwantes AM, Swenson JJ (2017) Trends in size of tropical deforestation events signal increasing dominance of industrial-scale drivers. Environ Res Lett 12(5):054009

Beck H (2006) A review of peccary-palm interactions and their ecological ramifications across the Neotropics. J Mammal 87:519–530

Beck H, Thebpanya P, Filiaggi M (2010) Do Neotropical peccary species (Tayassuidae) function as ecosystem engineers for anurans? J Trop Ecol 26:407–414

Beck H, Snodgrass J, Thebpanya P (2013) Long-term exclosure of large terrestrial vertebrates: implications of defaunation for seedling demographics in the Amazon rainforest. Biol Conserv 163:115–121

Butchart SHM, Walpole M, Collen B, van Strien A, Scharlemann JPW, Almond REA, Baillie JEM, Bomhard B, Brown C, Bruno J, Carpenter KE, Carr GM, Chanson J, Chenery AM, Csirke J, Davidson NC, Dentener F, Foster M, Galli A, Galloway JN, Genovesi P, Gregory RD, Hockings M, Kapos V, Lamarque J-F, Leverington F, Loh J, McGeoch MA, McRae L, Minasyan A, Morcillo MH, Oldfield TEE, Pauly D, Quader S, Revenga C, Sauer JR, Skolnik B, Spear D, Stanwell-Smith D, Stuart SN, Symes A, Tierney M, Tyrrell TD, Vie J-C, Watson R (2010) Global biodiversity: indicators of recent declines. Science 328(5982):1164–1168

Calenge C (2006) The package "adehabitat" for the R software: a tool for the analysis of space and habitat use by animals. Ecol Model 197(3–4):516–519

Carrillo E, Saenz JC, Fuller TK (2002) Movements and activities of white-lipped peccaries in Corcovado National Park, Costa Rica. Biol Conserv 108:317–324

Cullen L Jr, Bodmer RE, Valladares-Pádua C (2001) Ecological consequences of hunting in Atlantic forest patches, Sao Paulo, Brazil. Oryx 35:137–144

de Azevedo FCC, Conforti VA (2008) Decline of peccaries in a protected subtropical forest of Brazil: toward conservation issues. Mammalia 72(2):82–88

Desbiez ALJ, Santos SA, Keuroghlian A, Bodmer RE (2009) Niche partitioning between sympatric populations of native white-lipped peccary (*Tayassu pecari*), collared peccary (*Tayassu tajacu*), and introduced feral pigs (*Sus Scrofa*). J Mammal 90(1):119–128

Fragoso JM (1998) Home range and movement patterns of white-lipped peccary (*Tayassu pecari*) herds in the Northern Brazilian Amazon. Biotropica 30:458–469

Fragoso JM (2004) A long-term study of white-lipped peccary (*Tayassu pecari*) population fluctuation in Northern Amazonia. In: Silvius K, Bodmer RE, Fragoso JMV (eds) People in nature, wildlife conservation in South and Central America. Columbia University Press, New York, pp 286–296

Françoso RD, Brandao R, Nogueira CC, Salmona YB, Machado RB, Colli GR (2015) Habitat loss and the effectiveness of protected areas in the Cerrado Biodiversity Hotspot. Nat Conservação 13:35–40

Galetti M, Guevara R, Neves CL, Rodarte RR, Bovendorp RS, Moreira M, Hopkins JB III, Yeakel J (2015a) Defaunation affect populations and diet of rodents in Neotropical rainforests. Biol Conserv 190:2–7

Galetti M, Bovendorp RS, Guevara R (2015b) Defaunation of large mammals leads to an increase in seed predation in the Atlantic forest. Glob Ecol Conserv 3:824–830

Getz WM, Saltz D (2008) A framework for generating and analyzing movement paths on ecological landscapes. Proc Natl Acad Sci 105(49):19066–19071

Hofmann MPG, Signer J, Hayward MW, Balkenhol N (2016) Spatial ecology of a herd of white-lipped peccaries (*Tayassu pecari*) using GPS telemetry: challenges and preliminary results. Therya 7:21–37

Hansen MC, Potapov PV, Moore R, Hancher M, Turubanova SA, Tyukavina A, Thau D, Stehman SV, Goetz SJ, Loveland TR, Kommareddy A, Egorov A, Chini L, Justice CO, Townshend JRG (2013) High-resolution global maps of 21st-century forest cover change. Science 342(6160):850–853

Hayne DW (1949) Calculation of size of home range. J Mammal 30(1):1–18

Jacomo ATA, Furtado MM, Kashivakura CK (2013) White-lipped peccary home-range size in a protected area and farmland in the central Brazilian grasslands. J Mammal 94:137–145

Jorge MLSP, Galetti M, Ribeiro M, Ferraz KMPMB (2013) Mammal defaunation as surrogate of trophic cascades in a biodiversity hotspot. Biol Conserv 163:49–57

Keuroghlian A (2003) The response of peccaries to seasonal fluctuations in an isolated patch of tropical forest. Ph.D. dissertation. University of Nevada, Reno, Reno, Nevada

Keuroghlian A, Desbiez ALJ (2010) Biometric and age estimation of live peccaries in the Southern Pantanal, Brazil. Suiform Soundings 9:24–35

Keuroghlian A, Eaton DP (2008a) Fruit availability and peccary frugivory in an isolated Atlantic forest fragment: effects on peccary ranging behavior and habitat use. Biotropica 40:62–70

Keuroghlian A, Eaton DP (2008b) Importance of rare habitats and riparian zones in a tropical forest fragment: preferential use by *Tayassu pecari*, a wide-ranging frugivore. J Zool 275:283–293

Keuroghlian A, Eaton DP (2009) Removal of palm fruits and ecosystem engineering in palm stands by white-lipped peccaries (*Tayassu pecari*) and other frugivores in an isolated Atlantic Forest fragment. Biodivers Conserv 18:1733–1750

Keuroghlian A, Eaton DP, Longland WS (2004) Area use by white-lipped and collared peccaries (*Tayassu pecari* and *Tayassu tajacu*) in a tropical forest fragment. Biol Conserv 120:411–425

Keuroghlian A, Eaton D, Desbiez ALJ (2009) The response of a landscape species, white-lipped peccaries, to seasonal resource fluctuations in a tropical wetland, the Brazilian Pantanal. Int J Biodivers Conserv 1(4):87–97

Keuroghlian A, Santos MCA, Eaton DP (2015) The effects of deforestation on white-lipped peccary (*Tayassu pecari*) home range in the southern Pantanal. Mammalia 70:491–497

Keuroghlian A, Reyna-Hurtado R, Meijaard E, Altrichter M, Beck H, Gongora JR (2017) Conservation of wild pigs and peccaries. In: Melletti M, Meijaard E (eds) Ecology, conservation and management of wild pigs and peccaries. Cambridge University Press, Cambridge, UK, pp 277–290

Klink CA, Machado RB (2005) Conservation of the Brazilian cerrado. Conserv Biol 19(3):707–713

Köppen W (1884) Translated by Volken, E.; Brönnimann, S. "Die Wärmezonen der Erde, nach der Dauer der heissen, gemässigten und kalten Zeit und nach der Wirkung der Wärme auf die organische Welt betrachtet" [The thermal zones of the earth according to the duration of hot, moderate and cold periods and to the impact of heat on the organic world)]. Meteorol Z (published 2011) 20(3):351–360

Kiltie RA, Terborgh J (1983) Observations on the behavior of rain Forest peccaries in Perú: why do white-lipped peccaries form herds? Z Tierpsychol 62(3):241–255

Leite CC, Costa MH, Soares-Filho BS, Hissa LBV (2012) Historical land use change and associated carbon emissions in Brazil from 1940 to 1995. Glob Biogeochem Cycles 26:GB2011

Mohr CO (1947) Table of equivalent populations of north American small mammals. Am Midl Nat 37(1):223–249

Painter LRE (1998) Gardeners of the forest: plant–animal interactions in a Neotropical forest ungulate community. PhD dissertation. Liverpool: University of Liverpool

Paviolo A, De Angelo C, Di Blanco Y, Di Bitetti M (2008) Jaguar population decline in the Upper Paraná Atlantic Forest of Argentina and Brazil. Oryx 42:554–561

Powell RA (2000) Animal home ranges and territories and home range estimators. In: Boitani L, Fuller TK (eds) Research techniques in animal ecology: controversies and consequences. Columbia University Press, New York, pp 65–110

Peres CA (2000) Effects of subsistence hunting on vertebrate community structure in Amazonian forests. Conserv Biol 14(1):240–253

Pott A, Pott VJ (2003) Espécies de Fragmentos Florestais em Mato Grosso do Sul. In: Costa RB (ed) Fragmentação Florestal e Alternativas de Desenvolvimento Rural na Região Centro-Oeste. MS, UCDB, Campo Grande, pp 26–52

Reyna-Hurtado R, Rojas-Flores E, Tanner GW (2009) Home range and habitat preferences of white-lipped peccaries (*Tayassu Pecari*) in Calakmul, Campeche, Mexico. J Mammal 90:1199–1209

Reyna-Hurtado R, Chapman CA, Calme S, Pedersen E (2012) Searching in heterogeneous environments: foraging strategies in the white-lipped peccary (*Tayassu pecari*). J Mammal 93:124–133
Reynolds J, Wesson K, Desbiez A, Ochoa-Quintero J, Leimgruber P (2016) Using remote sensing and random forest to assess the conservation status of critical Cerrado habitats in Mato Grosso do Sul, Brazil. Land 5(2):1–12
Santana WSC (2015) Análise-multitemporal da paisagem, potencialidades e fragilidades sob a ótica do uso e ocupação: destaque para as terras da Alta Bacia do Rio Taboco-MS. Ph.D thesis. State University of Sao Paulo, Rio Claro (UNESP). 152 pp
SEMAC (2011) Secretaria de Estado de Meio Ambiente,do Planejamento, da Ciência e Tecnologia. Caderno Geoambiental das Regiões de Planejamento de MS – 5. Região do Pantanal. Governo do Estado de Mato Grosso do Sul, Campo Grande
Silman M, Terborgh JW, Kiltie RA (2003) Population regulation of a dominant rain forest tree by a major seed predator. Ecology 84:431–438
Silva NM (2008) Dinâmica de uso das terras nos municípios de. Bonito, Jardim e Bodoquena (MS) e o estado de conservação dos recursos biológicos do Parque. Nacional da Serra da Bodoquena e de sua zona de amortecimento. Ph.D thesis, University of Sao Paulo, S.P. 235 pp
Sowls LK (1997) Javelinas and other peccaries: their biology, management, and use, 2nd edn. Texas A&M University Press, College Station
Terborgh J, Nuñez-Iturri G, Pitman NCA, Valverde FHC, Alvarez P, Swamy V, Pringle EG, Paine CET (2008) Tree recruitment in an empty forest. Ecology 89:1757–1768
Tucker MA, Böhning-Gaese K, Fagan WF, Fryxell JM, Van Moorter B, Alberts SC, Ali AH, Allen AM, Attias N, Avgar T, Bartlam-Brooks H, Bayarbaatar B, Belant JL, Bertassoni A, Beyer D, Bidner L, van Beest FM, Blake S, Blaum N, Bracis C, Brown D, Nico de Bruyn PJ, Cagnacci F, Calabrese JM, Camilo-Alves C, Chamaillé-Jammes S, Chiaradia A, Davidson SC, Dennis T, DeStefano S, Diefenbach D, Hamilton ID, Fennessy J, Fichtel C, Fiedler W, Fischer C, Fischhoff I, Fleming CH, Ford AT, Fritz SA, Gehr B, Goheen JR, Gurarie E, Hebblewhite M, Heurich M, Mark Hewison AJ, Hof C, Hurme E, Isbell LA, Janssen R, Jeltsch F, Kaczensky P, Kane A, Kappeler PM, Kauffman M, Kays R, Kimuyu D, Koch F, Kranstauber B, LaPoint S, Leimgruber P, Linnell JDC, López-López P, Catherine Markham A, Mattisson J, Medici EP, Mellone U, Merrill E, Mourão G d M, Morato RG, Morellet N, Morrison TA, Díaz-Muñoz SL, Mysterud A, Nandintsetseg D, Nathan R, Niamir A, Odden J, O'Hara RB, Oliveira-Santos LGR, Olson KA, Patterson BD, de Paula RC, Pedrotti L, Reineking B, Rimmler M, Rogers TL, Rolandsen CM, Rosenberry CS, Rubenstein DI, Safi K, Saïd S, Sapir N, Sawyer H, Schmidt NM, Selva N, Sergiel A, Shiilegdamba E, Silva JP, Singh N, Solberg EJ, Spiegel O, Strand O, Sundaresan S, Ullmann W, Voigt U, Wall J, Wattles D, Wikelski M, Wilmers CC, Wilson JW, Wittemyer G, Zięba F, Zwijacz-Kozica T, Mueller T (2018) Moving in the Anthropocene: global reductions in terrestrial mammalian movements. Science 359(6374):466–469
Watson JEM, Iwamura T, Butt N (2013) Mapping vulnerability and conservation adaptation strategies under climate change. Nat Clim Chang 3:989–994

Chapter 5
Movements of White-Lipped Peccary in French Guiana

Cécile Richard-Hansen, Rachel Berzins, Matthis Petit, Ondine Rux, Bertrand Goguillon, and Luc Clément

5.1 Introduction

White-lipped peccary is unique among large- to medium-sized forest mammals, living in the largest-known social cohesive groups in such a close environment (Reyna-Hurtado 2009). They are also known as keystone ecosystem engineers, deeply structuring the forest by their rooting, trampling, feeding, wallowing, and ranging activities (Beck et al. 2010, 2017; Keuroghlian et al. 2013; Taber et al. 2016) and have crucial role for subsistence hunters (Altrichter 2005) and large predators (Altrichter et al. 2011; Foster et al. 2013, 2016; Mendes Pontes and Chivers 2007). Although they are well-known forest dwellers across South and Central American tropical forests, many unanswered questions remain on their movement ecology. Given the difficulties inherent both to the hard environmental conditions (remote areas and continuous canopy) and their morphological and ecological specificities (no neck-to-collar, large-scale ranging patterns), tracking studies are very scarce and recent comparing with other species of large mammals. Before the first radio-tracking monitoring (VHF) in northern Brazil (Fragoso 1998), white-lipped peccaries (white lips) were sometimes considered as nomadic or migratory

C. Richard-Hansen (✉)
ONCFS – Campus agronomique, Kourou, France

UMR EcofoG – Campus agronomique, Kourou, France
e-mail: cecile.richard-hansen@ecofog.gf

R. Berzins · M. Petit · O. Rux
ONCFS – Campus agronomique, Kourou, France

B. Goguillon
Parc amazonien de Guyane, Remire-Montjoly, France

L. Clément
ONCFS – Campus agronomique, Kourou, France

Parc amazonien de Guyane, Remire-Montjoly, France

R. Reyna-Hurtado, C. A. Chapman (eds.), *Movement Ecology of Neotropical Forest Mammals*, https://doi.org/10.1007/978-3-030-03463-4_5

(Bodmer 1990b; Kiltie and Terborgh 1983). Since the last decade, some VHF and GPS monitoring of the species (Hofman et al. 2016; Keuroghlian et al. 2004; Reyna-Hurtado et al. 2009b; Tobler 2009) unveiled a more stable spatial behavior than previously thought but also many local variations according to the country and the environmental conditions.

Moreover, many regions or countries across the distribution area experienced temporary disappearances of the species, at local, medium or large spatial and temporal scales (Altrichter et al. 2011; Azevedo and Conforti 2008; Carrillo et al. 2002; Cullen et al. 2001; Fang et al. 2008; Fragoso 1997, 2004; Keuroghlian et al. 2012, 2013; Kiltie and Terborgh 1983; Peres 1996; Reyna-Hurtado 2009; Reyna-Hurtado et al. 2009a), which are probably at the origin of the migratory/nomad behavior hypothesis. Those occasional disappearances are well-known by indigenous people, and, for example, Wayãpi Amerindians in French Guiana consider the white lips as the most unpredictable game and often use magical practices to attract them back (Grenand 1993). In some cases, the species can disappear for several months, which could be related to seasonal movements (Carrillo et al. 2002; Keuroghlian and Eaton 2008a; Keuroghlian et al. 2009; Reyna-Hurtado et al. 2009b), but in other ones, the decline occurs during several years, and over very large areas, as proven by long-term monitoring in French Guiana (Richard-Hansen et al. 2014).

No one to date is able to explain these extreme population fluctuations. Overhunting is sometimes invoked (Azevedo and Conforti 2008; Carrillo et al. 2002; Chiarello 2000; Cullen et al. 2000; Peres 1996, 2000), but cannot be the only cause, given the scale of the phenomenon and that it has occurred also in very isolated areas with no or very low hunting pressure. Other hypotheses are currently debated. Do the herds occasionally migrate over long distances, because of local conditions of food limitation, overpopulation, and habitat alteration such as exceptional flooding (Bodmer 2012; Fang et al. 2008)? Or could it be related to demographic parameters such as diseases, inducing mass dying or reproductive failures (de Freitas et al. 2010; Fragoso 2004; Nava and Cullen 2003; Richard-Hansen et al. 2014)? Observations of large herds apparently moving unusually long distances, crossing large rivers, are reported, as well as occurrences of skinny and sick animals in other places with water scarcity (Reyna-Hurtado et al. 2014), but no clear link with demography and/or population collapse exists.

Regardless of regular seasonal movements or occasional long-distance moves, understanding the movement behavior is crucial for the management and conservation of this key species in forest ecology (Keuroghlian et al. 2017; Taber et al. 2016). Previous studies in other countries showed that home range size can vary from 20 to 200 km^2 (Carrillo et al. 2002; Fragoso 1998, 2004; Keuroghlian et al. 2004; Reyna-Hurtado et al. 2009b) according to the country and environmental conditions and that conservation efforts need to focus on landscape scale, in relation with dispersal of both sexes and low genetic differentiation on more than 80 km distances (Biondo et al. 2011). These examples illustrate how spatial ecology and movement behavior are central issues for management and conservation. Movement ecology encompasses several approaches and methodology to better understand the underlying mechanisms and the emergent spatiotemporal patterns of individuals' movements

(Nathan 2008). French Guiana experienced a strong decline of white-lipped peccaries' populations, which currently seem to progressively recover. During the same time, new wildlife management policies and hunting regulations are progressively set up. Developing adapted management tools for this species means being able to understand *where*, *when*, *how*, or even *why* white-lipped peccaries are moving (Nathan et al. 2008), to infer on which scale do we have to consider hunting management or which are the key habitats or climatic limitations for the species. Considering the high variability of the species' ecology across its range, we initiated this study primarily to acquire the baseline data on its spatial ecology in local French Guianan environmental conditions, concerning home range sizes, habitat use, and seasonal movements. We also began to have some insights on more detailed ranging behaviors, to more precisely approach underlying activities, and the patterns and drivers of observed movements.

5.2 Material and Methods

5.2.1 Study Area

French Guiana is an 84,000 km^2 French overseas department, with a pristine tropical humid forest cover of 93% (Gond et al. 2011). The region is part of the Guiana shield, and most of the forest is considered "terra firme" on poor nutrient soils, with a dense hydrographic network and small patches of tropical freshwater swamp forests, with periodically saturated hydromorphic soils, embedded between hills (Koponen et al. 2004; Sabatier et al. 1997). Forest types and spatial patterns of floristic composition across the country have been described on the basis of geomorphological landscape features (Guitet et al. 2013, 2015b). Small patches of open vegetation types as dry and temporarily inundated savannas occur on a 20-km-wide strip along the coast, representing only around 0.3% of the total French Guiana superficies. Annual rainfall ranges between 3600 mm in the northeast and 2000 mm in the south and west. A dry season occurs from August to November, and the rainiest periods from December to June, with a short lull around March. In this study, we consider two main seasons, with mid of July as transition. Overall human density in French Guiana is below three inhabitants km^{-2}, and 75% of the population is restricted to the five major towns, in the northern coastal area. The remaining population lives in a few small villages and settlements mainly along the two main rivers that form the borders with Suriname and Brazil. In that part, a National Park (*Parc amazonien de Guyane*) covers 34 million ha, 20 million ha of which comprise the core area where only the resident population is allowed to hunt for subsistence. Hunting regulations are currently very embryonic, but needs for better wildlife management are becoming urgent, both for the northern populated areas and southern autochthonous populations.

5.2.2 Study Sites

The initial objective was to undertake the study within the National Park area, as the white lips' hunting management was identified as a crucial issue regarding subsistence needs of local human populations. However, the white-lipped peccary populations' level was really too low at this time to give any reasonable capture opportunities. Conversely, two small protected areas on the coastal region showed evidence of regular presence of white-lipped peccaries. The first study site (*Centre Spatial Guyanais*, CSG) is located in the area surrounding the European spaceport (5°12′39″N, 52°48′51″W), where access is highly regulated, and hunting is prohibited for security reasons on a ≈ 700 km^2 area. With the exception of scattered buildings and industrial places such as rocket launchpads, the surrounding environment is quite preserved and shelters an apparently healthy large vertebrate's community (Petit et al. 2017). The area includes continuous forest from coastal plain and low-valley type (Guitet et al. 2015b), almost contiguous to the large forest block of the rest of the country (from which it is split only by a road), and wet and dry savannas interspersed by small gallery forests on old cheniers. The second study site is located in the protected area *Réserve Nationale du Mont Grand Matoury* (RNMGM), near the largest city in French Guiana, Cayenne, and the international airport. The reserve was created in 2006, and the medium-to-large vertebrate community has been previously altered by hunting and has not yet recovered; however informal observations from local people ascertain the regular presence of white lips in the area. The area includes a hilly part (234 m asl) and a flatter area including wet savannas, mangrove, and swampy forest (ONF 2016). Although almost surrounded by anthropized areas, the southwestern part of the reserve is still connected to less disturbed environment and to the rest of the forest block.

After many prospections, four main capture sites were selected on the basis of frequent track observations and camera trap data. Four box traps (110 × 70 × 90 cm) were displayed, and five larger corral traps (±20 m^2) were built on those places. Traps were deployed almost continuously since June 2013. Traps were regularly baited with local fruits (mainly *Astrocaryum vulgare* palm tree), and a trap transmitter (Vectronics) alarmed us via satellite transmission when the trap triggered. Once trapped, we selected one or two individuals to be immobilized with 8 mg/kg of Zoletil 100 (Tiletamine 250 mg, Zolazepam 250 mg) delivered by a Dan-Inject dart gun. They were then fitted with GPS collars, respectively, from Followit (Tellus Medium and Small, six individuals), Vectronics (GPS-PLUS-2, one individual), and Lotek (Track M 2D and 1D, eight individuals) fitted with a drop-off (Telonics CR-2A) if not already integrated into the collar design. Collars were programmed to try a fix attempt every 1, 2, or 4 h and to transfer data via Iridium satellite transmission when 10–12 fixes were stored (theoretically every 2 days).

To characterize the home ranges, we used the autocorrelated kernel density estimation (AKDE) (Fleming et al. 2015) which better fitted data from GPS-tracking technology, as the reduced time between successive fixes strongly increases the autocorrelation, violating the basic assumption of independence needed in conventional KDE analysis. We used the ctmm package (Calabrese et al. 2016) for the R statistical environment (R Development Core Team 2017) to calculate the empirical variogram and select the best fitting semi-variance function (SVF) on the basis of the Akaike information criterion. The minimum convex polygons (MCP) were estimated using Qgis (Qgis Development Team 2017), for comparison with other studies and investigation of habitat use.

To analyze the ranging patterns, we first estimated the cumulative daily distances walked by animals from the most complete daily dataset. We calculated the cumulative distance between each successive fixes ("movement steps") on a daily basis. For greater accuracy, we only took into account "validated" fixes that were calculated with more than four satellites and that had a low dilution of precision (DOP<10). As it is rare to obtain 100% accurate fixes each day, we decided to consider all days that recorded at least 80% of "validated" fixes to calculate daily distance. We also estimated straight-line distances (not cumulative) between 2 and 3 successive days, based on the locations of the animal at midnight. Those two distance measurements are complementary, the last one detecting if animals are moving from one place of the home range to another one in a quick period (medium or long-distance move), while the first one mainly reflects the motion capacity on the species (Nathan 2008). Both movements probably have different internal or external drivers and reflect different movement phases (Nathan et al. 2008).

To investigate habitat use and selection within the home range, we used a recent forest habitat classification for French Guianese forests (Guitet et al. 2015a; ONF 2015) and a land use cover for the coastal area (ONF 2016). We compared the proportion of each habitat type in the total MCP surface as available and the proportion of fixes in each habitat type as the proportion of habitat used (Hofman et al. 2016). We also calculated the Jacobs index (Jacobs 1974) to detect preferences and avoidances of habitat within the home range's availability. It is calculated as $D = (r - p)/(r + p - 2rp)$, where r is the proportion of habitat used and p is the proportional habitat availability, and varies from −1 for strong avoidance to +1 for strong selectivity.

5.3 Results

During *ca* 4 years of study, 44 animals were captured on 10 trapping occasions, and 15 animals were fitted with GPS/iridium collars (4 captures, 19 individuals of which 7 GPS-collared on study site 1; 6 captures, 25 animals trapped and 8 collared on study site 2). Individual monitoring lasted 51 days in average (2–242 days), ended in most cases by the signal lost without explanation (nine cases) and the drop-off triggering in only two cases (Table 5.1). The collar fell a few days after being attached to the animal in three cases (without breaking), and one animal died. The collars were recovered on six occasions, when iridium transmission of data was still operational and allowed us to find it. In all other cases, no iridium signal was received, and systematic prospections by foot or by plane did not allow finding the

Table 5.1 Details of captures and monitoring results of all individuals over the two study sites

Animal (site)	Date beginning	Date end	No. of fixes	No. days	Collar brand	End cause
Clementine (2)	19/05/2014	11/06/2014	115	22	Followit	Signal loss
Desire (2)	18/06/2014	08/07/2014	501	16	Lotek	Drop off
Ariane (1)	16/12/2014	29/01/2015	264	43	Followit	Signal loss
Seb (1)	09/02/2015	08/08/2015	4328	179	Lotek	Drop off
Oliver (1)	09/02/2015	11/02/2015	2	2	Lotek	Collar fell
Pim (2)	29/07/2015	25/08/2015	155	26	Followit	Signal loss
Pam (2)	29/07/2015	14/08/2015	90	15	Followit	Signal loss
Poum (2)	29/07/2015	19/08/2015	238	20	Followit	Signal loss
Mimi (1)	17/05/2016	25/06/2016	385	38	Vectronics	Signal loss
Rustine (2)	19/07/2016	21/03/2017	1064	242	Lotek	Signal loss
Vegas (1)	30/07/2016	17/08/2016	150	18	Lotek	Collar fell
Malma (1)	16/05/2017	18/05/2017	12	2	Lotek	Death
Noury (1)	16/05/2017	22/08/2017	369	96	Lotek	Signal loss
Stone (2)	09/07/2017	27/07/2017	146	18	Lotek	Collar fell
Chardene (2)	09/07/2017	30/08/2017	84	28	Lotek	Signal loss

collars with VHF. When recovering the collar, we found that the percentage of fix success/fix attempts was rather good, around 86%, indicating the first step of fix acquisition performed correctly. The major issue seemed therefore to be the satellite data transfer. Finally, two animals gave satisfactory results with, respectively, 6 and 8 months of continuous data, one on each study site. The size of the group in site 1 could be evaluated on a video when crossing the road in single line, to more than 200 individuals. We have no indication if this group periodically splits, as noted in some other places (Keuroghlian et al. 2004). In site 2, more than 86 individuals were counted once.

5.3.1 *Home Ranges*

Home ranges were estimated with minimum convex polygons (MCP) for the two animals with longest surveys and for each whole group, cumulating all fixes from all tracked animals in each group (Table 5.2). The MCP of the whole groups matches quite well with the MCP of individuals tracked over 6 and 8 months, with the exception of a small area in site 2 (Fig. 5.1), which means that these may be very cohesive groups. Moreover, the locations of all animals monitored over 3 years were all within the same global polygon, indicating a spatial stability of the two groups over the study time (2014–2017).

For Seb (adult male, 6 months monitoring in site 1) and Rustine (adult female, 8 months monitoring in site 2), home ranges were estimated with the autocorrelated kernel density estimator (AKDE) and with minimum convex polygon (MCP) for comparison with other studies (Fig. 5.1, Table 5.2). For both animals, the best model selected was the Ornstein–Uhlenbeck-F (OUF, (Calabrese et al. 2016)). The estima-

Table 5.2 Annual and seasonal home range size estimates, using minimum convex polygon (MCP) and autocorrelated kernel density estimator (AKDE) methods, for each group and two best-monitored individuals (Seb and Rustine) in the two study sites

	MCP 100% km^2			AKDE 95% km^2	AKDE 50% km^2
	Year	Dry season	Rainy season	(95% CI)	(95% CI)
Site 1 – Seb (♂)	70	–	–	49 (32–69)	12 (8–17)
Site 1 – all group	73	64	73	–	–
Site 2 – Rustine (♀)	39.7	–	–	34 (25–45)	7.7 (5.6–10)
Site 2 – all group	55	41	49	–	–

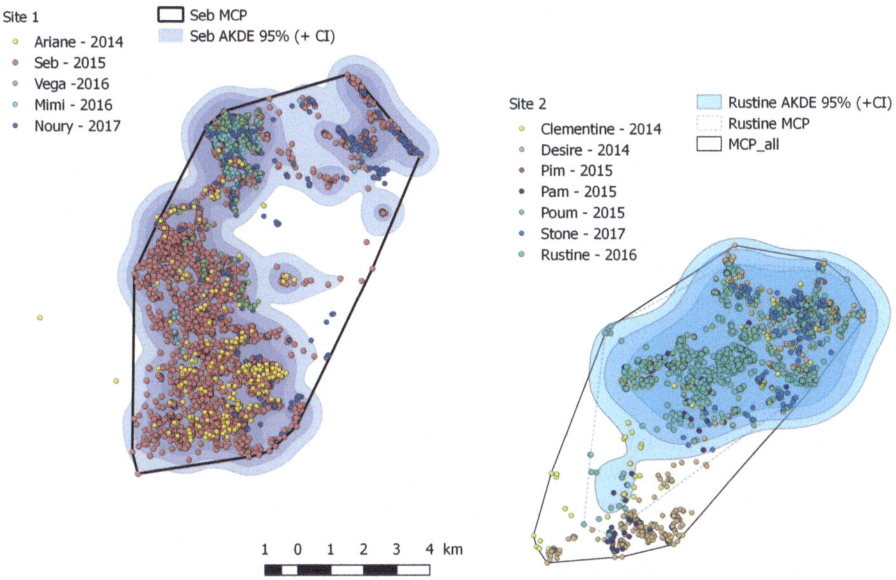

Fig. 5.1 Location of all monitored animals (one color for each individual) in the two study sites of French Guiana and home ranges of the two best-monitored ones. Home range is figured by minimum convex polygon (100% MCP) and by autocorrelated kernel density estimator (AKDE 95%), with 95% confidence interval

tions were slightly lower than a preliminary estimate by 95% KDE estimator (respectively 60 km^2 for Seb and 57 km^2 for Rustine, made with Hawth's tools ArcGis software after selection of independent fixes).

The cumulative area used by the animals reached 90% of the maximum recorded home range area (Fig. 5.2) after, respectively, 84 (Seb, site 1) and 204 days (Rustine, site 2). Comparing with the global area used by all the animals in each site, it is likely that the cumulative area is representative of the annual home range. The difference in the cumulative curves between the two surveys may be related to the study area characteristics, to the period of the year in which the survey took place, and/or to different behaviors. Indeed, Rustine's survey began a few times before she moved and stayed in a restricted area for the dry season, from which she moved again 2 months later, while Seb was captured at the beginning of the rainy season.

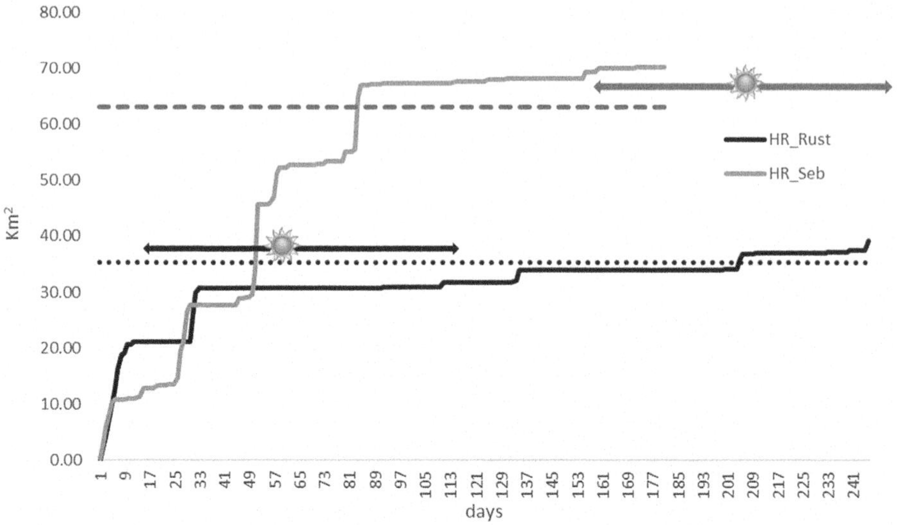

Fig. 5.2 Cumulative area used by two GPS-tracked white-lipped peccaries over the time. The dashed line indicates the 90% of the whole home range estimate for each animal. The sunny symbol and arrow indicates the dry season period

5.3.2 *Seasonality*

There are no apparent large seasonal movements and little difference between the global area prospected by white lips during the wet and the rainy seasons (Fig. 5.3). In site 1, an area presenting forest strips grown on Pleistocene beach ridges (see Sect. 5.3.4) seems to be used more often during the rainy season. Unfortunately, too few locations were collected during the dry season because of collar failures. However, within the group home range, a higher proportion of the dry season fixes was recorded within the wettest forest types, compared to the rainy season (32% vs. 25%). In site 2, a small wet area located in the core area of the home range has been extensively used during the drier month (September).

5.3.3 *Daily Movements*

Based on best (satellite accuracy and full-day data, see "Methods" section) data on hourly moves, white-lipped peccaries could travel up to 8.3 km per day (maximum cumulative daily distance, ranging from 3.6 to 8.3 km). Mean values of cumulative daily distances range from 1.9 to 4.2 km/day, depending on the animal. During daytime (6:00–19:00), an animal travels an average of 240 m/h (109–310 m, based on Seb's data, which collar was scheduled to 1 fix/h). On the overall dataset, the mean daily range seems higher during the wet season 3.6 km/day (3.2–4.1 IC95%) than

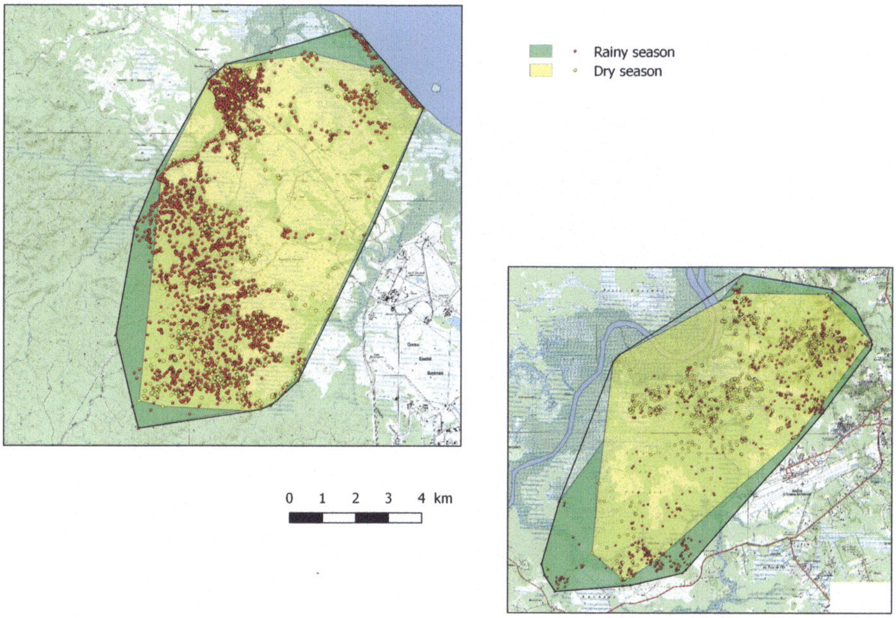

Fig. 5.3 Seasonal locations and seasonal home ranges of two white-lipped peccary groups in two study sites of French Guiana

during the dry one 2.4 km/day (2.2–2.6 IC95%) (Wilcoxon test, $p = 8.561e{-}06$), but the situation may be different between the two study areas, depending on the spatial distribution of habitat.

The highest cumulative daily ranges (>4 km for Seb in site 1; >3 km for Rustine in site 2) occurred only four times during each survey (in March, April, May, and July for Seb; January, October, November, and July for Rustine) and revealed different movement patterns. During these days, the movement appeared more directional and the peccaries quickly traveled to a different place within their home range and even entirely crossed it in 2–3 days (Fig. 5.4). The straight-line distance between three consecutive days was 6–9 km at this time, against 1.5–2 km for the rest of the survey. The map (Fig. 5.4) also shows that Seb seemed to follow the streams and borders of the home range for that large-scale movement, which could give some indications on their navigation capacities.

5.3.4 Habitat Use

In both study sites, the MCP includes wide savanna areas, which are consistently avoided by animals (Fig. 5.5). Savannas represent 47% of the MCP area in site 1 and around 20% in site 2, whereas, respectively, only 6% and 5% of fixes were located

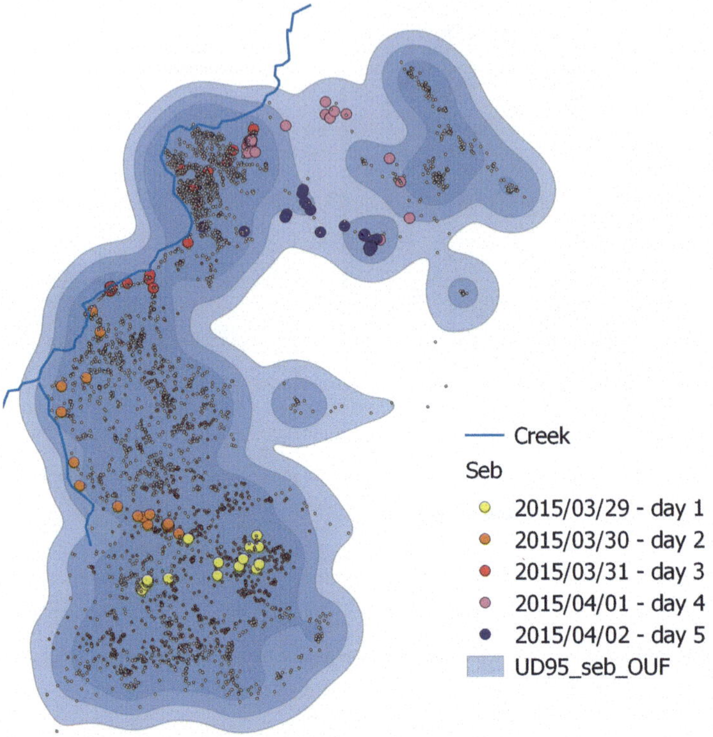

Fig. 5.4 Long-distance directional move a white-lipped peccary living in the study site of Centre Spatial Guyanais in French Guiana, in 5 days of March–April 2015

in this habitat ($p < 0.001$, Chi-square test). The proportion is lower when considering the AKDE estimates of home ranges, but some embedded savannas are still included in the area, although not used by animals. The Jacobs index shows the avoidance ($D = -0.9$).

The proportion of fixes in various habitat types is statistically different from random use based on availability in the MCP ($p < 0.001$, Chi-square test, for two sites and groups, Fig. 5.6). In site 1, 34% of fixes were located in *coastal dryland wet forest* (representing 23% of the MCP area) and 33% of fixes in *lowland wet forest* which represent only 14% of the available area in the MCP home range. This forest type forms narrow strips across the savannas, partly grown on old sandy barriers, and seems particularly used during the rainy (and fruiting) season (Fig. 5.5). The Jacobs index ($D = 0.5$) shows the preference for that habitat. During the driest season, there is a tendency to use more the riparian and wettest habitats, but these habitats are interlaced within the dryland forest, so the global area remains the same.

In site 2, there is no *dryland wet forest* type, as in site 1, and the majority of forest is the *lowland wet forest* (42% of MCP), in which 70% of fixes were located. Many fixes recorded during the dry season (yellow dots, Fig. 5.5) appear as if they were located in flooded savannas. However, this is due to low accurate mapping, and field

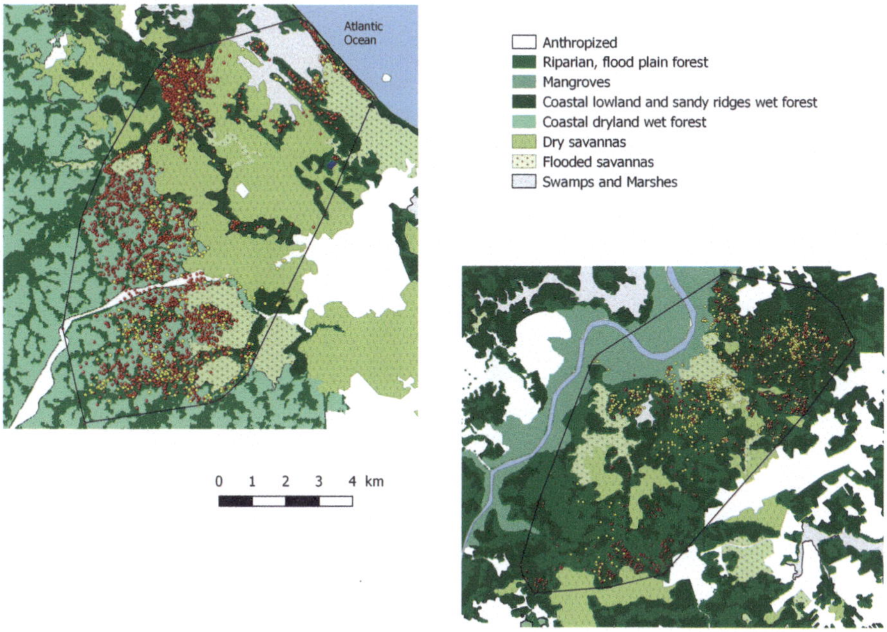

Fig. 5.5 Locations of two white-lipped peccary groups in two study sites in French Guiana, according to various habitats and seasons. Habitats are characterized following ONF (2015, 2016) typology

verification indicated that they are actually located in small floodplain forest patches, remaining wet all year long. Open savanna habitats are avoided in this site also (Jacobs index = −0.7), and the forest connectivity allows the full extension of the home range, despite the high anthropization level of this area. The mangrove is not used by white-lipped peccaries ($D = -0.4$).

5.4 Discussion

5.4.1 Technology Limitations

Tracking white-lipped peccaries is challenging everywhere. Their wide home ranges and the inaccessibility of their habitat, particularly in French Guiana with very few road infrastructure, make the classic VHF tracking impossible. During the study, we faced several issues with GPS tracking, probably in relation with the close and high canopy forest existing in their French Guianese habitat. Several GPS collar brands were tested, and at present time, only one gave some satisfactory results (Table 5.1), although the signal has been lost after few weeks in three cases out of nine. For the other brands, 6/6 surveys ended after 15–35 days, because of unexplained satellite signal reception loss. The Lotek drop-off system also failed very quickly, probably

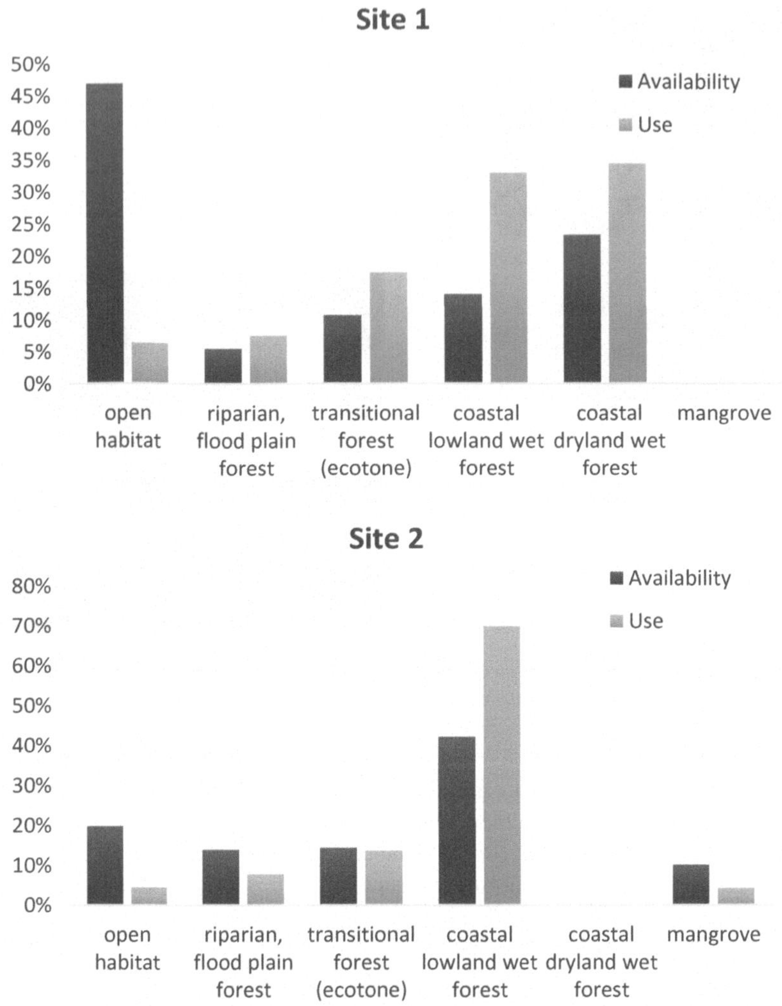

Fig. 5.6 Habitat use by two white-lipped peccary groups in two study sites in French Guiana. Forest habitats are characterized following ONF (2015) typology. Available habitat is the proportion of the habitat in the total home range, and use is the proportion of location recorded in this habitat

because it was too protruding for this social and rough species (rubbing collars against trees or mutual grooming; Fig. 5.7). The same collars on jaguars, in the same environment, worked much better, lasting 1-year long and with scheduled drop-off release (Berzins unpubl. data). As we could check on several occasions that the limit does not come from the GPS acquisition system (>85% of success/fix attempt), our hypothesis is that the remotely retrieve transmission system is the main limitation in our case, as also noted by other white lips trackers (Moreira et al. 2018; this book). Jaguars use more open habitats, providing more opportunities to transmit the data.

Fig. 5.7 White-lipped peccary group resting in a very cohesive way. French Guiana. (Photo: ONCFS/CNES, camera trapping)

Another hypothesis to test is the influence of the scheduled acquisition rate of fixes on the transmission success rate. The incorrect attachment of the collar on white lips also caused three early losses, because of the species' triangle-shaped head.

Given the large and remote area prospected by the animals, and the high probability of technological failures, we have to rely on the remote retrieval of data, and not on collar recovery. However, the precision of remotely retrieved data (validation parameter) is not always provided in detail, and we found here that it might cause bias in detailed movement analysis. For the same animal, the daily distances calculated from the remotely received data were sometimes higher than those calculated on the collar-retrieved data, after filtering for "validated" data only, because the remote-retrieved data included more location errors (even considering the 3D data only). This has to be taken into account for further movement analyses from GPS data. Moreover, detailed analyses in movement ecology need to parse the movement path in elementary units (Nathan et al. 2008), and only data directly retrieved from the collar are regular enough to allow such a spatiotemporal high resolution.

5.4.2 Ranging Behavior and Habitat Use

Our dataset showed stable home ranges during the study length, with several animals from a same herd using the same area from year to year. No very long-distance moves were observed during this period, as noted in other studies (Reyna-Hurtado et al. 2012). Across their distribution area, mean home range sizes of white-lipped peccaries vary from 21 to 200 km^2 according to the study site and the ecosystem, climate, and habitat fragmentation (Beck et al. 2017). The size of home ranges in our study is smaller than previously found in Northern Amazonia (up to 200 km^2 for largest herd (Fragoso 1998, 2004), Mesoamerica (Reyna-Hurtado et al. 2009b), or central Brazilian farmland area (Jacomo et al. 2013), higher than in tropical forest fragment of Atlantic forest in southeastern Brazil (Keuroghlian et al. 2004), and similar to Southern Pantanal estimates (Keuroghlian et al. 2015). As underlined by Keuroghlian et al. (2015) and Jacomo et al. (2013), the white lips' home ranges are significantly larger and may be polynuclear in artificially fragmented landscapes compared to continuous forest. Although French Guiana suffers from limited deforestation compared to other countries, and is almost entirely covered by continuous forest, our two study sites strangely take place in the only part of the country with natural open savannas. In site 1, peccaries' home range encompasses, without using, large open areas and is larger than the home range of the herd in site 2, where savannas are less present (70 km^2 vs 55 km^2, 100% MCP). In site 1, there is no apparent limitation in the southern part of the home range (except probable higher hunting pressure), and we assume that utilization of forest patches is not related to artificial constraints but to natural behavior. These forest strips are grown on coastal lowland and old Pleistocene sandy ridges and are characterized by high abundances of palm trees (Guitet et al. 2015a), particularly *Mauritia flexuosa*, *Astrocaryum vulgare*, or *Maximiliana maripa*, key feeding resources for peccaries (Beck 2006). This particular habitat, although making up a small proportion of the home range, contributes to habitat heterogeneity and may provide unique resources for this species. In the Pantanal Keuroghlian et al. also reported the importance of rare and heterogeneous habitats and the regular use of gallery and "cordilheira" forests (Keuroghlian and Eaton 2008b; Keuroghlian et al. 2009, 2015).

Although the habitat fragmentation may apparently increase the home range of herd 1, including non-used areas, a highly nutritious forest may provide concentrated food availability, which does not exist in the inner continuous *terra firme* forest of French Guiana. Except for some small *Euterpe oleracea* patches on hydromorphic soils, palm trees are not very abundant and do not form large patches as noted for *Mauritia flexuosa* in the Peruvian Amazon (Bodmer 1990a), and food resources may be more scattered. Indeed, in site 2, this *lowland forest* type is dominant and also highly used by white-lipped peccaries. In this site, these forest types cover the hilliest parts, while the rest (southern part) is mostly covered by floodplain forests. The high proportion of nutritive lowland forest and the lower proportion of open savannas could explain the smaller home range size in site 2. Our next step will be to compare home ranges of herds living in continuous forest, to assess the effects of habitat and forest types on home range sizes of white-lipped peccaries in French Guiana.

Seasonal movements were limited, and the seasonal home ranges overlap almost entirely. During the dryer season, the two herds seem to prefer wet and riparian habitats, downhill, along the stream network, as previously noted in other studies (Keuroghlian and Eaton 2008b). However, water shortage is clearly not an issue in French Guiana, even during the dry season, comparing to drier habitats of Mexico, for example (Reyna-Hurtado et al. 2012, 2009b). In site 1, these wet areas are tightly interlaced within small low elevated hill systems, distributed all over the home range, and animals moved from one to the other without needing long-distance movements. In site 2, the animals spent almost 2 months in a reduced and central area, of lower floodplain forest, explaining the smaller home range and daily movements in this season, in this study site. The larger floodplain forests located in the southwestern part were not used.

The palm tree *Astrocaryum vulgare* is mainly present on small *coastal lowland forest* patches and savanna edges, and its seasonal and mass fructification (beginning of the rainy season, December–April) could explain the higher frequentation of this habitat (most northern, coastal, and fragmented part of home range in site 1) during the rainy season.

5.4.3 Management and Conservation

Our results also highlight that healthy populations of white-lipped peccaries can live very near important human activities, as long as habitat and hunting are correctly managed. Indeed, the herd in site 1 lives just between two international rocket launchpads of the European Space Center and herd 2 between the largest city and the international airport. However, both sites are protected against hunting, and habitat is quite preserved. Our results highlight in particular the role of small forest patches in forest continuity, a potential preferred habitat when the species gets protection from hunters. In both sites, white-lipped peccaries are using very narrow forest strips and are crossing the open areas (savannas, roads) at the narrowest point. This is important and should be addressed and taken into account for land management.

5.4.4 Movement Ecology

In more heterogeneous and limiting environments, larger-scale movements up to 16 km were observed, inducing a complete temporary change of home range and alternating with shorter-scale movements (Reyna-Hurtado et al. 2012). Distances traveled in our study are smaller (maximum 9 km), but we observed the same pattern of alternating small-scale and occasional directional larger movements during the year. In both cases, the mean daily traveled distance remains comparable, but results in local movements in one case and directional moves in the other one. In French Guiana, the need for water was not necessarily the main driver, as water is accessible all year long in almost the whole home range. Looking for fruiting resources may explain the frequentation of certain habitat types, but the motivation underlying a sudden movement to the other side of the range remains unclear. Shifting areas could be a foraging strategy to manage renewable feeding resources, such as fallen fruits, for example.

When resources become scarce, straighter and faster movements become more efficient than tortuous ones, as they result in high net displacement, thereby minimizing the chance of revisiting an already visited resource, increasing the chance of finding new resources, and giving time for stock to replenish (de Knegt et al. 2007; Viswanathan et al. 2008). Elucidating the proximate and ultimate (evolutionary) causes responsible for the observed movement paths is one of the central challenging questions in the movement ecology (Nathan et al. 2008). However, very regular data acquisition and more reliable transfer system or collar recovering will be needed to achieve this goal. Present data are still too fragmented to give good-enough temporal series both on short and long term, but once the technical issues are fixed, movement analyses will allow to better understand not only where but how, when, and why white-lipped peccaries travel in their heterogeneous habitat, to better fit management tools to their ecological requirements.

Acknowledgments This study was funded by European funds BEST and by Centre National d'Etudes Spatiales (CNES). We thank the Réserve naturelle nationale du Mont Grand Matoury – co-managed by la Mairie de Matoury and l'Office National des Forêts – for collaboration and authorization to capture and work within this protected site and the Centre Spatial Guyanais for collaboration, access to the protected site, and technical assistance. We gratefully acknowledge the Zoo de Guyane and "SOS Faune Sauvage" for veterinary assistance in anesthetic procedures and animal welfare, and particularly Olivier Bongard for his enthusiast, friendly, and competent help in the mud and the night. Many thanks to all people who helped in the field work, especially Laetitia Proux, Gaëlle Jaouen and Thomas Denis, and to Elisabeth Berzins for language editing.

References

Altrichter M (2005) The sustainability of subsistence hunting of peccaries in the Argentine Chaco. Biol Conserv 126:351–362

Altrichter M, Taber A, Beck H, Reyna-Hurtado R, Lizzaragua L, Keuroghlian A, Sanderson EW (2011) Range-wide declines of a key Neotropical ecosystem architect, the Near Threatened white-lipped peccary *Tayassu pecari*. Oryx 46(1):87–98

Azevedo FCC, Conforti VA (2008) Decline of peccaries in a protected subtropical forest of Brazil: toward conservation issues. Mammalia 72:82–88

Beck H (2006) A review of peccary-palm interactions and their ecological ramifications across the neotropics. J Mammal 87(3):519–530

Beck H, Thebpanya P, Filiaggi M (2010) Do Neotropical peccary species (Tayassuidae) function as ecosystem engineers for anurans? J Trop Ecol 26:407–414

Beck H, Keuroghlian A, Reyna-Hurtado R, Altricher M, Gongora J (2017) White-lipped peccary *Tayassu pecari* (Link, 1795). In: Melletti M, Meijaard E (eds) Ecology, conservation and management of wild pigs and peccaries. Cambridge University Press, Cambridge, pp 265–276

Biondo C, Keuroghlian A, Gongora J, Miyaki CY (2011) Population genetic structure and dispersal in white-lipped peccaries (*Tayassu pecari*) from the Brazilian Pantanal. J Mammal 92(2):267–274

Bodmer RE (1990a) Fruit patch size and frugivory in the lowland tapir (*Tapirus terrestris*). J Zool (Lond) 222:121–128

Bodmer RE (1990b) Response of ungulates to seasonal inundations in the Amazon floodplain. J Trop Ecol 6:191–201

Bodmer RE (2012) Impacto del cambio climático sobre la fauna silvestre de la Amazonia. In: X Congreso Internacional de Manejo de Fauna Silvestre en la Amazonía y Latinoamérica – Cimfauna, Salta, Argentina. p 19

Calabrese JM, Fleming CH, Gurarie E (2016) ctmm: anR package for analyzing animal relocation data as a continuous-time stochastic process. Methods Ecol Evol 7:1124–1132

Carrillo E, Saenz J, Fuller TK (2002) Movements and activities of white-lipped peccaries in Corcovado National Park, Costa Rica. Biol Conserv 108:317–324

Chiarello AG (2000) Density and population size of mammals in remnants of Brazilian Atlantic forest. Conserv Biol 14(6):1649–1657

Cullen LJ, Bodmer R, Valladares-Padua C (2000) Effects of hunting in habitat fragments of the Atlantic forests, Brazil. Biol Conserv 95:49–56

Cullen LJ, Bodmer R, Valladares-Padua C (2001) Ecological consequences of hunting in Atlantic forest patches, São Paulo, Brazil. Oryx 35(2):137–144

de Freitas TPT, Keuroghlian A, Eaton DP, Freitas EB, Figueiredo A, Nakazato L, de Oliveira JM, Miranda F, Paes RCS, Carneiro Monteiro LAR, Lima JVB, da C. Neto AA, Dutra V, de Freitas JC (2010) Prevalence of *Leptospira interrogans* antibodies in free-ranging *Tayassu pecari* of the Southern Pantanal, Brazil, an ecosystem where wildlife and cattle interact. Trop Anim Health Prod 42:1695–1703

de Knegt HJ, Hengeveld GM, van Langevelde F, de Boer WF, Kirkman KP (2007) Patch density determines movement patterns and foraging efficiency of large herbivores. Behav Ecol 18(6):1065–1072

Fang T, Bodmer R, Puertas P, Mayor P, Perez P, Acero R, Hayman D (2008) Certificación de pieles de pecaries (*Tayassu tajacu* y *T. pecari*) en la Amazonía peruana: Una estrategia para la conservación y manejo de fauna Silvestre en la Amazonia peruana. Wust Editions-Darwin Institute, Lima

Fleming CH, Fagan WF, Mueller T, Olson KA, Leimgruber P, Calabrese JM (2015) Rigorous home range estimation with movement data: a new autocorrelated kernel density estimator. Ecology 96(5):1182–1188

Foster VC, Sarmento P, Sollmann R, Torres N, Jacomo ATA, Negroes N, Fonseca C, Silveira L (2013) Jaguar and Puma activity patterns and predator-prey interactions in four Brazilian biomes. Biotropica 0(0):1–7. https://doi.org/10.1111/btp.12021

Foster RJ, Harmsen BJ, Macdonald DW, Collins J, Urbina Y, Garcia R, Doncaster CP (2016) Wild meat: a shared resource amongst people and predators. Oryx 50(1):63-75

Fragoso JMV (1997) Desapariciones locales del baquiro labiado (*Tayassu pecari*) en la Amazonia: migracion, sobre-cosecha, o epidemia? In: Fang TG, Bodmer RE, Aquino R, Valqui MH (eds) Manejo de fauna Silvestre en la Amazonia. OFAVIM, La Paz, pp 309–312

Fragoso JMV (1998) Home range and movement patterns of white-lipped peccary (*Tayassu pecari*) herds in the Northern Brazilian Amazon. Biotropica 30(3):458–469

Fragoso JMV (2004) A long-term study of white-lipped peccary (*Tayassu pecari*) population fluctuations in Northern Amazonia. Anthropogenic vs. "natural" causes. In: Silvius KM, Bodmer RE, Fragoso JMV (eds) People in nature. Wildlife conservation in south and central America. Columbia Univeristy Press, New York, pp 286–296

Gond V, Freycon V, Molino J-F, Brunaux O, Ingrassia F, Joubert P, Pekel J-F, Prévost MF, Thierron V, Trombe PJ, Sabatier D (2011) Broad scale pattern of forest landscape types in Guiana Shield. Int J Appl Earth Obs Geoinf 13:357–367

Grenand P (1993) Fruits, animals and people: hunting and fishing strategies of the Wayãpi of Amazonia. In: Tropical forests, people and food: biocultural interactions and applications, vol 13. Unesco and the Parthenon Publishing group, New York, pp 425–434

Guitet S, Cornu JF, Brunaux O, Betbeder J, Carozza JM, Richard-Hansen C (2013) Landform and landscape mapping, French Guiana (South America). J Maps 9(3):325–335. https://doi.org/10.1080/17445647.2013.785371

Guitet S, Brunaux O, de Granville J-J, Gonzalez S, Richard-Hansen C (2015a) Catalogue des habitats forestiers de Guyane. DEAL, Cayenne.

Guitet S, Pelissier R, Brunaux O, Jaouen G, Sabatier D (2015b) Geomorphological landscape features explain floristic patterns in French Guiana rainforest. J Biogeogr 24:1215–1237. https://doi.org/10.1007/s10531-014-0854-8

Hofman MPG, Signer J, Hayward MW, Balkenhol N (2016) Spatial ecology of a herd of white-lipped peccaries (*Tayassu pecari*) in Belize using GPS telemetry: challenges and preliminary results. THERYA 7(1):21–37

Jacobs J (1974) Quantitative measurement of food selection. Oecologia 14:413–417

Jacomo ATA, Malzoni Furtado M, Kayo Kashivakura C, Marinho-Filho J, Sollman R, Mundim Torres N, Sillveira L (2013) White-lipped peccary home-range size in a protected area and farmland in the central Brazilian grasslands. J Mammal 94(1):137–145

Keuroghlian A, Eaton DP (2008a) Fruit availability and peccary frugivory in an isolated Atlantic forest fragment: effects on peccary ranging behavior and habitat use. Biotropica 40(1):62–70

Keuroghlian A, Eaton DP (2008b) Importance of rare habitats and riparian zones in a tropical forest fragment: preferential use by *Tayassu pecari*, a wide-ranging frugivore. J Zool 275:283–293

Keuroghlian A, Eaton DP, Longland WS (2004) Area used by white-lipped and collared peccaries (*Tayassu pecari* and *Tayassu tajacu*) in a tropical forest fragment. Biol Conserv 120:411–425

Keuroghlian A, Eaton DP, Desbiez ALJ (2009) The response of a landscape species, white-lipped peccaries, to seasonal resource fluctuations in a tropical wetland, the Braz Pantanal. Int J Biodivers Conserv 1(4):087–097

Keuroghlian A, Desbiez A, de Mello Beisiegel B, Medici P, Gatti A, Mendes Pontes AR, Bueno de Campos C, de Tófoli CF, Moraes EAJ, Cavalcanti de Azevedo F, Medeiros de Pinho G, Passos Cordeiro JL, da Silva Santos TJ, Aparecida de Morais A, Mangini PR, Flesher K, Rodrigues LF, Bonjorne de Almeida L (2012) Avaliação do Risco de Extinção do Queixada *Tayassu pecari* Link, 1795, no Brasil. Biodivers Bras 3 (Número Temático: Avaliação do Estado de Conservação dos Ungulados):84–102

Keuroghlian A, Desbiez A, Reyna-Hurtado R, Altrichter M, Beck H, Taber A, Fragoso J (2013) *Tayassu pecari*. The IUCN red list of threatened species version 20131

Keuroghlian A, Santos MCA, Eaton DP (2015) The effects of deforestation on white-lipped peccary (*Tayassu pecari*) home range in the southern Pantanal. Mammalia 70:491–497

Keuroghlian A, Reyna-Hurtado R, Meijaard E, Altricher M, Beck H, Gongora J (2017) Conservation of Wild Pigs and peccaries. In: Melletti M, Meijaard E (eds) Ecology, conservation and management of wild pigs and peccaries. Cambridge University Press, Cambridge, pp 277–290

Kiltie RA, Terborgh J (1983) Observations on the behavior of rain forest peccaries in Perú: why do white-lipped peccaries form herds? Z Tierpsychol 62:241–255

Koponen P, Nygren P, Sabatier D, Rousteau A, Saur E (2004) Tree species diversity and forest structure in relation to microtopography in a tropical freshwater swamp forest in French Guiana. Plant Ecology (formerly Vegetatio) 173(1):17–32

Mendes Pontes AR, Chivers DJ (2007) Peccary movements as determinants of the movements of large cats in Brazilian Amazonia. J Zool 273:257–265

Moreira-Ramírez JF, Reyna-Hurtado R, Hidalgo-Mihart M, Naranjo E, Ribeiro M, García-Anleu R, McNab R, Radachowsky R, Mérida M, Briceño-Méndez M, Ponce-Santizo G (2018) White-lipped peccary home-range size in the Maya Forest of Guatemala and México. Chapter 2. In: Reyna-Hurtado R, Chapman CA (eds) Movement ecology of Neotropical forest mammals. Cham: Springer

Nathan R (2008) An emerging movement ecology paradigm. Proc Natl Acad Sci U S A 105(49):19050–19051

Nathan R, Getz WM, revilla E, Holyoak M, Kadmon R, Saltz D, Smouse PE (2008) A movement ecology paradigm for unifying organismal movement research. Proc Natl Acad Sci U S A 105(49):19052–19059

Nava A, Cullen L (2003) Peccaries as sentinel species: conservation, health and training in Atlantic Forest Fragments, Brazil. Suiform Soundings 3(2):15–16

ONF (2015) Habitats forestiers de Guyane. GeoGuyane. https://www.geoguyane.fr/geonetwork/srv/0906df56-3bf2-4df3-8ad8-645df6f03308. Accessed Mar 2018

ONF (2016) Carte d'occupation des sol sur la bande littorale guyanaise en 2015. GeoGuyane. https://www.geoguyane.fr/geonetwork/srv/0906df56-3bf2-4df3-8ad8-645df6f03308. Accessed Mar 2018

Peres CA (1996) Population status of white lipped *Tayassu pecari* and collared pecaries *T. tajacu* in hunted and unhunted Amazonian forests. Biol Conserv 77:115–123

Peres CA (2000) Effects of subsistence hunting on vertebrate community structure in Amazonian forests. Biol Conserv 14(1):240–253

Petit M, Denis T, Rux O, Richard-Hansen C, Berzins R (2017) Estimating jaguar (*Panthera onca*) density in a preserved coastal area of French Guiana. Mammalia. https://doi.org/10.1515/mammalia-2016-0150

Qgis Development Team (2017) QGIS geographic information system. Open Source Geospatial Foundation Project

R Development Core Team (2017) R: a language and environment for statistical computing, 3.01 edn. R Foundation for Statistical Computing, Vienna

Reyna-Hurtado R (2009) Conservation status of the white-lipped peccary (*Tayassu pecari*) outside the Calakmul Biosphere Reserve in Campeche, Mexico: a synthesis. Trop Conserv Sci 22(2):159–172

Reyna-Hurtado R, Naranjo E, Chapman CA, Tanner GW (2009a) Hunting and the conservation of a social ungulate: the white-lipped peccary *Tayassu pecari* in Calakmul, Mexico. Oryx 44(1):89–96

Reyna-Hurtado R, Rojas-Flores E, Tanner GW (2009b) Home range and habitat preferences of white–lipped peccary groups (*Tayassu pecari*) in a seasonally tropical forest of the Yucatan Peninsula, Mexico. J Mammal 90(5):1199–1209

Reyna-Hurtado R, Chapman CA, Calme S, Pedersen EJ (2012) Searching in heterogeneous and limiting environments: foraging strategies of white-lipped peccaries (*Tayassu pecari*). J Mammal 93(1):124–133

Reyna-Hurtado R, Moreira-Ramirez JF, Briceño-Méndez M, Sanvincente-Lopez M, McNab RB, Garcia-Anleu R, Mérida M, Sandoval-Seres E, Ponce-Santizo G, Hyeroba D (2014) White-lipped peccaries with skin problems in the Maya Forest. Suiform Soundings 13(1):29–31

Richard-Hansen C, Khazraie K, Surugue N, LeNoc M, Grenand F (2014) Long-term fluctuations of white-lipped peccary populations in French Guiana. Mammalia 218(3):291–301. https://doi.org/10.1515/mammalia-2013-0087

Sabatier D, Grimaldi M, Prévost M-F, Guillaume J, Godron M, Dosso M, Sabatier D (1997) The influence of soil cover organization on the floristic and structural heterogeneity of a Guianan rain forest. Plant Ecology formerly 'Vegetatio' 131(1):81–108

Taber A, Beck H, Gonzalez S, Altricher M, Barbanti Duarta JM, Reyna-Hurtado R (2016) Why neotropical forest ungulates matter. The case for deer, peccary, and tapir conservation. In: Aguirre AA, Sukumar R (eds) Tropical conservation: perspectives on local and global priorities. Oxford University Press, Oxford, pp 255–261

Tobler M (2009) New GPS technology improves fix success for large mammal collars in dense tropical forests. J Trop Ecol 25:217–221

Viswanathan GM, Raposo EP, da Luz MGE (2008) Lévy flights and superdiffusion in the context of biological encounters and random searches. Phys Life Rev 5:133–150

Chapter 6
Spatial Ecology of a Large and Endangered Tropical Mammal: The White-Lipped Peccary in Darién, Panama

Ninon F. V. Meyer, Ricardo Moreno, Miguel Angel Martínez-Morales, and Rafael Reyna-Hurtado

6.1 Introduction

Worldwide, many species are under threat because of habitat loss and fragmentation (Schipper et al. 2008). This is especially true in tropical ecosystems where native vegetation is being converted to agricultural land faster than in any other biome (Laurance et al. 2006). Habitat fragmentation hinders species dispersal, reduces gene flow, and isolates populations, which then face a higher risk of local extinction, ultimately threatening the overall species viability (Pardini et al. 2010). Large-bodied mammals are typically at a higher risk of extinction than smaller ones because of their large home ranges, low population densities, slow reproduction rates, and persecution by humans (Ripple et al. 2015). Consequently, many large mammal species are nowadays often restricted to protected areas or areas with low human density (Urquiza-Haas et al. 2011; Meyer et al. 2015; Reyna-Hurtado et al. 2017). This is problematic because these species often play a key role in ecosystem functions and their loss can have cascading effects on the environment (Ripple et al. 2015).

N. F. V. Meyer (✉)
El Colegio de la Frontera Sur, Lerma, Campeche, Mexico

Fundación Yaguará Panamá, Ciudad del Saber, Panama

R. Moreno
Fundación Yaguará Panamá, Ciudad del Saber, Panama

M. A. Martínez-Morales
El Colegio de la Frontera Sur, San Cristobal de las Casas, Chiapas, Mexico

R. Reyna-Hurtado
El Colegio de la Frontera Sur (ECOSUR), Department of Biodiversity Conservation, Lerma, Campeche, Mexico

The Wildlife Conservation Society (WCS), Bronx, NY, USA

R. Reyna-Hurtado, C. A. Chapman (eds.), *Movement Ecology of Neotropical Forest Mammals*, https://doi.org/10.1007/978-3-030-03463-4_6

To facilitate large mammals conservation in response to habitat loss and fragmentation, a common strategy consists in restoring connectivity between areas of suitable habitat (Cushman et al. 2013). It is increasingly recognized that considering animal movement and the amount of habitat needed is key to improve the design of such connectivity models (Cushman et al. 2013). Therefore, in the case of large mammals, assessing space use of large mammals by integrating empirical data is paramount both for effective conservation planning and for basic species ecological understanding.

The white-lipped peccary (WLP), *Tayassu pecari*, is a large frugivorous ungulate species from the Neotropics, ranging from Mexico to Argentina. It is considered a keystone species because it shapes the structure and function of the landscapes it inhabits via consumption of plants, seed predation, and trampling and also contributes to the nutrient cycling by being an important prey for carnivores (Beck 2006; Keuroghlian et al. 2004; Altrichter et al. 2012). The WLP is mostly a mature forest specialist, but it is also able to use fragmented landscapes (de Almeida Jácomo et al. 2013; Keuroghlian et al. 2004, 2014). However, since the WLP is a social animal that lives in large herds of up to more than 300 individuals, it usually requires large tracts of continued forests or other native vegetation types to survive (Altrichter et al. 2012; Reyna-Hurtado et al. 2016). As a result, it is among the first species to be affected by habitat loss and fragmentation (Altrichter et al. 2012; Moreno and Meyer 2014). Despite its wide distributional range, the WLP is under severe pressure mainly due to poaching and habitat loss (Reyna-Hurtado et al. 2010, 2017; Altrichter et al. 2012), particularly in Panama, where it is listed as "Endangered" (Moreno and Meyer 2014; Meyer et al. 2016; MiAmbiente 2016). The species was indeed extirpated from most of the country, including large protected forests that harbor a rich biodiversity (Goldman 1920; Enders 1935; Glanz 1982; Moreno and Meyer 2014; Meyer et al. 2015, 2016). As a result, the WLP has become one of the rarest mammal species in Panama, and its long-term survival is uncertain (Moreno and Meyer 2014). Moreover, it is most likely that WLP populations are no longer connected across Panama, thus neither between Mesoamerica and South America (Moreno and Meyer 2014; Meyer et al. 2016).

To promote the persistence of the WLP in Panama, conservation actions are urgently needed. Nevertheless, without reliable information on spatial requirements, developing a strategy for sustaining WLP populations will be challenging. An important first step in establishing areas of conservation is to know the size and composition of the area that fulfills critical life history requirements of a given species, as the disproportionate use of available resources may suggest where conservation efforts should be focused (Foerster and Vaughan 2002). Despite several studies that investigated the spatial ecology of the WLP throughout the Neotropics (e.g., Fragoso 1998; Keuroghlian et al. 2004, 2014; Reyna-Hurtado et al. 2009; de Almeida Jácomo et al. 2013; Hofman et al. 2016; Moreira Ramírez 2017), none has been conducted in Panama. Consequently, the herd dynamics, movement patterns, and seasonal home range were unknown in the moist tropical forests of Panama. Because animal behavior, ecology, and movement markedly depend on environmental conditions that are dynamic in space and time (Nathan et al. 2008; Kays et al. 2015), information extrapolated from studies carried out in other biomes that differ in habitat characteristics and seasonality may result in conservation strategies that are not appropriate. Therefore, the aim of this study was to describe the basic movement ecology of wild WLPs in the forests of the Darién National Park (NP), which was identified as one of the last strongholds for the WLP in Panama and in Mesoamerica (Meyer et al. 2016; Reyna-Hurtado et al. 2017).

6.2 Methods

6.2.1 Study Area

The Darién NP, located in eastern Panama at the border with Colombia, was created in 1980 and is listed as a UNESCO World Heritage site and Biosphere Reserve (Fig. 6.1). It is the second largest protected area in Mesoamerica with 5970 km^2 of forests. In addition to the Darién forest, this region also includes adjacent conserved areas in Panama (Comarca Guna) and Colombia (Los Katios NP). Its emergence several millions years ago allowed the formation of a natural bridge between North and South America, which probably enabled the Tayassuidae family to disperse to South America (Webb 2006). Besides its exceptional biodiversity, the Darién NP also exhibits a cultural dimension since it includes the Comarca Emberá-Wounaan (Sambu and Cemaco), a semiautonomous indigenous land.

We conducted our study at the Serranía de Pirre (8.03535 N, 77.7586 W) within the Darién NP, which includes both the core and buffer areas of the NP (Fig. 6.2). It is covered by wet lowland, highland, and cloud forests. Annual temperature averages 25 °C and annual rainfall reach 4000 mm. There is a relatively short dry season from mid-December to early May, while the rest of the year corresponds to

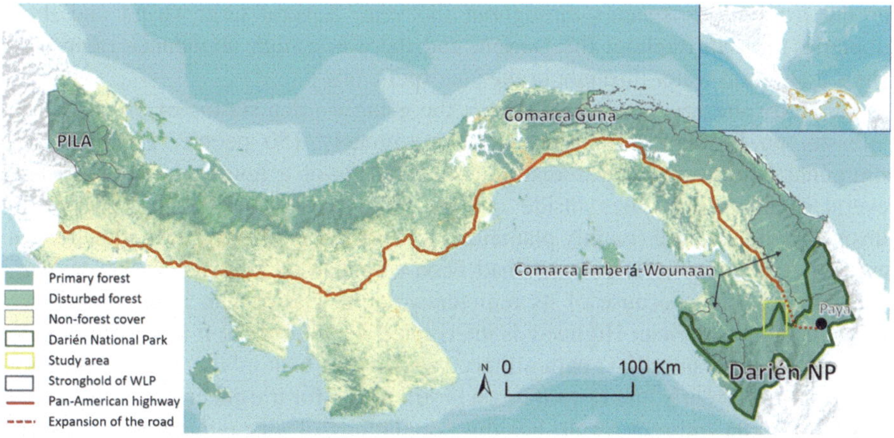

Fig. 6.1 Land cover of Panama and our study area. The Darién National Park together with the La Amistad International Park (PILA), the Comarca Emberá-Wounaan, and the Comarca Guna are the strongholds for white-lipped peccaries in Panama

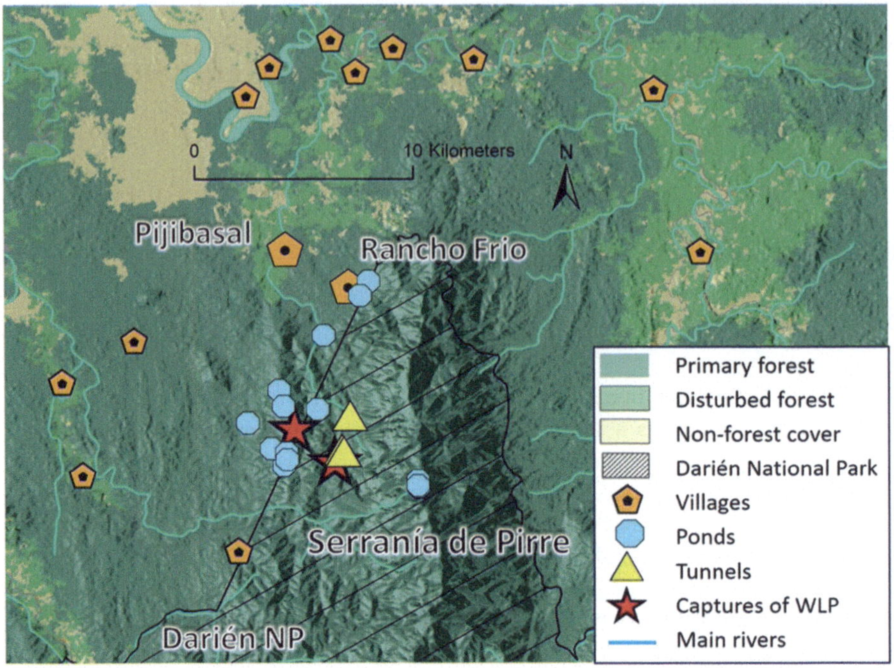

Fig. 6.2 Study area at the Serranía de Pirre depicting land cover, main rivers, villages, capture sites, known locations of ponds and mud baths, and tunnels used by white-lipped peccaries

the wet season. The terrain is rugged in some parts, with steep and abrupt slopes to the ridge at 1450 m of altitude. The lowland area is characterized by abundant palm tree species (i.e., *Astrocaryum* sp., *Attalea* sp., and *Oenocarpus* sp.), almond trees (*Dipteryx* sp.), and jobos (*Spondias* sp.), which provide a substantial amount of food for the frugivorous species throughout the year. Several large rivers and their tributaries flow throughout the Darién, and there is a high abundance of muddy ponds that retain water throughout the year (pers. obs.).

Our study site was near Rancho Frio, the game warden station of the park, and Pijibasal, an Emberá community of approximately 150 people. Pijibasal is a communal land covered by forest and small farms of a few hectares. There are several other communities inside and near the park. Most people engage in subsistence agriculture (mainly plantains, rice, maize, and beans), and some have a few cattle. So far, the Darién NP had been safeguarded from excessive human-induced disruption because of its remoteness – the Darién gap is the only place where the Pan-American Highway is interrupted – and because it is a region where conflicts between guerillas, paramilitary groups, and Colombian army took place for decades. Currently, the Darién forest is still a path for traffickers of all kinds and migrants. As an indirect result, healthy populations of all mammal species were maintained, including the most endangered and sensitive ones such as the WLP. Nowadays however, and despite its size and remoteness, the Darién NP is facing increasing pressure from an expanding agricultural frontier, human invasion, deforestation inside its core area for cattle ranching, and uncontrolled timber extraction. Moreover, apart from subsistence hunting by indigenous communities which is tolerated, poaching pressure is relatively high in the buffer area and moderate inside the park (but its exact magnitude is difficult to assess; Meyer and Moreno pers. obs.). A network of trails used by poachers and "gente del monte" (traffickers of all kinds) traverses remote portions of the Darién NP.

6.2.2 Capture and Collaring of WLPs

We used direct darting to capture WLPs from different groups. We intensively searched for fresh signs of WLPs (tracks or rooted soils from the same day or the day before) in areas with a high probability of WLP presence, which were identified based on both camera-trap data and the knowledge of Emberá hunters, who were also our guides. When presence was detected, we rapidly followed the tracks to get close to the group (30–60 m). This usually took from a half day to 2 days after finding the fresh tracks. Once the group was almost reached, one of our highly experienced guides discretely approached the herd in order to evaluate its behavior, while the rest of the team would remain at a safety distance.

The WLP in Darién usually displays two types of behavior: aggressive or fugitive. If the animals were aggressive, we would not attempt any capture for safety reasons. If they were fugitive, we would isolate an individual – occasionally two – from the rest of the group that would run away. We selected a single adult individual and anesthetize it with a combination of ketamine (5 mg kg) and xylazine (2 mg kg) administered via a CO_2-powered pistol (DanInject, Inc., TX) from a distance of 4–8 m. It took approximately 20 min for the WLP to be asleep and ready to be handled with low risk. Because animals were not completely tranquilized by the first injection, we administered an additional 1 mg kg of ketamine. We examined each individual for general body condition; gathered information on sex, age, and weight; collected ectoparasites; and fitted a global positioning system (GPS) collar unit (TGW-4570-4 GPS/Iridium) equipped with a CR-2A automatic release mechanism (Telonics, AZ, USA). Anesthesia was reversed with atipamezole (0.8 mg kg) as soon as the animal handling was finished, about 60 min after the initial dose. The animals then woke up in approximately 10 min and walked away. We always made capture attempts before 15:00 h in order to allow sufficient time for the animal to fully recover before dark, when predation risk increases. All procedures followed standard protocols approved by the Ministry of Environment of Panama (permit No. SE/A-104-15) and the Research Ethics Committee of El Colegio de la Frontera Sur, Mexico (CEI-O-068/16).

The GPS collars were programmed to get an hourly fix for 14 months, and data were scheduled to be transferred to the satellite every 2 days. Data remained on the sending list for 10 days, and if not successfully sent, it was stored onboard. Due to the lack of signal from the collars after the release date, and the rugged terrain, we could not recover the collars to extract all the data stored onboard.

6.2.3 Data Analysis and Home Range

A movement path usually represents a complex mixture of different movement behaviors that occur on different spatiotemporal scales, e.g., small-scale foraging versus long-distance migration or dispersal (Fleming et al. 2014). Reliably decomposing this mix of multiple movement modes is challenging albeit necessary to understand the movement pattern of an individual. A common tool used to analyze multiple movement modes is the set of discrete-time correlated random walk models (Calabrese et al. 2016). Nevertheless, these models depend heavily on the sampling schedule, i.e., they assume that the relocation data are independent and identically distributed, and consequently, they are not able to detect and disentangle the underlying movement behaviors. This is especially true if the data feature irregular gaps, as is frequent when using GPS collars in dense tropical forests (Fleming et al. 2014; Hofman et al. 2016). As a result, the risk of underestimating the home range is higher when analyzing sparse, autocorrelated data in conventional frameworks (Fleming et al. 2015). As an alternative, the recent development in modeling movement as a continuous-time stochastic process allows to account for the inherent autocorrelated structure of the animal tracking data to identify the multiple movement modes. Most importantly, this method is able to accommodate data with uneven sampling intervals, gaps, and short periods of data collection (Fleming et al. 2014, 2015). As this was our case, we favored this approach over more traditional methods because it would allow yielding better home range estimates.

Following the workflow provided by Calabrese et al. (2016), we used the semi-variance approach developed by Fleming et al. (2014) to generate continuous-space and continuous-time stochastic movement models. We preliminary examined whether the herds of WLPs displayed home ranging behavior. For each marked individual, we inspected the empirical semi-variogram which is a plot of the semi-variance in relocations as a function of the time lag among observations. In other words, it measures the distance between two relocations and computes the variability of distances among all relocation pairs with the same time lag. The herds were considered range residents as the semi-variograms displayed an asymptote. It is indeed unlikely that animals travel more than their home range diameter even in longer periods (Fleming et al. 2014). In addition, the inspection of the semi-variograms at short lags showed an upward curve suggesting directional persistence (or velocity autocorrelation) in the data. Subsequently, we fitted three candidate movement models that featured a home range with the R package *ctmm* (Fleming and Calabrese 2015). (1) The independent identically distributed (IID) process features the home range, but not the position autocorrelation, whereas (2) the Ornstein-Uhlenbeck process (OU – Brownian motion within a home range, i.e., random, undirected movement) features autocorrelated positions that are described by two parameters, i.e., home range crossing time (days) and variance (km^2). (3) The Ornstein-Uhlenbeck motion model with foraging included (OUF) features both position and velocity autocorrelation that are described by three parameters, i.e., home range crossing time (days), variance (km^2), and velocity autocorrelation time scale (min). OUF assumes regular Brownian motion while foraging on a very short time scale and Ornstein-Uhlenbeck motion on longer time scales. Models were fitted via maximum likelihood and ranked based on the AICc (Akaike 1974).

Once an appropriate continuous-time model was selected, we estimated the home range conditional to that model with an autocorrelated kernel density estimation (AKDE), which can accommodate sparse data. For the herd that yielded sufficient data, we tested the effect of seasonality by analyzing subsets of fixes in the movement path. Relocations from 15 December to 15 May were considered as dry season data, while data from the rest of the year were included in the wet season subset. Home range overlap was estimated based on selected models.

In order to reliably compare the behavior of the WLPs in the Darién forest to that in other biomes (Signer et al. 2015), we also estimated the home range using tools that have been used more frequently, i.e., the minimum convex polygon (MCP) and the kernel density estimators (KDE), with the *rhr* package in R (Signer and Balkenhol 2015). We estimated a core home range area as well, using the method of Seaman and Powell (1990) as provided in the *rhr* package. MCP and KDE tools are derived from discrete-time models, making them less robust when dealing with irregular and sparse relocation data (Fleming et al. 2015). However, Signer et al. (2015) found that although the choice of the home range estimator can change considerably the absolute size of the home range, it was negligible for assessing relative differences. They argued that home range estimators should be seen as an index of an animal's requirements, and as a mean to explore changes in space-use patterns over time or space, which is what we wanted to test here (i.e., difference between seasons).

6.3 Results

6.3.1 Captures, GPS Fix Rate, and Data Period Transmission

On 10 May and 14 July 2016, we successfully captured a female, "Serafina," and a male, "Cachito," WLPs (both were named after our Emberá assistants) from herds of about 100 and 70 individuals, respectively. It took three and two expeditions of 2 days each to successfully capture these animals. Cachito weighted 36 kg, and although we were not able to weight Serafina, we estimated a weight of *ca.* 40 kg because she was bigger than Cachito. Based on the telemetry data from the days right after the capture, both individuals managed to rapidly rejoin their herd.

Serafina's collar transmitted 2127 fixes during 466 days (11 May 2016–20 August 2017) hence a successful fixing rate of 19% (based on an expected fix schedule of 1 h). This allowed us to calculate several home ranges, i.e., total, dry season, and wet season, the latter one by combining the relocations of the 2016 and 2017 wet seasons, as they spatially overlapped (Fig. 6.3). Cachito's collar had a slightly higher successful fixing rate of 35%, but for unknown reasons only worked for 33 days (17 July–19 August 2017), during which time it transmitted 280 fixes.

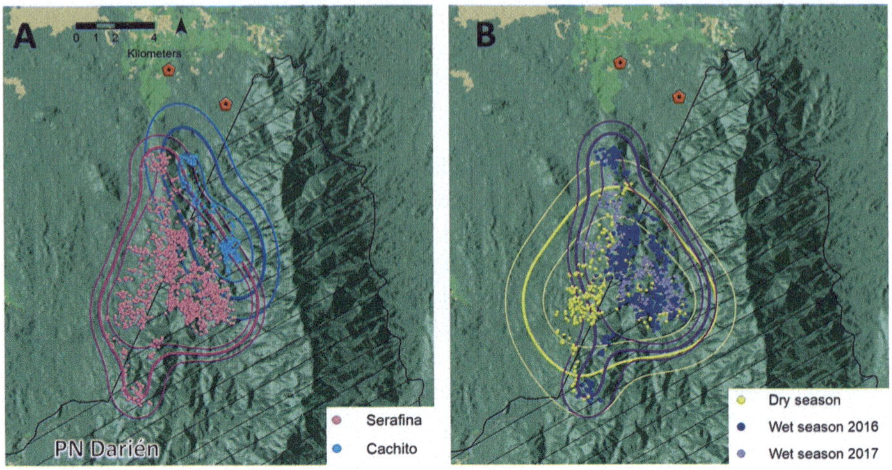

Fig. 6.3 Autocorrelated kernel density estimate home ranges of the marked white-lipped peccaries in the Darién National Park (dashed area) and its buffer area. (**a**) All relocations and home ranges (bold contour lines) with its 95% confidence intervals (light contour lines) of Serafina and Cachito. (**b**) Seasonal relocations of Serafina and home ranges (bold contour lines) with its 95% confidence intervals (light contour lines) during the dry season and the wet season. Land cover as in Figs. 6.1 and 6.2. Pijibasal and Rancho Frio are depicted in orange

6.3.2 Home Range

The scatterplots of the WLPs relocations suggested that both herds were range residents with no obvious migrations during the time they were monitored (Fig. 6.3). This behavior was supported by the asymptotes in the semi-variograms (Fig. 6.4). The empirical semi-variograms of each WLP were best approximated by the semi-variance function of the OUF that included all the main features of the semi-variograms (i.e., home range, position autocorrelation, and velocity autocorrelation; Table 6.1). The selection of this model gave evidence of the WLP home ranging behavior both during the wet and the dry seasons, and even for Cachito's herd that was only monitored for 33 days. At very short time scales, the animals displayed random, undirected movement, indicative of foraging.

Conditioning on the selected OUF models, Serafina's home range using 15 months of relocations data was estimated at *ca.* 58 (95% CI: 42.7–76.1) km^2 that she could traverse in 5.5 (95% CI: 3.9–7.9) days, and the average distance traveled per day reached 8.6 (95% CI: 8.4–8.9) km. During the wet season the estimated home range of Serafina's group was *ca.* 52 (95% CI: 34.9–70.3) km^2 that she could cross in approximately 6 (95% CI: 3.8–9.2) days as the average distance traveled per day was 8 (95% CI: 7.7–8.3) km. In contrast, during the dry season, home range increased by *ca.* 20%, to reach 64 (95% CI: 36.1–99.6) km^2 that she could traverse in fewer days, i.e., 4.5 (95% CI: 2.2–9.1) days because the average distance traveled per day was 9.7 (95% CI: 9.3–10.1) km. The overlap of the confidence intervals

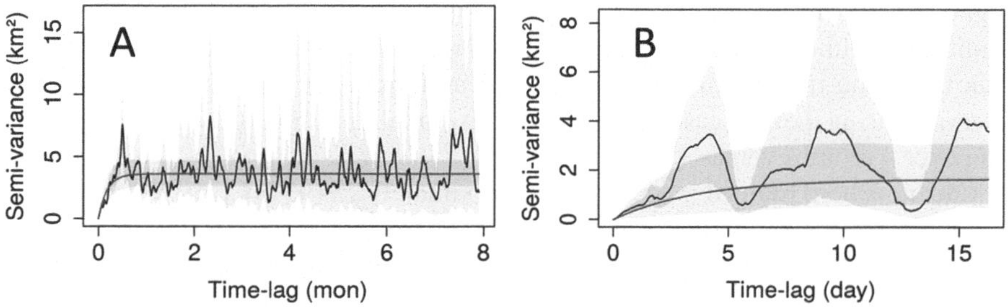

Fig. 6.4 Semi-variograms of the GPS fixes of two resident white-lipped peccaries from different herds (**a**, Serafina's, and **b**, Cachito's) in the Darién National Park. In both cases, the semi-variance reached an asymptote after a few days which represents the time to traverse their home range. The Ornstein-Uhlenbeck foraging model was fitted (purple line) with its 95% CI (purple shading)

Table 6.1 Movement parameters and home range sizes with their 95% confidence intervals in brackets for two herds of white-lipped peccaries in Darién, Panama. Home range sizes were estimated through different levels via autocorrelated kernel density estimate (AKDE); kernel density estimate (KDE) with 50, 95, and 100% isopleths; and 50, 95, and 100% minimum convex polygon (MCP)

		Serafina		Cachito
Group size		80–100		70
Period	Total	Wet season	Dry season	
No. of fixes/days tracked	2127/466	1404/315	723/151	280/33
Movement model				
OUF anisotropic	0.00	0.00	0.00	0.00
OU	361.48	194.30	153.30	39.70
IID	19057.79	12684.45	5934.90	1850.42
Movement parameters				
Home range crossing time (day)	5.5 (3.9–7.9)	5.9 (3.8–9.2)	4.5 (2.2–9.1)	3.7 (1.0–13.6)
Velocity autocorrelation time scale (min)	25.5 (23.0–28.2)	23.0 (20.1–26.3)	28.8 (24.5–34.1)	24.9 (17.9–34.8)
Average distance traveled (km/day)	8.6 (8.4–8.9)	8.0 (7.7–8.3)	9.7 (9.3–10.1)	7.2 (6.6–7.9)
Home range (km^2)				
AKDE[a]	58.2 (42.7–76.1)	51.1 (34.9–70.3)	63.9 (36.1–99.6)	24.9 (10.5–45. 47)
KDE 50	10.1	8.8	9.6	4.3
95	45.9	32.6	34.4	19.4
100	118.1	101.7	90.9	54.3
MCP 50	10.9	8.6	9.4	3.5
95	44.2	32.4	22.7	8.1
100	51.6	34.1	26.9	8.3
Core area (level)	29.1 (84.4)	26.6 (90.7)	28.9 (90.9)	12.5 (83.6)

[a]Autocorrelated kernel density estimation conditioned to the best supported OUF model

suggested that the difference in home range size between the two seasons was not significant. Although the estimated home range during the dry season was larger than the home range estimated over the 15 months, there were larger confidence intervals during the dry season, which is due to a smaller sample size (i.e., relocations data over 4 months versus 15 months).

Cachito's herd roamed in an area of approximately 25 (95% CI, 10.5–45.47) km^2 during the 33 days this herd was monitored, and it would take 3.7 (95% CI, 1.0–13.6) days to cover it (Table 6.1). The estimated home ranges in Darién were smaller than those in most other areas in the Neotropics, but group size was larger (Table 6.2).

When using MCP and 95% KDE estimates, the home range estimated was smaller (8-51 km^2 and 19-46 km^2 using the MCP and 95% KDE respectively, Table 6.1). The slightly larger home range during the dry season was also observed when using 95% KDE and the core area. However, when using the MCP, the home range estimated was larger during the wet than the dry season (32 versus 23 km^2).

Although the home range boundaries (based on the AKDE) of the two herds overlapped, the relocations never did, i.e., Serafina was never recorded in the area that Cachito used at the same time (Fig. 6.3a). The two herds wandered in the core and buffer areas of the Darién NP, oblivious to the manmade park boundaries, but they did not venture to areas that were not forested or disturbed. They always remained in areas of primary forest. They also used both lowlands and rugged terrain, but they did not cross the Pirre mountain range.

The overlap index of Serafina's home ranges between the wet seasons of 2016 and 2017 was 0.95 (95% CI: 0.81–0.99), which justifies our decision to pool the data of both wet seasons. Between the wet and dry seasons, Serafina home ranges

Table 6.2 Home range estimates (km^2) and group size of white-lipped peccary herds in several localities in the Neotropics

	Home range			Group	
Locality	AKDE	95% KDE	100% MCP	size	Source
Darién	25–59	19–46	8–51	70–100	This study
Calakmul	–	39–98	23–121	20–31	Reyna-Hurtado et al. (2009)
Calakmul	62–312	40–140	47–154	25–30	Moreira Ramírez (2017)
Maya Forest	156	99	138	46	Moreira Ramírez (2017)
Southern Belize	55	70	34	60	Hofman et al. (2016)
Maraca Island	–	–	22–110	40–124	Fragoso (1998)
Manu, Amazon	–	–	–	90–138	Kiltie and Terborgh (1983)
Atlantic forest[a]	–	–	21	41 (203)	Keuroghlian et al. (2004)
Cerrado	–	91–277	–	65–110	De Almeida Jácomo et al. (2013)
Pantanal	–	46–70	76–80 (95%)	56	Keuroghlian et al. (2014)

[a]In their study, Keuroghlian et al. (2004) monitored three subherds that sometimes merged into a single large herd

overlapped by 86% (95% CI: 0.68–0.97), while the home ranges of Cachito and Serafina overlapped by 41% (95% CI: 0.19–0.69) during the 2016 wet season (when the two collars were working at the same time).

6.4 Discussion

This is the first study on the spatial ecology of the endangered WLP in the moist tropical forests of Panama. We focused on estimating the home range of two herds of WLPs in the Darién NP, an area that was identified as one of the most important strongholds for the species in Mesoamerica (Reyna-Hurtado et al. 2017).

Our results corroborate the annual home ranging behavior of the WLP. Whichever estimator we used, the home range size in Darién was within the lower ranges reported in other studies from different biomes (Table 6.2), and particularly similar to the Maya Mountains in southern Belize (i.e., 55 km^2; Hofman et al. 2016). However, it is important to consider that given the short period of time the data was collected for Cachito's herd, its actual home range is likely to be larger than the estimated 25 km^2. The group of Serafina (followed during 15 months) did not show any pattern of nomadic or migratory movement. Although it slightly expanded its home range during the dry season, it did not suggest a seasonal movement into a new range, because the home ranges of the dry and wet seasons largely overlapped. The nonsignificant larger home range during the dry season contrasts with the Maya Forest in northern Guatemala and southeastern Mexico, where WLPs remain in a relatively small area during the dry season and considerably expand their home range in the wet season (Reyna-Hurtado et al. 2012; Moreira Ramírez 2017). The Maya Forest covered by the seasonal tropical dry forest receives less rainfall (900–2000 mm annually) and has a marked dry season during which animals rely on a few ponds of water that gradually dry up (Reyna-Hurtado et al. 2009; Moreira Ramírez 2017). Consequently, water is the primary limiting factor in the Maya Forest (Reyna-Hurtado et al. 2009), whereas in Darién water is widely available all year long, even during the dry season. However, while backtracking the WLP herds, we noticed that at fine scales, the WLPs would systematically search for water, as they daily visited two to three large muddy ponds and crossed several streams and larger rivers. Similarly, when following them in the field, we could clearly notice how their movement pattern alternated between short undirected movement with frequent sharp turns and climbing up steep hills and directional movement straight to the ponds or rivers, usually using ridges. Besides, it was very interesting to uncover several tunnels and large burrow-like holes that were used by WLPs (based on fresh tracks found) and that were most likely excavated by them (Plate 6.1). Photos from camera traps and relocation data confirmed the frequent visits to these sites (Plate 6.1). Although, the exact reason for their use has yet to be determined, they probably have the same function than clay licks in the Amazon (Tobler et al. 2009). Our guides also mentioned that the WLPs seek refuge in these tunnels and holes when they are being hunted.

Plate 6.1 Tunnels used by white-lipped peccaries in the Darién National Park, Panama

Besides water, food availability has been identified as a primary driver in herbivore space use, with both food density and high-quality food having a negative effect on home range size (van Beest et al. 2011). Several studies offer clear examples of the impact of fruit variability on the WLP movement pattern in many biomes, e.g., Keuroghlian et al. (2004) in the Atlantic forest, De Almeida Jácomo et al. (2013) in the Cerrado, Keuroghlian et al. (2014) in the Pantanal, Fragoso (1998) in the Amazon, and Carrillo et al. (2002) in the Pacific wet forest. In our study, the slight increase in home range size during the dry season might be associated with lower fruit availability in the drier months. Nevertheless, the seasonal difference in relocations was not significant – as shown by the seasonal home range overlap – which suggests that the WLP did not experience severe food scarcity over the 15 months surveyed.

In Panama, although fruit production and availability peaks at the beginning of the wet season, heavy rainfall coupled with the warm temperature all year long favors a flourishing productivity throughout the year. Our results suggest that habitat conditions in the forests of Pirre, Darién, provide an excellent balance of water and food: (1) to fulfill the requirements of the resident WLP population within a relatively small home range and (2) for the groups to remain in the area.

Our home range estimate is relatively small given the size of the herds in Darién, assuming that home range increases with herd size (Fragoso 1998). We estimated the two groups we monitored to be composed of at least 70–100 individuals. According to personal observations and local hunters and based on camera-trap data, this seems to be common in Darién, even though we are aware of the presence of one or two groups that are smaller (i.e., *ca.* 30–40 individuals) probably because of poachers who regularly decimate the herds as they roam closer to human settlements on the edge of the park. In general, herd size in our study site is quite large in comparison with other sites throughout the Neotropics (reviewed in Reyna-Hurtado et al. 2016; Table 6.2). As far as we are aware, WLP herds reach such size or larger only in the Amazonia (Kiltie and Terborgh 1983; Fragoso 1998). Reyna-Hurtado et al. (2016) found that WLP group size changes as a function of climatic, ecological, and human factors. Larger WLP groups usually occur in areas that are further away from anthropogenic disturbance with low hunting pressure, and in wet areas, probably due to the positive relationship between rainfall and primary productivity – hence fruit availability (Reyna-Hurtado et al. 2016). Therefore, the large herd size in our study site is probably indicative of the good health of the forest ecosystem in Darién.

Our data did not allow us to estimate the density of individuals, but our camera-trap data suggest a relatively high abundance of WLP in our study area (Meyer et al. 2015, 2016). Moreover, several studies report an overlap between groups (Fragoso 1998; Reyna-Hurtado et al. 2009), and although we cannot confirm this pattern based on our telemetry data, we often saw tracks from different groups on the same day. We also came across tracks in many areas that did not coincide with the herds we monitored. Such evidence indicate that there are many large herds that still roam the area of Pirre and suggest a good conservation status of the WLP in Darién.

Nevertheless, the apparent healthy WLP population, the protection status, and remoteness of the Darién NP do not mean that the species is safe from menace. Hunting, whether for subsistence or illegal, remains a major threat for the WLP, which is also the favorite game species in Panama (Moreno and Meyer 2014). We found many trails – most likely made and used by both traffickers and poachers – that were 2- to 3-day walk from the closest communities, deep inside the core area of the NP. Although less common than a few decades ago, Darién is still inhabited by many experienced hunters who are willing to travel far and for several days to specifically hunt WLPs (pers. obs. and pers. com. with local people). For example, once that we were following the tracks of Serafina's group, two poachers who were ahead of us by an hour (as later discovered on our camera trap) reached the herd before us, killed several individuals, and later sold the meat for US$2 per pound in the nearby communities. Unlike other species such as the Baird's tapir, *Tapirus bairdii*, for which poaching is severely punished, WLP hunting often occurs with poor control and usually leads to no consequences (pers. obs.). Subsistence hunting by indigenous is tolerated, but reliable information about WLP hunting is not easy to collect due to a lack of trust and because a large proportion of the inhabitants of Darién appreciate and consume that meat. Anytime a herd is known to be in the area, people pass the message to hunt them together. Reports of culling of a large proportion of the herd at a single hunting event are not unusual, both in the Darién province (e.g., Yavisa, Boca de Cupe, Canglón Forest Reserve, and Serranía de Majé until 1989 when WLPs were seen for the last time), including deep inside the Darien NP (e.g., Cana, Moreno 2006), and in other regions (e.g., Comarca Guna Yala). As a result, WLPs are nowadays hardly recorded outside the boundaries of the NP and rarely venture close to human communities, as shown by our telemetry data. More troubling is that such harvest practices can have deleterious effects on the WLP population and have contributed to drive the species to local extinctions in several sites throughout Panama (Enders 1935; Moreno and Meyer 2014; Meyer et al. 2015, 2016).

It is therefore urgent to prioritize conservation actions in order to improve WLP population viability in Darién and guarantee its persistence in Panama. Altrichter et al. (2012) identified the size of conservation units and the level of deforestation and hunting as critical factors for the survival of the WLP. Despite several studies that showed that the WLP can persist in fragmented landscapes (de Almeida Jácomo et al. 2013), and survive in forest patches of about 20 km^2 (Keuroghlian et al. 2004), our study in Darién showed that, whichever home range estimate we used, a substantial area of intact forest is needed for the herds to fulfill their requirements. Given these results, restoration of suitable habitat and connectivity will be challenging in some areas of Panama, particularly in Central Panama. There, a large proportion of the landscape consists in a mosaic of small forest patches, sometimes as small as a few hectares, in a matrix of agricultural lands and cattle pastures, with a high human density and a dense road network (Meyer et al. 2015).

In the light of our findings, conservation efforts should be made at the landscape level by focusing on the preservation of large and continuous forested areas such as the Darién, the Comarca Guna, and the transboundary NP of La Amistad near Costa Rica. All were identified as the last strongholds for the WLP in Panama (Moreno and Meyer 2014). Nevertheless, as some development projects threaten the pristine condition of the Darién NP, we stress the importance to assess the impacts of such projects and take adequate measures to mitigate them. Of particular concern are (1) the power transmission corridors from Central Panama to Colombia and (2) the possible expansion of the Pan-American Highway from Yavisa inside the Darién NP (shown in Fig. 6.1), which will most likely result in drastic changes in the forests of the park and its wildlife. Furthermore, while focusing on protecting their habitat, the survival of the WLP could not be adequately achieved without a tighter hunting control. Fostering greater awareness toward the WLP conservation in key areas may contribute to reducing poaching pressure.

Last but not least, the WLP remains largely overlooked in spite of its critical conservation status. This reflects the general lack of awareness toward the relevance of conserving the species and its habitat in Panama, which is problematic since the WLP usually co-occurs with other important species such as the jaguar *Panthera onca*, the tapir, and the giant anteater *Myrmecophaga tridactyla*. Therefore, including the WLP on the species conservation agenda of the government of Panama will be a first step to help for the survival of this and others imperiled yet significant species in tropical forests.

Globally, this study highlights the potential of modeling movement as a continuous-time stochastic process coupled with statistical methods for autocorrelated data to offer insights into animal space use in time, even when relocation data are irregular and sparse. We were therefore able to estimate in a rigorous manner the home range of WLP in Darién. Our results increased our understanding of the WLP movement and habitat requirements in the moist tropical forests and provided much needed information to inform effectively on its management and conservation in Panama.

Acknowledgments We thank the Secretaria Nacional de Ciencia, Tecnología e Innovación de Panamá (SENACYT proyecto FID 14-145), the Rufford Foundation, IDEA WILD, GEMAS/Fondo Darién, Fundación Natura, D. Anderson from Cat Haven, N. Correa from Asociación Panamericana para la Conservación, A. Perez from Green Rainforest, and R. Morales for funding, equipment, and logistics, especially A. Lu and L. Hernandez. We are extremely grateful to our guides from Pijibasal A. Chami, T. Contreras, U. Contreras, M. Manyoma, and C. Contreras; our wildlife veterinarian L. Pretelt, as well as A. Artavia and E. Sanchez for assistance during the captures. Thanks to the Ministry of Environment of Panama and C. Sanchez, director of the Darién NP, for permits; the game wardens at Rancho Frio, N. Correa, Y. Yanguez, and people of Pijibasal for housing and welcoming us. R. Kays and J. Signer provided valuable advice about the analysis. NM received a scholarship from the Consejo Nacional para la Ciencia y Tecnología de México (scholar # 576309). Data were uploaded on Movebank (www.movebank.org).

References

Akaike H (1974) A new look at the statistical model identification. IEEE Trans Autom Control 19:716–723

Altrichter M, Taber A, Beck H et al (2012) Range-wide declines of a key Neotropical ecosystem architect, the Near Threatened white-lipped peccary *Tayassu pecari*. Oryx 46:87–98

Beck A (2006) A review of peccary-pal interactions and their ecological ramifications across the Neotropics. J Mammal 87:519–530

Calabrese JM, Fleming CH, Gurarie E (2016) ctmm: an R package for analyzing animal relocation data as a continuous-time stochastic process. Methods Ecol Evol 7:1124–1132

Carrillo E, Saenz JC, Fuller T (2002) Movements and activities of white-lipped peccaries in Corcovado National Park, Costa Rica. Biol Conserv 108(3):317–324. https://doi.org/10.1016/S0006-3207(02)00118-0

Cushman SA, McRae B, Adriaensen F, Beier P, Shirley M, Zeller K (2013) Biological corridors and connectivity. In: Macdonald DW, Willis KJ (eds) Key topics in conservation biology 2. John Wiley and Sons, Oxford. https://doi.org/10.1002/9781118520178.ch21

De Almeida Jácomo AT, Furtado MM, Kashivakua CK et al (2013) White-lipped peccary home-range size in a protected area and farmland in the central Brazilian grasslands. J Mammal 94:137–145

Enders RK (1935) Mammalian life histories from Barro Colorado Island, Panama. Bull Mus Comp Zool 78:383–502

Fleming CH, Calabrese JM (2015) ctmm: Continuous-time movement modeling. R package version 0.4.1

Fleming CH, Calabrese JM, Mueller T et al (2014) From fine-scale foraging to home ranges: a semivariance approach to identifying movement modes across spatiotemporal scales. Am Nat 183:154–167

Fleming CH, Fagan WF, Mueller T, Olson KA, Leimgruber P, Calabrese JM (2015) Rigorous home range estimation with movement data: a new autocorrelated kernel density estimator. Ecology 96:1182–1188
Foerster R, Vaughan C (2002) Home range, habitat use, and activity of Baird's Tapir in Costa Rica. Biotropica 34:345–353
Fragoso JMV (1998) Home range and movement patterns of white-lipped peccary (*Tayassu pecari*) herds in the Northern Brazilian Amazon. Biotropica 30:458–469
Glanz WE (1982) The terrestrial mammal fauna of Barro Colorado Island: censuses and long-term changes. In: Leigh EG, Rand AS, Windsor DM (eds) The ecology of a tropical forest: seasonal rhythms and long-term changes. Smithsonian Institution Press, Washington, DC
Goldman EA (1920) Mammals of Panama (with thirty-nine plates). Smithson Misc Collect 69:5. https://doi.org/10.5962/bhl.title.15684
Hofman M, Signer J, Hayward MW et al (2016) Spatial ecology of a herd of white-lipped peccaries (*Tayassu pecari*) in Belize using GPS telemetry: challenges and preliminary results. Therya 7:21–37
Kays R, Crofoot MC, Jetz W, Wikelski M (2015) Terrestrial animal tracking as an eye on life and planet. Science 348:aaa2478
Keuroghlian A, Eaton DP, Longland WS (2004) Area use by white-lipped and collared peccaries (*Tayassu pecari* and *Tayassu tajacu*) in a tropical forest fragment. Biol Conserv 120:411–425
Keuroghlian A, Andrade Santos MC, Eaton DP (2014) The effects of deforestation on white-lipped peccary (*Tayassu pecari*) home range in the southern Pantanal. Mammalia. https://doi.org/10.1515/mammalia-2014-0094
Kiltie RA, Terborgh J (1983) Observations on the behavior of rain forest peccaries in Peru: why do white-lipped peccaries form herds? Z Tierpsychol 62:241–255
Laurance WF, Croes BM, Tchignoumba L et al (2006) Impacts of roads and hunting on central African rainforest mammals. Conserv Biol 20:1251–1261
Meyer NFV, Esser HJ, Moreno R et al (2015) An assessment of the terrestrial mamma communities in forests of Central Panama using camera-trap surveys. J Nat Conserv 26:28–35
Meyer NFV, Moreno R, Sanches E et al (2016) Do protected areas in Panama support intact assemblages of ungulates? Therya 7:65–76
Ministerio de Ambiente de Panamá (2016) Resolución N. DM-0657-2016. Gaceta Oficial Digital, jueves 29 de diciembre de 2016, N. 28187-A
Moreira Ramírez JF (2017) Movimientos del pecarí de labios blancos en relación con la disponibilidad de agua y cacería en la Selva Maya de Guatemala y México. Tesis de Doctorado, El Colegio de la Frontera Sur, Mexico
Moreno R (2006) Parámetros poblacionales y aspectos ecológicos de los felinos y sus presas en Cana, Parque Nacional Darién, Panamá. Tesis de Maestría. Heredia, Costa Rica
Moreno R, Meyer NFV (2014) Distribution and conservation status of the white-lipped peccary (*Tayassu pecari*) in Panama. Suiform Soundings 13:32–35
Nathan R, Getz WM, Revilla E et al (2008) A movement ecology paradigm for unifying organismal movement research. Proc Natl Acad Sci U S A 1051:9052–19059
Pardini R, de Arruda Bueno A, Gardner TA et al (2010) Beyond the fragmentation threshold hypothesis: regime shifts in biodiversity across fragmented landscapes. PLoS One 5(10):e13666
Reyna-Hurtado R, Rojas-Flores E, Tanner GW (2009) Home range and habitat preferences of white-lipped peccaries (*Tayassu pecari*) in Calakmul, Campeche, Mexico. J Mammal 90:1199–1209
Reyna-Hurtado R, Naranjo E, Chapman CA, Tanner GW (2010) Hunting and the conservation of a social ungulate: the white-lipped peccary *Tayassu pecari* in the Calakmul, Mexico. Oryx 44:88–96
Reyna-Hurtado R, Chapman CA, Calme S, Pedersen E (2012) Searching in heterogeneous environments: foraging strategies in the white-lipped peccary (*Tayassu pecari*). J Mammal 93:124–133
Reyna-Hurtado R, Beck H, Altrichter M, Chapman CA, Bonnell TR, Keuroghlian A, Desbiez AL, Moreira-Ramirez JF, O'Farrill G, Fragoso J, Naranjo EJ (2016) What ecological and

anthropogenic factors affect group size in white-lipped peccaries (*Tayassu pecari*)? Biotropica 48:246–254

Reyna-Hurtado R, Radachowsky J, Mcloughlin L et al (2017) White-lipped peccary in Mesoamerica: status, threats and conservation actions. Suiform Soundings 15:31–35

Ripple WJ, Newsome TM, Wolf C, Dirzo R, Everatt KT, Galetti M, Hayward MW, Kerley GIH, Levi T, Lindsey PA, Macdonald DW, Malhi Y, Painter LE, Sandom CJ, Terborgh J, Van Valkenburgh B (2015) Collapse of the world's largest herbivores. Sci Adv 1:e1400103

Schipper J, Chanson JS, Chiozza F et al (2008) The status of the world's land and marine mammals: diversity, threat, and knowledge. Science 322:225–230

Seaman DE, Powell RA (1990) Identifying patterns and intensity of home range use. Bears Biol Manag 8:243–249

Signer J, Balkenhol N (2015) Reproducible home ranges (rhr): a new, user friendly R package for analyses of wildlife telemetry data. Wildl Soc Bull 39:358–363

Signer J, Balkenhol N, Ditmer M, Fieberg J (2015) Does estimator choice influence our ability to detect changes in home range size? Anim Biotelemetry. https://doi.org/10.1186/s40317-015-0051-x

Tobler MW, Carillo-Percastegui, Powell G (2009) Habitat use, activity patterns and use of mineral licks by five species of ungulate in south-eastern Peru. J Trop Ecol 25:261–270

Urquiza-Haas T, Perez CA, Dolman PM (2011) Large vertebrate responses to forest cover and hunting pressure in communal landholdings and protected areas of the Yucatan Peninsula, Mexico. Anim Conserv 14:271–282

Van Beest FM, Rivrud IM, Loe LE et al (2011) What determines variation in home range size across spatiotemporal scales in a large browsing herbivore? J Anim Ecol 80:771–785

Webb SD (2006) The Great American Biotic Interchange: patterns and processes. Ann Mo Bot Gard 93:245–257

Chapter 7
Movements of Neotropical Forest Deer: What Do We Know?

Francisco Grotta-Neto and José Maurício Barbanti Duarte

7.1 Introduction

The deer family, Cervidae (clade Cetartiodactyla, suborder Ruminantia), comprises ruminants with a divided, fermenting stomach, in association with the grazing nature of their diet of grasses, forbs, and leaves (Duarte and Merino 1997; Eisenberg and Redford 1999; Oliveira and Duarte 2006). Antlers are deciduous and found only in males (with the exception of being found in both sexes in the reindeer, *Rangifer tarandus*) (Duarte and Merino 1997). Deer are found on all continents except Australia and Antarctica (Eisenberg and Redford 1999), and today there are 56 recognized extant species (IUCN 2018).

In the Americas, deer are most diverse in the Neotropical region, from Mexico to southern South America (Fig. 7.1), where species range from 38 to 127 cm in height and weight from 5 to 150 kg (Eisenberg and Redford 1999; Duarte and González 2010). Morphological diversity in the neotropics is associated with their adaptations to a wide variety of habitat types and the many different ecoregions in which they are found (Merino and Rossi 2010). Due to that diversity, studies of free-ranging deer are unequally distributed among the different species because of the difficulty in observing or capturing some of those species in the wild. Thus, some species and

F. Grotta-Neto
Núcleo de Pesquisa e Conservação de Cervídeos (NUPECCE), Faculdade de Ciências Agrárias e Veterinárias da Universidade Estadual Paulista (UNESP), Campus de Jaboticabal, São Paulo, Brazil

Laboratório de Biodiversidade, Conservação e Ecologia de Animais Silvestres (LABCEAS), Universidade Federal do Paraná (UFPR), Curitiba, Brazil

J. M. B. Duarte (✉)
Núcleo de Pesquisa e Conservação de Cervídeos (NUPECCE), Faculdade de Ciências Agrárias e Veterinárias da Universidade Estadual Paulista (UNESP), Campus de Jaboticabal, São Paulo, Brazil
e-mail: barbanti@fcav.unesp.br

R. Reyna-Hurtado, C. A. Chapman (eds.), *Movement Ecology of Neotropical Forest Mammals*, https://doi.org/10.1007/978-3-030-03463-4_7

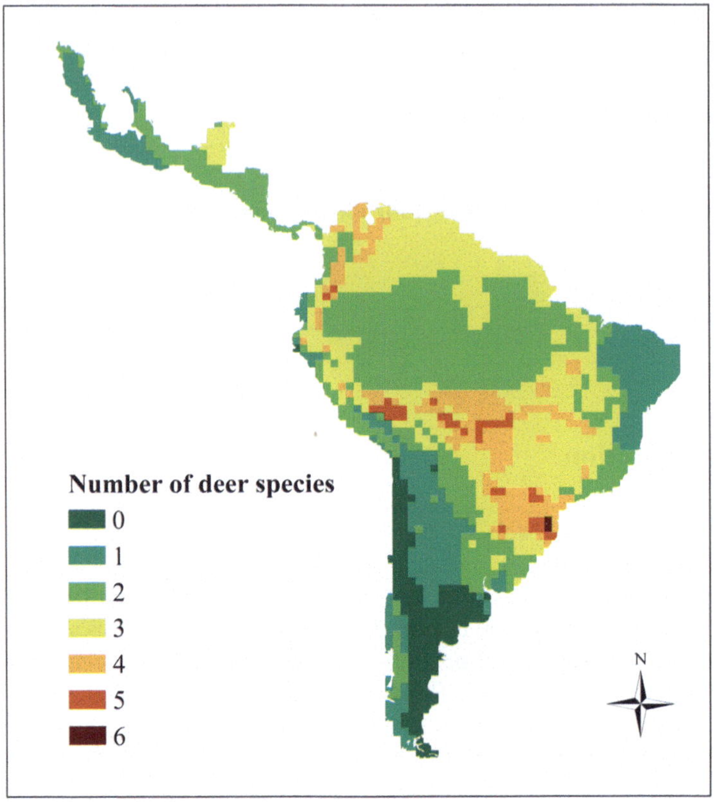

Fig. 7.1 Deer species richness in the neotropics

some regions where they are found have been neglected, and very little is known of their biology in the wild (Taber et al. 2016).

Similarly, behavior in general is poorly studied in Neotropical deer due to a variety of reasons, from difficulty of study in the field (due to the nature of the species of interest) to the high costs associated with the equipment necessary to follow animals. Thus, studies with Neotropical terrestrial mammals using radiotelemetry have largely been ignored (Zucco 2014), and movement patterns studies with deer species are not exception. Documenting animal movement can be key to understanding how an animal interacts with its environment and in communities wherever they are found. Movements of free-ranging animal are related to foraging, reproduction, dispersal, and spatial structure of the landscape and can be influenced by internal and/or external factors such as sex, reproductive phase, food availability, anthropic factors, climate, and interspecific interaction (Morales and Ellner 2002; Lusseau 2004; Nathan et al. 2008; Webb et al. 2010; Baguette et al. 2014). In this chapter, we provide a review of what is known about Neotropical forest deer movement patterns along with summary of information about basic biology of some of the genera and species. Additionally, we discuss those factors that limit or impede research on deer species in the wild.

7.2 A Brief History of Neotropical Deer

The place of origin of the family Cervidae would be Eurasia, where some of the earliest forms have been recorded. Evolutionary history for the Neotropical deer began in the late Pliocene (around 2.5 million years ago) when North American deer reached South America by crossing the Isthmus of Panama (Eisenberg and Redford 1999; Webb 2000). In the absence of large herbivore competitors, deer spread widely and diversified, occupying a variety of habitats from closed forests to open savanna and humid lowlands to high mountains (Gilbert et al. 2006; Duarte et al. 2008). This diversification generated an uneven distribution of species richness over the various bioregions of the neotropics (Webb 2000; Gilbert et al. 2006; Duarte et al. 2008, Fig. 7.1). Today there are 17 Neotropical deer species in 6 genera (Duarte and González 2010).

The marsh deer (*Blastocerus dichotomus*, Illiger 1815) is the largest Neotropical deer and is found in seasonally flooded regions of Bolivia, Peru, Paraguay, Argentina, Uruguay, and Brazil. The pampas deer (*Ozotoceros bezoarticus*, Linnaeus 1758) is only found in open areas, including pampas, savannas, and grasslands of Argentina, Paraguay, Uruguay, Bolivia, and Brazil. Five subspecies have been described, the most recent of which is from Uruguay (González et al. 2002).

The white-tailed deer (*Odocoileus virginianus*, Zimmermann 1780) is the most widespread, found in most habitat types from forests to savannas to semi-arid deserts, from southern Canada to northern South America (including Colombia, Venezuela, Guyana, Surinam, French Guiana, Ecuador, Peru, Bolivia, and northern Brazil). The white-tailed deer is among the most often studied species of deer, especially in North America and neotropics. The white-tailed deer has adapted well to different environments, which reflect in a diversity of characteristics, like body weight, external dimensions, coat coloration, antler growth, and behavioral distinctions (Gallina et al. 2010). The genus *Hippocamelus* comprises two species: the taruka, or North Andean deer (*Hippocamelus antisensis*, d'Orbigny 1834), and the guemal or huemul, or South Andean deer (*Hippocamelus bisulcus*, Molina 1784). These are both medium-sized deer and are found in non-overlapping ranges along the cordillera of the Andes, from Peru to northern Chile and Argentina, and in southern Chile and Argentina, respectively (Eisenberg and Redford 1999). They use commonly high-altitude regions that are also rocky and steep. The genus *Pudu* also comprises two species with non-overlapping (parapatric) distributions: the Northern Pudu (*Pudu mephistopheles*, De Winton 1896) and the Southern Pudu (*Pudu puda*, Molina 1782). Pudus are the smallest deer in South America, being *P. mephistopheles* even smaller than *P. puda* (Escamilo et al. 2010). The Northern Pudu is found in the Andes of Colombia, Ecuador, and Peru, while the Southern is found in Chile and Argentina (Eisenberg and Redford 1999). The Northern Pudu is associated with temperate forests in Colombia to cloud forests in the eastern Andes of Peru (Escamilo et al. 2010). The Southern Pudu is found in temperate forests as well as secondary forests and in association with dense stands of bamboo (Jiménez 2010).

Finally, the brocket deer, in the genus *Mazama* (Rafinesque, 1817), includes the greatest number of species of Neotropical deer (ten), found in a wide variety of habitats (forests to savannas) from southern Mexico to Argentina (Eisenberg and Redford 1999). Due to a complex and poorly studied taxonomy and morphological similarities, species identification is difficult (Duarte and Merino 1997). Thus, clearly a literature review will find more information of the easier species to study.

Of all species of Neotropical deer, only *O. bezoarticus* and *B. dichotomus* are not associated with forest habitats. These two species are found in open habitats, with some differences between their habitat selection and behavior. For example, *O. bezoarticus* is often found in pastures and they tend to tolerate areas with people, and thus this species is more easy to study, and, as a consequence, it is the best-studied Neotropical deer species in South America. On the other hand, *B. dichotomus* is found in swampy, seasonally flooded areas, farther from people. To study the marsh deer requires the use of helicopters which makes their study more expensive, difficult, and dangerous.

7.3 Movement Patterns of Neotropical Forest Deer

7.3.1 **Hippocamelus bisulcus,** *the Huemul, Guemal, Güemul, or South Andean Deer*

The huemul is the second largest Neotropical deer found only in the Andes along the border of southern Chile and Argentina. It is found up to 2000 m elevation, in habitats dominated by highland steppe and rocky outcrops (Vila et al. 2010). The few studies of movement patterns tend to examine altitudinal movements at different seasons of the year and in association with reproduction. In Aysén, in Chile, these altitudinal movements have been studied in three nature reserves: Tamango Forest Reserve, La Bangala Forest Reserve, and the Candonga (Gill et al. 2008). Studies over intervals of 19–36 months followed movements of 16 adult huemuls (9 male and 7 female) using visual observations and radiotelemetry triangulation (Gill et al. 2008). Fourteen animals were fitted with VHF collars and with GPS collar. The animals with VHF collars yielded data with irregular time intervals and those with GPS collars provided fixes every 6 h. During their travels, the deer ranged from 160 to 1100 m elevation.

In general, huemul selected mid-slope habitats while avoiding the highest elevations (Gill et al. 2008). The mean distance moved for both sexes between mean summer and winter was 552 m (range 44–1219 m), while the median daily distance moved was 706 m (Gill et al. 2008). In general, during movement, huemuls tend to move over a wider elevational distance at two higher elevations sites, yet without migratory movements per se, but with a tendency to descend during the winter and ascend again during summer (Gill et al. 2008; Díaz et al. 2013). It has been suggested that these movement patterns are responses to adverse conditions as well as to find the best areas for better quality foraging. Also, appropriate habitat begins to decrease at higher elevations simply due to altitude and at lower elevations due to changing climate and the presence of forests and human activities (such as farming). Hence, the deer is relegated to these intermediate elevations.

There was a clear relationship between altitude and movement (more movement at higher altitudes), because two reserves were at higher elevations (Candonga and La Bangala), so deer in those reserves showed more seasonal movement than in the other, lower reserve (Gill et al. 2008). Also, as a consequence of the relatively small distances the deer move, it has been proposed that populations may easily become isolated and unsustainable. If so, conservation recommendations need to include maintaining corridors between appropriate habitats to assure gene flow and that habitat will always be available (Saucedo and Gill 2004; Gill et al. 2008).

Similar movement patterns were also found in Torres del Paine National Park, in southern Patagonian Chile, where snow and ice can often overlay grasses and impede grazing. There, the huemul was found to move as a response to seasonality and as a consequence of reproductive cycles and social interactions (Garay et al. 2016). For example, reproductive movements were associated with birthing (fawning), when females that lived on the slopes of hills moved down to the valleys (Garay et al. 2016). Social movements were related to the time of the fawns became yearlings, when they moved out of the family groups during the fawning period and during their dispersion at the end of winter (Garay et al. 2016).

The huemul has been observed (by park guards) to travel, on average, at about 2 km h^{-1} and nearly 4 km day^{-1} in the Torres del Paine National Park, southern of Patagonia (Rau 1980). The population of this park was studied during 10 years (Garay et al. 2016). In this study, the animals were identified individually by natural marks or scars on their bodies, coloring of body and face, and shape and length of antler in males. Moreover, 16 fawns (10 females and 6 males) were captured and marked with ear tags. The population density of Torres del Paine National Park has also remained relatively constant during the study and was suggested to be at its carrying capacity and maintained by keeping transient individuals out of their home ranges and by the dispersal of yearlings into other areas (Garay et al. 2016). While these suggestions may be true, continued study will require more sophisticated analyses of mortality and birth to test whether the population is at carrying capacity, for example. This population also did not appear to migrate, which suggests that there is plenty of space and food to make migration unnecessary. Nonetheless, huemuls can move altitudinally over short distance from areas under snow cover to areas without (Garay et al. 2016). More complex and dynamic analyses were not possible due to budget restrictions to use radios or GPS collars.

The most recent study of the huemul was about six individuals (three of each sex) in Shoonem Protected Park in southern Chubut province, Argentina (Smith-Flueck et al. 2018). In that study they found that the maximum distance moved, albeit short, from where they were caught in winter to where they spent their summer tended to be greater in females (1.82–3.85 km) than males (1.39–1.5 km) (Smith-Flueck et al. 2018). Because the distances were so small, the authors considered the deer to be nonmigratory.

7.3.2 Odocoileus virginianus

In a tropical deciduous forest at the Chamela Biological Station in Jalisco, Mexico, the behavioral and habitat use patterns of two white-tailed deer (one male and one female) were described (Sánchez-Rojas et al. 1997). Both individuals were monitored with VHF collar during wet and dry seasons (male just in dry) between 1992 and 1994. The home range of female in dry was smaller (11 ha) than wet (21 and 34 ha) season (Sánchez-Rojas et al. 1997). Daily moving distance of female in wet was greater (1.98 and 2.58 km) than dry (1.44 and 1.48 km) season, which could be related to the intense search for best foraging places in wet season (Sánchez-Rojas et al. 1997). For male, home range and daily moving distance in dry season were, respectively, 26 ha and 2.5 km. The greater distance traveled by male could be related to his energetics requirements (Sánchez-Rojas et al. 1997).

In northeastern of Mexico, home range size and distance traveled of *O. virginianus* were analyzed in relation to the amount and distribution of rainfall, environmental temperature, and other factors such as sex, physiological stage, and year (Bello et al. 2004). For this, 18 (9 males and 9 females) deer were captured and monitored with radiotelemetry from October 1994 to September 1998. In both sexes, the mean of home range size varied from 174 ± 20 to 258 ± 18 ha, with significant differences between years (Bello et al. 2004). Female ranges were smaller (0 = 193 ± 13 ha) than those of males (0 = 234 ± 14 ha), with differences among sexes just in 1997 (females 0 = 200 ± 25 and males 0 = 316 ± 25 ha), when the highest rainfall occurred during the dry period. In this way, an increase of food availability is expected with a more precipitation, allowing the animals spend more time searching for and selecting food that is high in nutrients. No significant differences were found in distances covered per day between sex and seasons. Some seasonal movement differences were in reproductive phases between years. A significant relationship was found between precipitation and distance traveled per day, but not for maximum temperature (Bello et al. 2004).

7.3.3 Mazama *spp.*

The genus *Mazama* has been described as small, solitary forest deer (Barrette 1987). The ten species comprising the genus are found in a variety of habitat types, including lowland tropical forests and savannas (Duarte and Merino 1997; Eisenberg and Redford 1999). Morphologically, all species are very similar and so field identification is difficult (Duarte and Merino 1997; Duarte et al. 2008). Additionally, these deer are shy, often nocturnal, living usually in very dense vegetation that is often difficult to reach; therefore, of the few studies of this genus, even fewer address movement patterns. For example, to use telemetry they must be captured and fitted with radio collars; however, to date, no efficient method of capture has been developed (Duarte et al. 2010). Indeed, captures have usually been opportunistic, such as during the filling of reservoirs or after being tracked by dogs (Marques and Santos-Junior 2002; Vogliotti 2003). As a consequence, most studies of these deer have been through the use of trail cameras or limited to a very few individuals by radiotelemetry (Barrientos and Maffei 1999; Leeuwenberg et al. 1999; Vogliotti 2003; Rivero et al. 2005; Oliveira-Santos 2009; Tobler et al. 2009; Antunes 2012; Rodrigues et al. 2014, 2017; Ferreguetti et al. 2015; Oliveira et al. 2016).

Movement patterns in *Mazama*, therefore, have been inferred from these trail camera and radiotelemetry studies. For example, in the Atlantic Forest of eastern Brazil, *M. bororo* has been photographed most frequently along streams as opposed to trails, and that behavior may be because traveling there is easy or it avoids predators (Vogliotti 2003). In Intervales State Park, in the Atlantic Forest of the state of São Paulo, one *M. gouazoubira* and two *M. bororo* were radio-tracked for home range and habitat use studies (Vogliotti 2003). In this study, the home ranges (estimated with minimum convex polygon – MCP) of the two *M. bororo* were 48.5 and 11.5 ha and for *M. gouazoubira* was 348.2 ha. Also, the home range of *M. gouazoubira* varied from 290.6 ha in the winter to 173.8 and 179.6 ha in the two summers. The author discusses that this difference may be related to the seasonality in food availability or reproduction, although none of these aspects were analyzed (Vogliotti 2003).

In Bolivia, home range and activity were studied in two *M. gouazoubira* (one of each sex) that were followed by radio collars, and their positions were obtained by triangulation (Barrientos and Maffei 1999). The animals were tracked from June 1998 to July 1999 (male) and from January to May 1999 (female), and their home ranges were estimated in 33.6 and 24.9 ha, respectively. Similarly, an adult female of *M. gouazoubira* was radio-tagged between September and December 1990 in the Brazilian *cerrado* (savanna) (Leeuwenberg et al. 1999). During this time, its home range varied monthly between 28.4 and 41.9 ha. Also in Brazil, five *M. gouazoubira* were followed in the Pantanal using radio collars and GPS between 2 and 17 weeks during 2010 and 2011 (Antunes 2012). The home ranges estimated with MCP of these animals varied between 6.4 and 26.9 ha. In general, these studies discussed aspects related to home range, habitat use, and activity period, and none of them explained movement patterns of these species in detail.

In southern Mexico efforts were done to study brown Yucatan brocket deer *Mazama pandora* and Central American red brocket deer *Mazama temama* with very limited results as after 2 intensive years of trapping effort, only four animals were captured (Weber and Medellin 2010). Only one *M. pandora* survived enough to exhibit a remarkably small home range of near 50 ha for a period of 3 months (Weber and Medellin 2010).

To date, the most detailed study of a Neotropical forest deer was carried out with gray brocket deer (*Mazama gouazoubira*) in the Brazilian Pantanal during 2011–2012 (Grotta-Neto et al. submitted (a)). Despite the challenge of capturing the species, the heterogeneous habitat mosaic from Pantanal (Rodela 2006) and the open-area explorations provide capture chances to radio-collars studies. Four males and two females (Table 7.1) were collared with GPS radio transmitters (LOTEK Wireless® 6000SL, Ontario, Canada) and fitted with ear tags after being darted and

Table 7.1 Details of the six gray brocket deer (*Mazama gouazoubira*) that we monitored using telemetry to study movement patterns in the Pantanal of Brazil (Table from Grotta-Neto et al. submitted (a))

Sex	Age (months)*	Antler condition	Weight (kg)
Female	14	–	16
Female	24	–	16
Male	8	Cast	17
Male	36	Velvet	18
Male	60	Hard antler	22
Male	72	Hard antler	20

*Explains the way of estimating deer age

sedated. Collars collected coordinates every 15 min for 72 h during two different monitoring intervals (15–17 September 2011, 24–26 March 2012). Based on distance and angles between points, movements were classified as either encamped or exploratory, which were characterized, respectively, by short step lengths with large turning angles and long step lengths with small turning angles (Morales et al. 2004; Fryxell et al. 2008). The coordinates collected were placed in context with habitat type (open and closed) and were compared over time as diurnal, crepuscular, and nocturnal. Also, the two 72 h intervals were associated with highest flood level (September) and the lowest water levels (March).

Movement patterns of these deer were shown to be associated with habitat type and time of day. Significant differences of turning angles were observed between habitats (Watson-Williams test: $F_{1,2762} = 1628.5$, $p < 0.001$) and between time of day (Watson-Williams test: $F_{2,3309} = 543.32$, $p < 0.001$). Deer turned much more in closed habitats (average turn angle was 164°) than open habitats (9°). Deer also turned at smaller angles when moving during sunrise and sunset (crepuscular, 2°), followed by day (115°) and night (179°). Also, deer travel at different speeds (thereby covering different distances) in different habitats and times of day (Mann-Whitney-Wilcoxon test, $W = 610{,}810$, $p < 0.001$; Kruskal-Wallis test, $\chi^2 = 132.31$, gl = 2, $p < 0.001$, respectively). Distance traveled (mean ± standard deviation) was greatest in open habitats (27.5 ± 45.7 m) and during sunrise and sunset (48.4 ± 64.2 m), as compared to closed habitats (13.3 ± 17.4 m), day (33.7 ± 52.2 m) and night (21.0 ± 32.8 m). Thus, encamped movements were more common in closed habitats during day and night, while exploratory movements were in open habitats during crepuscular time periods (Figs. 7.2 and 7.3, Grotta-Neto et al. submitted (a)).

Encamped behaviors are to be expected when and where food is more available (Morales et al. 2004). Thus, closed, dense habitat is important for *M. gouazoubira* as places where they can rest and forage and where they found the protection against predators (Pinder 1997; Pinder and Leeuwenberg 1997; Black-Décima et al. 2010; Rodrigues et al. 2014). In the Pantanal, more than 80% of the diet of *M. gouazoubira* comprises of forbs and browse (Pinder 1997), and these are most abundant in dense vegetation. Clearly dense vegetation affords protection against predators as well, and thus, food, rest, and refuge together explain the more encamped state of the deer in dense vegetation.

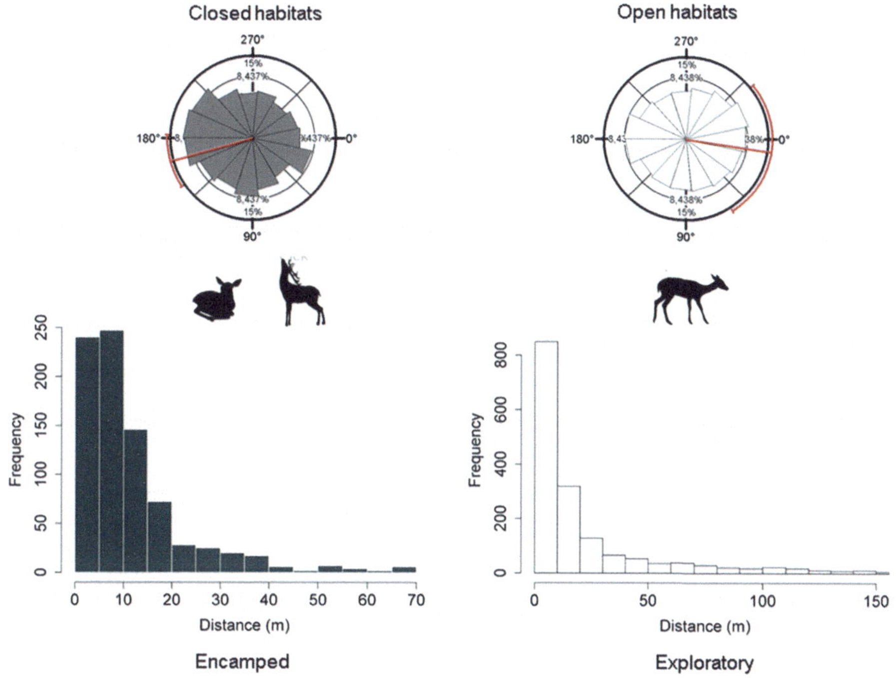

Fig. 7.2 Circular graphs illustrating how movement patterns depend on behavior (turn angles and 95% confidence intervals in red) and step lengths in closed and open habitats in encamped and exploratory behaviors of *Mazama gouazoubira* monitored in the Pantanal of Brazil. (Figure from Grotta-Neto et al. submitted (a))

Exploratory movements occur most during dawn and dusk and in open regions, such as grasslands and savannas, through which the deer travel in search of other foraging areas with denser vegetation. Brocket deer (*Mazama* spp.) are considered strict forest species that suggests that exposure in open fields places them at greater risk and therefore is likely to occur only for limited time and when necessary – such as in search of new foraging grounds. In the Pantanal, *M. gouazoubira* tends to leave dense vegetation and cross open fields during dawn and dusk and travel along lake shores, perhaps for water and forage (Antunes 2012). Thus, activity patterns suggest that movement occurs more frequent and more extensive during dawn and dusk when animals tend to cross open fields that are probably very risky habitat types for the species.

The greatest distances moved in Pantanal tended to be during the flooding season (Fig. 7.4, Grotta-Neto et al. submitted (a)). Period flooding of the Pantanal is an essential part of seasonal changes that is fundamentally important in maintaining habitat structure and, consequentially, foraging resources for all wildlife in the region throughout the year (Rodela et al. 2007). Movement is likely to be greater at this time simply because areas that are not underwater are separated by greater distances.

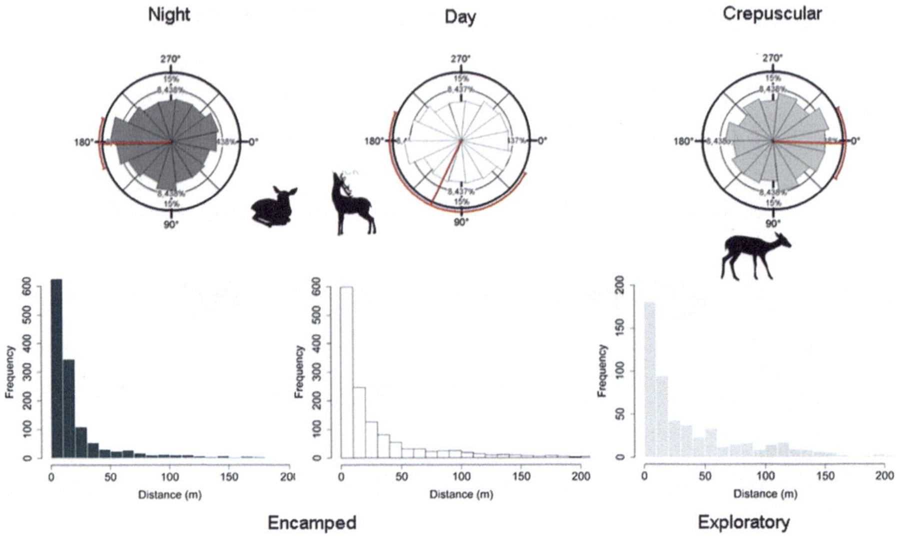

Fig. 7.3 Time of day and turn frequency and angle of turn (with 95% confidence interval in red) and step length of *Mazama gouazoubira* monitored in the Pantanal of Brazil. (Figure from Grotta-Neto et al. submitted (a))

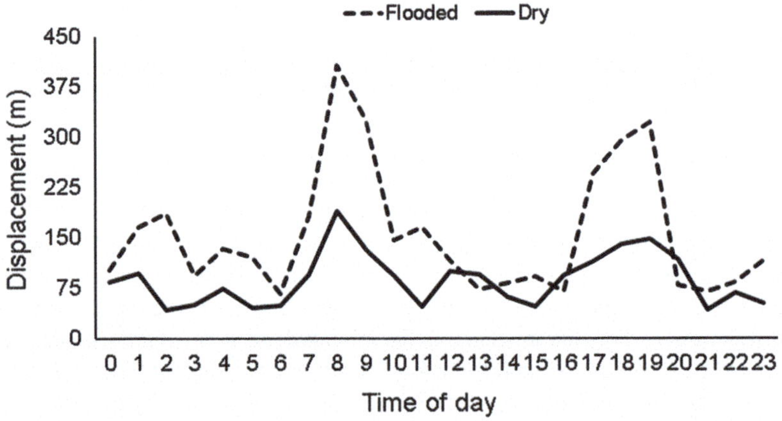

Fig. 7.4 Comparison between the wet and dry season of the distances traveled during 24 h of the six deer (*Mazama gouazoubira*) monitored in the Pantanal of Brazil. (Figure from Grotta-Neto et al. submitted (a))

Also in the Pantanal, six gray brocket deer (four males and two females) were monitored with GPS collars during 1 year, between October 2011 and September 2012 (Grotta-Neto et al. submitted (b)). Using kernel's estimator with 95% contour, the home range sizes varied from 41 to 112 ha for males and from 68.1 to 90.6 ha for females (Grotta-Neto et al. submitted (b)).

7.4 Final Considerations

Neotropical deer, especially forest species, have been neglected in studies, especially in comparison to the amount of scientific attention deer receive in northern hemisphere countries. Movement patterns of Neotropical forest deer over space and time have received even less scientific attention, as demonstrated by the literature in this chapter. In the relatively few studies of these deer, most focus on basic details of movement, such as distance traveled and seasonal, altitudinal movements. In terms of spatial ecology, most studies are limited to the home range size and habitat use. Also, movement patterns in Neotropical terrestrial mammals are limited by observations over relatively short time intervals (Zucco 2014). The scarcity of studies and their limitations allow us to explain some of the causes for these lacunae.

Telemetry is the method of choice for movement studies, because it permits gathering the basic data that allow us to describe and analyze movement patterns, including locations, trajectories, speed, and so on. To use telemetry, individuals must be captured and fitted with equipment. This is the first hurdle for Neotropical deer, in that capturing free-living animals is extremely difficult. Deer are notably more timid, solitary, elusive, and hard to find, especially in comparison to deer in the northern hemisphere that can easily be seen from rural to suburban areas. The genus *Mazama* provides a good example. While comprising ~10 species that are found throughout the neotropics, all of them are difficult to capture due to their secretive, elusive, and unpredictable behaviors. As a consequence, studies of free-living animals have had to resort almost exclusively to the use of trail cameras (Rivero et al. 2005; Azevedo 2008; Vogliotti 2008; Tobler et al. 2009; Ferreguetti et al. 2015; Oliveira et al. 2016; Rodrigues et al. 2017). Even when using trail cameras, deer usually represent a small percentage of the many pictures that are taken because of the diversity of other animals that can trigger the cameras. From 1998 to 2002, 11 *Mazama* deer capture techniques were compared to evaluate the positive and negative factors of each technique (Vogliotti 2003). Trapping was carried out through a variety of methods, including bait, where food and scents (feces, pheromones) were used to attract animals, lasso-traps, tracking with dogs, and helicopters. The efficiency of each method was evaluated based on capture opportunities observed (captures, escape or death due to capture) and on the approximate effort (in hours) employed in each one, generating an average interval of time between each opportunity. As a result, each and every method had problems, and no single method could be demonstrated to be recommended for use in most situations when studying forest deer.

Elusive behavior is not the only obstacle to studying deer, especially when access is limited or forests are dense. Huemuls, for example, are found in very isolated regions in the Andes, with steep slopes and rocky terrain and few access roads. Dense vegetation where *Mazama* spp. are found is more problematic because they are difficult to see and follow in the dense vegetation, and also they are often difficult to reach because of bad roads and lack of trails in the forest.

In the relatively few situations in which capture is more feasible, extreme care must be taken in handling the animals once caught. Traumatic accidents have been known to occur to the detriment of the animals during capture (Vogliotti 2003). Dosages of anesthetics are also complicated because of individual variation in their reactions, as well as individual variation in stress and myopathy that accompany capture (Catão-Dias and Camargo 2010). Thus, few studies today have successfully captured and studied Neotropical deer due to these limitations (Duarte et al. 2010), and these limits have resulted in the few studies available today of Neotropical forest deer and especially movement patterns of free-living animals.

Particularly in Latin America, the cost (due to unfavorable exchange rates and limited financial support) of telemetry equipment is a major obstacle. To study most free-living mammals in the tropics, capture of a large number of individuals might generate expenses associated with large field effort and costly monitoring devices, which may limit the sample size (Zucco 2014). Telemetry equipment is much more expensive than other kinds of monitoring equipment, such as trail cameras, and so trail cameras become the tool of choice because all equipment must be imported, which adds costs to already costly equipment.

Understanding movement patterns of large animals is fundamental for their conservation and to understand their basic ecology and biology. Here, we discussed that the movement patterns have been ignored in the study of Neotropical terrestrial mammals through radiotelemetry (Zucco 2014) and that Neotropical forest deer are much more difficult to study for diverse reasons, including their evasive behavior, the often isolated regions in which they are found, and the difficulty of observing them in complex habitat (Ramos et al. 2010). These facts seem to be strongest for the small deer (*Mazama* and *Pudu* genera), in which only one movement study and some behavioral and habitat use aspects are found in literature. Thus, the knowledge and studies in free-ranging with these species remain rare. Although practically restricted to altitudinal distance traveled and basic descriptions, the majority of movement patterns studies are found in Chile and Argentina for the largest species (*Hippocamelus* genus), specially for huemul. In relation to small deer, the largest size and open habitat use of *Hippocamelus* sp. could be associated with the use of collars and successful capture. Therefore, the future of understanding the ecology and biology aspects of Neotropical forest deer is at risk due to the economics of studying these fascinating animals, as well as the continuing habitat loss due to human activities. Yet, from the little we know of deer movement patterns, they are likely to be fundamentally important to maintain viable populations. From the Pantanal of Brazil, Bolivia, and Paraguay, where floods are important constraints on deer movement patterns, to the Andes, where the lowlands are being increasingly converted to agriculture, for many species, we may lose deer habitat more rapidly than we can determine the importance of that habitat for their population dynamics.

We highlight here that the greatest challenges to studying Neotropical forest deer are associated with forest deer behavior and with costs, also with finding new and efficient methods for capture and access to less costly telemetry equipment. We hope that some of these hurdles can be overcome through collaborative (perhaps international) efforts and imaginative use of resources and to find funding, and through such efforts we may begin to understand these elusive and important animals that play vital ecological roles in communities wherever they are found.

References

Azevedo ADK (2008) Análise comparativa do período de atividade entre duas populações de *Mazama americana* (veado-mateiro). Universidade de São Paulo, Piracicaba, São Paulo, Brazil

Antunes VS (2012) Área de uso e seleção de habitats por veado-catingueiro (*Mazama gouazoubira*; Fisher, 1814) no Pantanal da Nhecolândia- MS. MS thesis, Universidade de São Paulo, Piracicaba, São Paulo, Brazil

Baguette M, Stevens VM, Clobert J (2014) The pros and cons of applying the movement ecology paradigm for studying animal dispersal. Mov Ecol 2:13

Barrette C (1987) The comparative behavior and ecology of chevrotains, musk deer, and morphologically conservative deer. In: Wemmer CM (ed) Biology and management of the Cervidae. Smithsonian Institution Press, Washington, DC, pp 200–213

Barrientos J, Maffei L (1999) Radio-telemetría en la urina Mazama gouazoubira en el campamento Cerro Cortado, Izozog, Santa Cruz, Bolivia. Resúmenes de investigaciones – Proyecto Kaa-Iya-CABI/WCS. pp 369–371

Black-Décima P et al (2010) Brown Brocket Deer *Mazama gouazoubira* (Fischer 1814). In: Duarte JMB, González S (eds) Neotropical cervidoloy: biology and medicine of Latin American Deer. FUNEP, Jaboticabal, pp 190–201

Bello J, Galinna S, Equihua M (2004) Movements of the white-tailed deer and their relationship with precipitation in northeastern Mexico. Interciencia 29(7):357–361

Catão-Dias JL, Camargo CMS (2010) Capture myopathy. In: Duarte JMB, González S (eds) Neotropical cervidoloy: biology and medicine of Latin American deer. FUNEP/IUCN, Jaboticabal, p 393

Díaz P, Marqués BI, Vila AR (2013) Seasonal habitat use and selection of the endangered huemul deer (*Hippocamelus bisulcus*) in Patagonian Andes. Mammalia 77:371–380

Duarte JMB, González S (2010) Neotropical cervidology: biology and medicine of Latin American Deer. FUNEP/IUCN, Jaboticabal

Duarte JMB, Merino ML (1997) Taxonomia e Evoluçao. In: Duarte JMB (ed) Biologia e Conservaçao de Cervídeos Sul-Americanos: *Blastocerus, Ozotoceros e Mazama*. FUNEP, Jaboticabal, pp 2–23

Duarte JMB, González S, Maldonado JE (2008) The surprising evolutionary history of South American deer. Mol Phylogenet Evol 49:17–22

Duarte JMB, Uhart MM, Galvez CES (2010) Capture and physical restrant. In: Duarte JMB, González S (eds) Neotropical cervidoloy: biology and medicine of Latin American deer. Funep/IUCN, Jaboticabal, pp 218–227

Eisenberg JF, Redford KH (1999) Order artiodactyla (even-toed ungulates). In: Mammals of the neotropics, 3rd edn. The University of Chicago Press, Chicago, pp 332–355

Escamilo LL, Barrio J, Benavides J, Tirira DG (2010) Northern pudu *Pudu mephistopheles* (De Winton 1896). In: Duarte JMB, Gonzalez S (eds) Neotropical cervidoloy: biology and medicine of Latin American Deer. Funep/IUCN, Jaboticabal, pp 133–139

Ferreguetti ÁC, Tomás WM, Bergallo HG (2015) Density, occupancy, and activity pattern of two sympatric deer (*Mazama*) in the Atlantic Forest, Brazil. J Mammal 96:1245–1254

Fryxell JM et al (2008) Multiple movement modes by large herbivores at multiple spatiotemporal scales. Proc Natl Acad Sci U S A 105:19114–19119

Gallina S, Mandujano S, Bello J, Arévalo HFL, Weber M (2010) White-tailed deer *Odocoileus virginianus* (Zimmermann 1780). In: Duarte JMB, González S (eds) Neotropical cervidoloy: biology and Medicine of Latin American deer. Funep/IUCN, Jaboticabal, pp 101–118

Garay G, Ortega IM, Guineo O (2016) Social ecology of the huemul at. An Inst Patagonia (Chile) 44:25–38

Gilbert C, Ropiquet A, Hassanin A (2006) Mitochondrial and nuclear phylogenies of Cervidae (Mammalia, Ruminantia): systematics, morphology, and biogeography. Mol Phylogenet Evol 40:101–117

Gill R, Saucedo Galvez C, Aldridge D, Morgan G (2008) Ranging behaviour of huemul in relation to habitat and landscape. J Zool 274:254–260

González S, Álvarez-Valin F, Maldonado JE (2002) Morphometric differentiation of endangered Pampas deer (*Ozotoceros bezoarticus*), with description of new subspecies from Uruguay. J Mammal 83:1127–1140

Grotta-Neto F, Peres PHF, Piovezan U, Passos FC, Duarte JMB (submitted_a) Influential factors on gray brocket deer (*Mazama gouazoubira*) activity and movement in Pantanal, Brazil

Grotta-Neto F, Peres PHF, Piovezan U, Passos FC, Duarte JMB (submitted_b) Camera trap feasibility in ecological studies of elusive forest deer

IUCN (2018) International union for conservation of nature. http://www.iucnredlist.org/technical-documents/spatial-data (30 Mar 2018)

Jiménez JE (2010) Southern pudu. In: Duarte JMB, Gonzalez S (eds) Neotropical cervidoloy: biology and medicine of Latin American Deer. Funep/IUCN, Jaboticabal, pp 140–150

Leeuwenberg F, Cabral IO, Resende SL (1999) Gray Brocket Deer (*Mazama gouazoubira*) in the Brazilian Savanna. Deer Specialist Group News 15:14–14

Lusseau D (2004) The energetic cost of path sinuosity related to road density in the wolf community of Jasper National Park. Ecol Soc 9(1):9–11

Marques SR, Santos-Junior TS (2002) Mamífeors terrestres de médio e grande porte. In: Alho CJR (ed) Monitoramento da fauna silvestre do APM Manso. Furnas Centrais Elétricas, São Carlos, pp 153–177

Merino ML, Rossi RV (2010) Origin, systematics, and morphological radiation. In: Duarte JMB, González S (eds) Neotropical cervidoloy: biology and medicine of Latin American Deer. Funep/IUCN, Jaboticabal, pp 2–11

Morales JM, Ellner SP (2002) Scaling up animal movements in heterogeneous landscapes: the importance of behavior. Ecology 83(8):2240–2247

Morales JM, Haydon DT, Frair J, Holsinger KE, Fryxell JM (2004) Extracting more out of relocation data: building movement models as mixtures of random walks. Ecology 85:2436–2445

Nathan R, Getz WM, Revilla E, Holyoak M, Kadmon R, Saltz D, Smouse PE (2008) A movement ecology paradigm for unifying organismal movement research. Proc Natl Acad Sci 49:19052–19059

Oliveira LD, Duarte JMB (2006) Gastro-intestinal transit time in South American Deer. Zoo Biol 25:47–57

Oliveira ML et al (2016) Phylogenetic signal in the circadian rhythm of morphologically convergent species of Neotropical deer. Mamm Biol 81:281–289

Oliveira-Santos LGR (2009) Ecologia e conservação de ungulados florestais em uma área do Pantanal. MS thesis, Universidade Federal do Mato Grosso do Sul, Campo Grande, Mato Grosso do Sul, Brazil

Pinder L (1997) Niche overlap among brown brocket deer, pampas deer, and cattle in the Pantanal of Brazil. Ph.D dissertation, University of Florida

Pinder L, Leeuwenberg F (1997) Veado-catingueiro (*Mazama gouazoubira*, Fischer 1814). In: Duarte JMB (ed) Biologia e Conservaçao de Cervídeos Sul-Americanos: *Blastocerus, Ozotoceros e Mazama*. FUNEP, Jaboticabal, pp 59–77

Ramos HGC, Mantovani JE, Vogliotti A, Piovezan U, dos Balta OS, Andriolo A (2010) Radiotelemetry. In: Duarte JMB, González S (eds) Neotropical cervidoloy: biology and medicine of Latin American deer. FUNEP/IUCN, Jaboticabal, p 393

Rau J (1980) Movimiento, habitat y velocidad del huemul del sul, (*Hippocamelus bisulcus*). Not Mensual Museo Nac Hist Nat 281–282:7–9

Rivero K, Rumiz DI, Taber AB (2005) Differential habitat use by two sympatric brocket deer species (*Mazama gouazoubira* and *Mazama americana*) in a seasonal Chiquitano forest of Bolívia. Mammalia 69:169–183

Rodela LG (2006) Unidades de Vegetação e Pastagens Nativas do Pantanal da Nhecolândia, Mato Grosso do Sul. Ph.D dissertation, Universidade de São Paulo. São Paulo, Brazil

Rodela LG, De Queiroz Neto JP, Santos SA (2007) Classificação das pastagens nativas do Pantanal da Nhecolândia, Mato Grosso do Sul, por meio de imagens de satélite. Anais XIII Simpósio Brasileiro de Sensoriamento Remoto. pp 4187–4194

Rodrigues TF, Cerveira JF, Duarte JMB (2014) Uso de áreas agrícolas por *Mazama gouazoubira* (Mammalia, Cervidae) no Estado de São Paulo. Iheringia. Sér Zool 104:439–445

Rodrigues TF et al (2017) Managed forest as habitat for gray brocket deer (*Mazama gouazoubira*) in agricultural landscapes of southeastern Brazil. J Mammal 98:1301–1309

Saucedo C, Gill R (2004) Huemul (*Hippocamelus bisulcus*) ecology research: conservation planning in Chilean Patagonia. Deer Specialist Group Newsl 22:84–87

Smith-Flueck JA, Flueck WT, Ruíz ME (2018) Preliminary data on movements and health condition of the first radio-collared huemul (*Hippocamelus bisulcus*) population study in Argentina. Deer Specialist Group News 30:4–14

Sánchez-Rojas G, Gallina S, Mandujano S (1997) Área de actividad y uso del hábitat de dos venados cola blanca (*Odocoileus virginianus*) en un bosque tropical de la costa de Jalisco, México. Acta Zool Mex 72:39–57

Taber A, Beck H, Gonzalez S, Altrichter M, Duarte JMB, Reyna-Hurtado R (2016) Why neotropical forest ungulates matter, the case for deer, peccary and tapir conservation. In: Aguirre AA, Sukumar R (eds) Tropical conservation: perspectives on local and global priorities. Oxford University Press, New York, pp 238–254

Tobler MW, Carrillo-Percastegui SE, Powell G (2009) Habitat use, activity patterns and use of mineral licks by five species of ungulate in south-eastern Peru. J Trop Ecol 25:261

Vila AR, Galvez CES, Aldridge D, Ramilo E, González PC (2010) South andean huemul *Hippocamelus bisulcus* (Molina 1987). In: Duarte JMB, González S (eds) Neotropical cervidoloy: biology and medicine of Latin American deer. Funep/IUCN, Jaboticabal, p 393

Vogliotti A (2003) História Natural de *Mazama bororo* (Artiodactyla; Cervidae) através da Etnozoologia, Monitoramento Fotográfico e Rádio-Telemetria. M.S. thesis, Universidade de São Paulo, Piracicaba, São Paulo, Brazil

Vogliotti A (2008) Partição de habitats entre os cervídeos do Parque Nacional do Iguaçu. Ph.D dissertation, Universidade de São Paulo. Piracicaba, São Paulo, Brazil

Webb SD (2000) Evolutionary history of new world deer. In: Vrba ES, Shaller GB (eds) Antelopes, deer, and relatives. Yale University Press, London, pp 38–64

Webb SL, Gee KL, Strickland BK, Demarais S, DeYoung RW (2010) Measuring fine-scale white-tailed deer movements and environmental influences using GPS collars. Int J Ecol 2010:1–12

Weber M, Medellin RA (2010) Yucatan brown brocket deer *Mazama pandora* (Merriam 1901). In: JMB D, Gonzalez S (eds) Neotropical Cervidoloy: biology and medicine of Latin American deer. FUNEP/IUCN, Jaboticabal, pp 166–170

Zucco CA (2014) Pampas deer on the move: spatial ecology and activity of *Ozotoceros bezoarticus* in the central Pantanal of Brazil. Ph.D dissertation, Universidade Federal do Rio de Janeiro, Rio de Janeiro, Brazil

Chapter 8
Daily Traveled Distances by the White-Tailed Deer in Relation to Seasonality and Reproductive Phenology in a Tropical Lowland of Southeastern Mexico

Fernando M. Contreras-Moreno, Mircea G. Hidalgo-Mihart, and Wilfrido M. Contreras-Sánchez

8.1 Introduction

Movement is the process by which animals make behavioral decisions to select resources in space and time (Turchin 1998). Animal movement reflects a wide range of phenomena, including habitat use, foraging and depredation (Gurarie et al. 2011), dispersal and migration (Delgado et al. 2010), social and territorial behaviors (Moorcroft et al. 1999), as well as coexistence with competitors (Keeling and Grenfell 1997) and community interactions (Hanski 1998). Animal movement usually is influenced by different factors, including the individual (age, sex, size, reproductive status), population (density and abundance, social interactions) as well as habitat characteristics (availability of resources, quality and quantity of resources, geographic barriers), and intensity of human activities (Webb et al. 2010; Salek et al. 2015).

Few studies have documented fine-scale temporal movements (i.e., daily traveled distances) of ungulate species (Pépin et al. 2004; Webb et al. 2010). In the case of the white-tailed deer (*Odocoileus virginianus*), the study of movement has focused on how movements of white-tailed deer vary at different spatial scales, from large-scale dispersal and migration to small-scale movements within home ranges and habitats (Webb et al. 2009). Movement of white-tailed deer (*Odocoileus virginianus*) could be influenced by many ecological, environmental, and behavioral variables such as hunger, reproduction, physiological condition, habitat, and predators or human activity (Ferguson et al. 1998; Phillips et al. 2004). Many studies of white-tailed deer have examined both large-scale movements (e.g., dispersal and migration) and movement within home ranges and among habitat types. These different types of movement occur at different hierarchical spatial scales (Webb

F. M. Contreras-Moreno · M. G. Hidalgo-Mihart (✉) · W. M. Contreras-Sánchez
División Académica de Ciencias Biológicas, Universidad Juárez Autónoma de Tabasco, Villahermosa, Tabasco, Mexico

R. Reyna-Hurtado, C. A. Chapman (eds.), *Movement Ecology of Neotropical Forest Mammals*, https://doi.org/10.1007/978-3-030-03463-4_8

et al. 2009). Small-scale movements, such as space-use patterns within a home range, may be influenced by physical habitat features, resource availability, or the distribution of conspecifics (Crist et al. 1992; McIntyre and Wiens 1999). Large-scale movements, such as dispersal, may be influenced by landscape structure or behavior as it relates to gene flow or population dynamics (Long et al. 2005).

The white-tailed deer movement ecology has been an important research subject in the United States and Canada (e.g., Webb et al. 2010; Long et al. 2005). However, in the southern portion of the distribution of the species, the information about the movement ecology is practically unknown, especially in tropical areas. It has been observed that in the northeastern arid lands of Mexico, the daily traveled distances of the white-tailed deer depended on the environmental temperature and precipitation (Delfín et al. 1998; Gallina et al. 2003), and deer moved longer distances during the fawning season (Bello et al. 2004). Also, in the tropical dry forests of the Pacific slope of the country, the daily traveled distances were longer during the wet season of the year (June–November) corresponding to the fawning season (Sánchez-Rojas et al. 1997).

The extensive wetland system located on western Campeche state in Mexico is subject to severe seasonal changes, with a marked dry season with temperatures above 40 °C (Instituto Nacional de Estadística y Geografía 2013), followed by long flooding season that can least for more than 8 months (Rivera-Monroy and Twilley 1996). These climatic conditions probably had a strong effect on the resource and habitat availability for the wildlife inhabiting the area (Hidalgo-Mihart et al. 2017) and are one of the possible causes that had produced that the fawning season of white-tailed deer synchronizes with the dry months of the year (February–June, Contreras-Moreno 2018).

Determining the effect of the environmental conditions and physiological status of individuals on the movement patterns of the white-tailed deer in this area of Mexico will help to understand how this species could be affected by the changes that are already occurring in the region, where it is anticipated a general decrease in mean precipitation and an increase in drought (Imbach et al. 2012; Chiabai 2015) with a strong effect on the wildlife (Reyna-Hurtado et al. 2010). Also, the sea level increase from global warming will produce longer periods of flooding as a result that most of this coastal area is in a very low terrain, with some areas even under the sea level (Yañez-Arancibia et al. 2014). This will probably have severe impact in the movement patterns of the white-tailed deer, as has already been detected in ungulates from the Amazon basin (Bodmer et al. 2018).

White-tailed deer is the most hunted species in Mexico (Mandujano et al. 2014). However, the study of movement patterns on the species is scarce in the country and specially on the tropical areas. The objective of this study was to determine the effect that environmental variables (rain seasonality and temperature) as well as reproductive status (fawning season) had on the daily traveled distances of the female white-tailed deer in a tropical lowland of southeastern Mexico.

8.2 Methods

Study Area Our research was conducted at the Nicte-Ha UMA (Wildlife Management and Conservation Unit per its Spanish initials; hereafter called Nicte-Ha), located in southwestern Campeche, México (91°43′56″ W, 18°19′56″N), and adjacent to the Laguna de Términos Flora and Fauna Protected Area (Fig. 8.1). The area is a 2300 ha tract of tropical lower coastal plain habitat situated between −1 and 3 m asl. The climate in the region is warm-humid, with a mean temperature of 27 °C and up to 2000 mm of precipitation per year (Instituto Nacional de Estadística y Geografía 2013). Rainy season in the area usually began at mid-June and finished by November–December and is followed by a dry season that least for around 5 months (January–May). Due to the intense rainy season and the plain terrain, most of the Nicte-Ha area is subject to seasonal floods that can be as high of 0.5 m and last for 7 months (July–January). The vegetation types of the area vary from hydrophilic vegetation, flooded savannas, mangroves, sub-evergreen flooded rain forests, tropical deciduous flooded forests, secondary growth forests, agricultural areas, and induced grasslands for cattle grazing (Ocaña and Lot 1996).

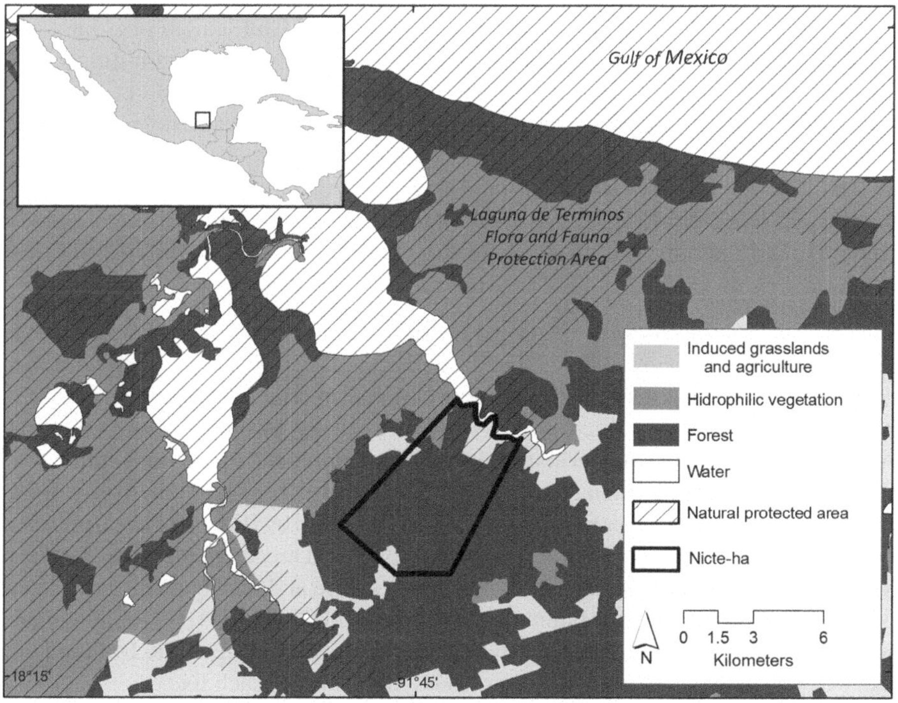

Fig. 8.1 Location of the study area in the tropical lowlands of Campeche, México. The Nicte-Ha area is located outside of the Laguna de Terminos Flora and Fauna Protected Area

Fig. 8.2 Photograph of white-tailed deer in Nicte-Ha during the dry season (**a**) and the wet season (**b**). Nicte-Ha is located in a tropical lowland area that is subject to a severe drought during the dry season (2016) followed by a long flooding season. The white-tailed deer in Nicte-Ha should adapt their movement behavior to the changing environmental conditions during the year

In 2010, the owners of Nicte-Ha remove all the cattle from the ranch, and since then, the area has been used exclusively for sustainable white-tailed deer hunting. As in other parts of the Yucatan Peninsula, water is probably the most limiting factor for wild ungulates in the area during the dry season (Reyna-Hurtado et al. 2010; Fig. 8.2). To improve the white-tailed deer habitat quality, six water holes were built along Nicte-Ha area, to offer water to the wildlife during the dry season. However, in occasions due to a very strong drought such as the one that occurred at the end of the dry season in 2017 (late May–early June), most of the artificial water sources were dry, and the only water source for the wildlife on Nicte-Ha was the Rio del Este.

Capture and Radio Tracking During May of 2016, five adult female white-tailed deer were captured using a 20 × 20 m drop net (Ramsey 1968). Once captured, the deer were physically restrained, and conventional measurements and mass were obtained. Each deer was fitted with a VERTEX Survey Iridium Collar (Vectronic Aerospace GmbH, Berlin, Germany) and released immediately on the capture site (Fig. 8.3). The capture, management, and collaring of the white-tailed deer were carried out under the collection permit SGPA/DGVS/01097/16 (10-Febrero-2016) granted to Mircea Gabriel Hidalgo Mihart by the Dirección General de Vida Silvestre, SEMARNAT, México, following the capture and management guidelines of the American Society of Mammalogists (Sikes and Gannon 2011). Radio collars were programmed to take fixes at 2-h intervals and transmitting to the Iridium system every 24 h the date, time, position, and temperature of each fix.

Data Analysis To determine the daily traveled distance by each deer, we selected only those 24-h periods that had 12 GPS fixes to derive estimates of daily travel

Fig. 8.3 Photograph of a radio-collared female deer in Nicte-Ha drinking from an artificial water hole. In the Nicte-Ha area, white-tailed deer are exposed to a severe drought during the dry season. Artificial water holes were built to improve the habitat quality for the deer inhabiting the area

distance from the five radio-collared deer. For each day, the distances between sequential GPS fixes were measured and added up to calculate total distance traveled. Distance moved between two successive locations was estimated by using HRT, Home Range Tools, in ArcGIS 9 (Rodgers et al. 2007).

We arranged the daily traveled distance data per deer in a bimester basis from May–June 2016 to May–June 2017. We obtained the mean values of daily traveled distances per deer at each bimester and use them to perform a one-way analysis of variance (ANOVA) for repeated measures to determine for differences in the daily traveled distance per bimester (Zar 1999). Tukey HSD tests were conducted a posteriori to determine which bimesters were different (Zar 1999). All statistical analyses were performed using the Statgraphics Centurion®, v XVIII, ($\alpha < 0.05$).

Environmental Variables To determine the daily temperature that each individual deer was exposed during the 24 h period when we estimated the daily travel distance, we average the temperature obtained from the radio collar on each fix, for each one of the complete 24 h periods. To determine if daily traveled distances were associated with average daily temperature, we performed a linear regression analysis between these two variables (Zar 1999), first relating all the daily traveled distances from all the deer with all the average daily temperatures and second with the mean values of daily traveled distances per deer at each bimester with the mean average temperature per deer per bimester.

We collected the precipitation data from the nearest meteorological stations to the study area (Toluca, Carmen, Campeche and Jonuta, Tabasco; Instituto Nacional de Investigaciones Forestales Agricolas y Pecuarias 2018) to determine the rainy and dry season during the studied period for the study area. We define that the rainy season began when we have a 15-day period with at least 10 days with an accumulated daily rain of 25 or more mm. The end of the rainy season was defined as the time when we had a 30-day period with and accumulated rain of less than 50 mm.

Fawning Season We follow Contreras-Moreno (2018) to determine the fawning season for white-tailed deer in the study area. Accordingly, we fix the beginning of the fawning season on February and at the end of June.

8.3 Results

All the radio-collared deer were alive by February 2018, 20 months after the initial deployment of the radio collars. Radio collars were originally projected to work for 24 months, but due to collar malfunction, two collars sent information to the Iridium system for 12 months, two for 14 months, and one for 16 months (Table 8.1). We obtained 461 daily traveled distances for the five radio-collared deer (36–129). The mean daily traveled distance along the year was 1044.06 m (SD ± 501.00 m) with a minimum of 302 m and a maximum 4831 m. The precipitation data showed that the rainy season began on the second half of June and finished by the first half of November.

In order to include only bimesters with data of more than one individual deer, we only include on the analysis 427 daily traveled distances, corresponding to the period May–June 2016 to May–June 2017 (Table 8.1). When we relate the observed seasonality of the study area to the average daily traveled distances per bimester for all deer (Fig. 8.4) and we observed the individual average distance per deer (Fig. 8.4b), we notice that the lowest values corresponded to the rainy season, while the highest values coincide with the dry season. We observed that the longest daily traveled distances were during the fawning season compared with the post-reproductive season (Fig. 8.4a, b).

There were differences between bimesters ($Fobs = 16.36$; $g.l. = 24, 6$; $p < 0.001$), while no differences were detected between individual deer ($Fobs = 0.91$, $g.l. = 34,4$; $p = 0.47$; Fig. 8.4a, b). The Tukey TSD showed that there were no differences between June–August 2016, September–October 2016, and November–December 2016 ($p > 0.05$) which had the lowest average observed deer daily traveled distances (± SD; 676.024 ± 160.053; 778.024 ± 240.003; and 876.136 ± 220.995 m, respectively; Fig. 8.4). We found differences between the bimesters with highest daily traveled distances May–June 2017 (1531.72 m ± 1151.71) followed by the bimesters May–June 2016 and March–April 2017 (1201.23 ± 460.547 and 1240.96 ± 377.787 m, respectively; Fig. 8.4).

Table 8.1 Bimonthly average daily distance (m) moved per each individual female white-tailed deer studied at Nicte-ha, Campeche, México, during the years 2016 and 2017.The number in parenthesis indicates the number of complete daily cycles during each bimester

Individual	2016				2017				
	May–Jun	Jul–Aug	Sep–Oct	Nov–Dec	Jan–Feb	Mar–Apr	May–Jun	Jul–Aug	Sep–Oct
17,796	1207 (20)	680 (12)	745 (16)	900 (13)	1104 (28)	978 (24)			
17,797	987 (23)	672 (20)	727 (7)	932 (12)	850 (10)	1389 (18)	1358 (9)		
17,798	1193 (8)	866 (12)	792 (3)	978 (6)	952 (7)	1242 (23)	1553 (14)	736 (8)	918 (3)
17,800	1307 (14)	918 (9)	554 (3)	1104 (2)	1357 (7)	1277 (1)			
17,801	1438 (16)	811 (29)	554 (12)	688 (11)	1037(24)	1355 (31)	1741 (6)		

Even the linear regression analysis between daily traveled distances and the average daily temperature was significant (Fobs = 12.48, g.l. = 426, $p < 0.001$); the explained variance of the model was low ($r^2 = 2.85\%$). We found that the linear regression between the mean values of daily traveled distances per deer at each bimester with the mean average temperature per deer per bimester was not significant (Fobs = 1.47, g.l. = 30, $p = 0.24$).

8.4 Discussion

The mean daily distances traveled by the white-tailed deer in Nicte-Ha (1044.06 ± 501 m) were shorter than those recorded in most of the studies in northern portions of the distribution of the species, where the usual daily traveled distances are above 1700 m (e.g., Sparrowe and Springer 1970; Ozoga et al. 1982; Schwede et al. 1993; Bertrand et al. 1996; Webb et al. 2009, 2010). Also, the mean daily traveled distances were shorter than those reported from the northwestern arid lands of Mexico (7016 ± 354 m; Gallina et al. 2003) and even in the tropical dry forest of the Pacific slope (1440 m in the dry season and 2580 m in the rainy season; Sánchez-Rojas et al. 1997).

Movement of white-tailed deer could be influenced by many ecological, environmental, and behavioral variables (Webb et al. 2010), and the differences we found between Nicte-Ha compared with studied sites should be reflecting the local adaptation of the deer in Nicte-Ha to the resource availability, as well the environmental conditions of the area. However, it is important to notice that to calculate the daily distance traveled by the deer, we are using a 2-h interval between fixes, which is a longer period than what has been used in other studies (e.g., Webb et al. 2010; Gallina et al. 2003) and could be underestimating the calculated movement distances

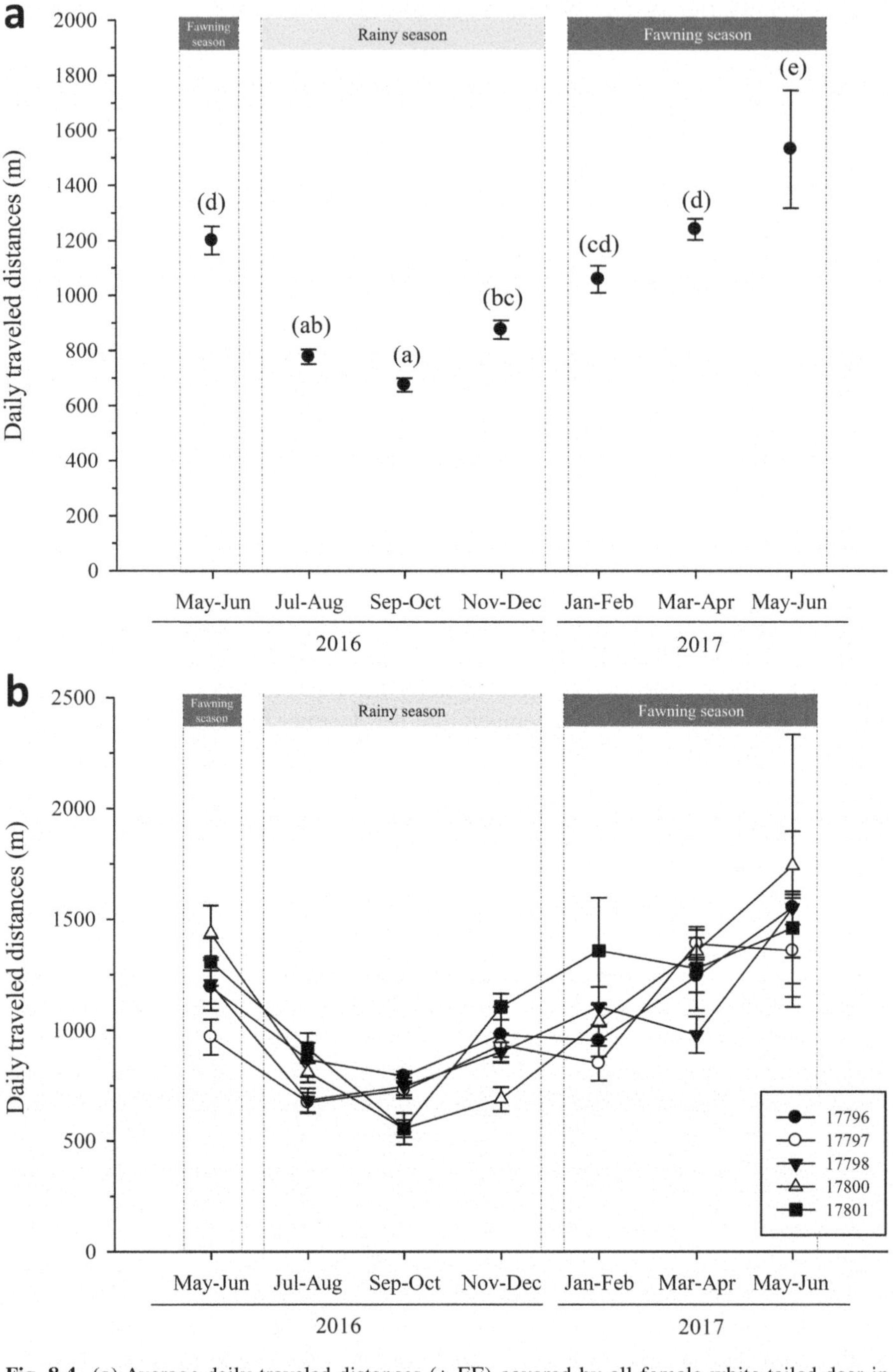

Fig. 8.4 (**a**) Average daily traveled distances (± EE) covered by all female white-tailed deer in Nicte-Ha per bimester during the study period (May 2006 and June 2017). Different letters over the values indicate statistically significant differences ($p < 0.05$). (**b**) Average daily traveled distances (± EE) covered by specific deer in Nicte-Ha per bimester during the study period (May 2006 and June 2017). The numbers in the label box indicate the identification number for each deer. The light-shaded area shows the approximate dates for the rainy season in 2016. The dark shaded areas indicate the fawning season in the Nicte-Ha area according to Contreras-Moreno (2018)

(Mills et al. 2006) because we are not capturing a more realistic movement distance produced by tortuous movement paths (Webb et al. 2010).

Several factors influence deer movements including physiology, which includes breeding and parturition, general changes in season and daylength, and environmental influences such as weather (Webb et al. 2010). In our study, the daily distances traveled by the deer showed differences throughout the study period. This suggests that environmental conditions or physiological traits of the deer that change over the course of the year influenced the behavior and the need of the deer to increase or decrease their activities. We found an increase in the traveled distances by the female deer during the dry season of 2016 and 2017, which corresponded also with the fawning season.

Delfín et al. (1998) and Sánchez-Rojas et al. (1997) found that in areas of semiarid and dry tropical forest, respectively, deer traveled shorter distances during the dry season, when temperatures were high, probably to reduce energy consumption. However, in both these areas (Northwest arid Mexico and tropical dry forest of the Pacific slope), the dry season corresponded to a post-reproductive season, and the females are not under the pressure to satisfy the fawn needs. Contrary, in Nicte-Ha, the dry season coincides with the fawning season (Contreras-Moreno 2018). In general, during the fawning season, the activity of the female white-tailed deer increases (e.g., Bello et al. 2004; Gallina et al. 2010; Gallina and López-Arévalo 2016), and the longer daily traveled distances of the females in Nicte-Ha could be related to the presence of fawns during the dry season, even taught it is the hottest time of the year.

In the Yucatan Peninsula, the dry season produced that herbivores like the white-lipped peccary (*Tayassu pecari*) and the tapir (*Tapirus bairdii*) concentrate their activities around the water ponds (Reyna-Hurtado et al. 2009; Pérez-Cortez et al. 2012). Due to the intense seasonality on Nicte-Ha, it was possible to expect that the white-tailed deer should concentrate their activities around the water sources reducing the daily traveled distance. During 2017, several of the artificial water holes in Nicte-Ha were dry because of the intense drought. This could cause that white-tailed deer increased the daily distance traveled during the May–June period, to travel to other water sources. However, in 2016 the drought was not as intense, and water holes do not get dry, but the deer still traveled longer distances than in the rainy season. This indicates that the reduced amount of water during the dry season is probably not the main cause of the longer traveled distances, supporting the hypothesis that the increase in the metabolic needs during the fawning produced that the female deer walk for longer distances.

The dry season brings about phenological changes that affect the diversity and availability of resources for herbivores, thereby prolonging natural periods of scarcity. As a result, herbivores must make decisions and modifications to their behavior, with some adjustments to their diets (Keuroghlian and Eaton 2008). In the region, diet of the white-tailed deer is comprised mostly by leaves and stems year-round and is richer in plant species during the wet season and much more restricted during the dry season (Weber 2005). Temporal variation in patch productivity has been observed to directly influence spatial use and deer movements (Van Beest et al. 2011). The reduction in available leaves and stems (the quality and quantity of natural fodder produced by the available biomass) in Nicte-Ha could be a factor that motivates deer females to travel longer distances, in search of food, to supplement their metabolic needs, and during the fawning season, there is an increase in energy needs due to the demands of breastfeeding.

The results of this study indicated that there is no relationship between temperature and the daily movements of the deer, although deer moved longest daily distances during the hottest months of the year (March–June). This is contrasting with other studies (e.g., Gallina et al. 2003; Webb et al. 2010) where there has been a correlation between activity and temperature. We still considered that the temperature could be affecting the white-tailed deer movement rate on Nicte-Ha. Rodríguez-Boza (2015) found that white tailed-deer was more active during the sunset and sunrise in Nicte-Ha and the activity was greatly reduced during the evening, the hottest time of the day. It is probable that the lack of relation between distance moved and the temperature could be because the use of average daily temperatures and daily distances moved by the deer is hiding the detailed movement pattern in relation to temperature that occurs during specific times of the day. Further studies should consider these problems to help to determine the relationship between deer movement and temperature in Nicte-Ha.

The flooding season in Nicte-Ha is coincident with the time when the deer in Nicte-Ha moved shorter distances. This behavior could be produced because it is the time when the fawning season is over (Contreras-Moreno 2018) and the metabolic needs for the female deer due to lactation are reduced. We did not detect that the white-tailed deer in Nicte-Ha migrate to higher ground areas during the flooding season, and even during the high flooding levels (July–October), the deer still stand on their home ranges. It is possible that the increased flooding levels reduced the mobility of the deer and they should concentrate their activities in areas of their home ranges where the flooding level is lower, as occurs in other wetland areas such as in the Florida Everglades (Fleming et a. 1994).

Nicte-Ha, as a most of the Campeche state coastal area, is in a high risk of flooding due to sea level rise, according to the current climate change scenarios (Ortiz and Méndez 1999; Yañez-Arancíbia et al. 2014). The reduced daily traveled distances during the flooding season by female white-tailed deer observed on this study indicate that the flooding season affected the movement of the individuals. The increase on flooding levels because of sea level rise in Nicte-Ha could have important effects on this species, as already happened in the Amazon where increase on water levels due to flooding had affected the mammal communities (Bodmer et al. 2018). Under these circumstances, the continuation of studies on the movement ecology of the species would help us to clarify how this species will adapt to the climate change scenarios in Nicte-Ha where longer flooding seasons with contrasting extreme drought are expected (Ortiz and Méndez 1999; Imbach et al. 2012; Chiabai 2015).

Although the number of radio-tracked individuals during this study was low, this is the first time that white-tailed deer movement is studied by satellite telemetry in the tropical portion of the distribution of the species. We hope that this study encourages other researchers to increment the knowledge of the species in the region, especially because it is probably the most hunted by subsistence and sport hunters and their importance for the natural communities.

Acknowledgments The UMA Nicte-Ha, Conservación Panthera Mexico A.C., and the División Académica de Ciencias Biológicas-UJAT (DACBiol-UJAT) provided logistic support for the project. The PFCE 2018 (Programa para el Mejoramiento de la Calidad Educativa) partially sponsored the acquisition of the radio collars used in this project through the DACBiol-UJAT. FCM received support from CONACYT (scholarship 271627) for their Ph. D. studies in Ecology and Management of Tropical Systems at the DACBiol-UJAT. We specially thank Alejandro Jesús, Yaribeth Bravata, Adrían Chahín, Ramón Sanz-Freeman, Martín Moreno, Ismael Sánchez, Diana Friedeberg, and Raul Valdez for their help on the fieldwork and their continuous support. Edwin Hernández-Pérez and Rafael Reyna-Hurtado, for their review of the manuscript.

References

Bello J, Gallina S, Equihua M (2004) Movements of the white-tailed deer and their relationship with precipitation in Northeastern Mexico. Interciencia 29(7):357–361

Bertrand MR, Denicola AJ, Beissinger SR, Swihart RK (1996) Effects of parturition on home ranges and social affiliations of female white-tailed deer. J. Wildl Manag 60:899–909

Bodmer R, Mayor P, Antunez M et al (2018) Major shifts in Amazon wildlife populations from recent intensification of floods and drought. Conserv Biol. https://doi.org/10.1111/cobi.12993

Chiabai A (2015) Climate change impacts on tropical forests in Central America: an ecosystem service perspective. The Earthscan Forest Library, Routledge

Contreras-Moreno FM (2018) Ecología del venado cola blanca (*Odoileus virginianus*) en hábitats tropicales del sureste de México. Dissertation, Universidad Juárez Autónoma de Tabasco

Crist TO, Guertin DS, Wiens JA, Milne BT et al (1992) Animal movement in heterogeneous landscapes: an experiment with eleodes beetles in shortgrass prairie. Funct Ecol. https://doi.org/10.2307/2390050

Delfín C, Mandujano S, Gallina S, Bello J, López N et al (1998) Patrones de desplazamiento del venado cola blanca en un rancho con manejo de agua en el Noreste de México In: Memorias VI Simposio sobre Venados de México. Universidad Nacional Autónoma de México. p 178–186

Delgado M, Penteriani V, Revilla E, Nams VO et al (2010) The effect of phenotypic traits and external cues on natal dispersal movements. J Anim Ecol. https://doi.org/10.1111/j.1365-2656.2009.01655.x

Ferguson SH, Rettie WJ, Messier F (1998) Fractal measures of female caribou movements. Rangifer 10:139–147

Fleming M, Schortemeyer J, Ault J (1994) Distribution, abundance, and demography of white-tailed deer in the Everglades. In: Dennis J (ed) Proceedings of the Florida panther conference. pp 247–274

Gallina S, Lopez-Arevalo H (2016) *Odocoileus virginianus*. The IUCN red list of threatened species 2016: e.T42394A22162580. Retrieved from https://doi.org/10.2305/IUCN.UK.2016-2.RLTS.T42394A22162580.en. Accessed 3 May 2018

Gallina S, Corona-Zárate P, Bello J (2003) El venado cola blanca: comportamiento en zonas semiáridas del Noreste de México. In: Manejo de Fauna Silvestre en Amazonía y Latinoamérica. Selección de trabajos V Congreso Internacional. (2003). CITES, Fundación Natura, Bogotá

Gallina S, Mandujano S, Bello J, López-Arevalo H, Weber M (2010) White-tailed deer *Odocoileus virginianus* (Zimmermann, 1780). In: Barbanti M, González S (eds) Neotropical cervidology. Funep/IUCN, Jaboticabal, pp 101–118

Gurarie E, Suutarinen J, Kojola I, Ovaskainen O (2011) Summer movements, predation and habitat use of wolves in human modified boreal forests. Oecologia. https://doi.org/10.1007/s00442-010-1883-y

Hanski I (1998) Metapopulation dynamics. Nature. https://doi.org/10.1016/0169-5347(89)90061-X

Hidalgo-Mihart MG, Contreras-Moreno FM, Jesús-De La Cruz A et al (2017) Inventory of medium-sized and large mammals in the wetlands of Laguna de Terminos and Pantanos de Centla, Mexico. Check List. https://doi.org/10.15560/13.6.711

Imbach P, Molina L, Locatelli B et al (2012) Modeling potential equilibrium states of vegetation and terrestrial water cycle of Mesoamerica under climate change scenarios. J Hydrometeorol 13(2):665–680

INEGI (Instituto Nacional de Estadística y Geografía) (2013) Anuario estadístico de los Estados Unidos Mexicanos 2013. Instituto Nacional de Estadística y Geografía, Aguascalientes Retrieved from http://internet.contenidos.inegi.org.mx/contenidos/productos/prod_serv/contenidos/espanol/bvinegi/productos/nueva_estruc/aegeum/702825063979.pdf

INIFAP (Instituto Nacional de Investigaciones Forestales Agrícolas y Pecuarias) (2018) Datos meteorológicos estaciones Meteorológicas de los estados. Laboratorio Nacional de Modelaje y Sensores Remotos. INIFAP/SAGARPA. http://clima.inifap.gob.mx/lnmysr/Estaciones. Consultado <21 junio 2018>

Keeling MJ, Grenfell BT (1997) Disease extinction and community size: modeling the persistence of measles. Science 275(5296):65–67

Keuroghlian A, Eaton P (2008) Fruit availability and peccary frugivory in an isolated Atlantic forest fragment: effects on peccary ranging behavior and habitat use. Biotropica 40:62–70

Long ES, Diefenbach DR, Rosenberry CS et al (2005) Forest cover influences dispersal distance of white-tailed deer. J Mammal 86:623–629

Mandujano M, Gallina S, Ortega JA (2014) Venado cola blanca en México. In: Valdéz R, Ortega-S. A (eds) Ecología y manejo de fauna silvestre en México. Colegio de Posgraduados, pp 399–420

Mcintyre NE, Wiens J (1999) Interactions between landscape structure and animal behavior: the roles of heterogeneously distributed resources and food deprivation on movement patterns. Landsc Ecol 14:437–447

Mills K, Patterson B, Murray D (2006) Effects of variable sampling frequencies on GPS transmitter efficiency and estimated wolf home range size and movement distance. Wildl Soc Bull 34(5):1463–1469

Moorcroft PR, Lewis MA, Crabtree RL (1999) Home range analysis using a mechanistic home range model. Ecology 80(5):1656–1665

Ocaña D, Lot A (1996) Estudio de la vegetación acuática vascular del sistema fluvio-lagunar-deltaico del río Palizada, en Campeche, Mexico. Anal Inst Biol 67(2):303–327

Ortiz M, Méndez A (1999) Escenarios de vulnerabilidad por ascenso del nivel del mar en la costa mexicana del Golfo de México y el Mar Caribe. Investig Geográficas 9:68–81

Ozoga JJ, Verme LJ, Bienz CS (1982) Parturition behavior and territoriality in white-tailed deer: impact on neonatal mortality. J Wildl Manag. https://doi.org/10.2307/3808402

Pépin D, Adrados C, Mann C, Janeau G (2004) Assessing real daily distance traveled by ungulates using differential GPS locations. J Mammal. https://doi.org/10.1644/BER-022

Pérez-Cortez S, Enríquez PL, Sima-Panti D et al (2012) Influencia de la disponibilidad de agua en la presencia y abundancia de *Tapirus bairdii* en la selva de Calakmul Campeche, Mexico. Rev Mex Biodivers. https://doi.org/10.7550/rmb.25095

Phillips ML, Clark WR, Nusser SM et al (2004) Analysis of predator movement in prairie landscapes with contrasting grassland composition. J Mammal. https://doi.org/10.1644/1545-1542(2004)085<0187:AOPMIP>2.0.CO;2

Ramsey CW (1968) A drop-net deer trap. J Wildl Manag 32:187–190

Reyna-Hurtado R, Rojas-Flores E, Tanner GW (2009) Home range and habitat preferences of white-lipped peccaries (*Tayassu pecari*) in Calakmul, Campeche, Mexico. J Mammal 90(5):1199–1209

Reyna-Hurtado R, O'Farril G, Sima D, Andrade M, Padilla A, Sosa L (2010) Las aguadas de Calakmul: Reservorios de vida silvestre y de la riqueza natural de México. CONABIO Biodiversitas 93:1–6

Rivera-Monroy VH, Twilley R (1996) The relative role of denitrification and immobilization in the fate of inorganic nitrogen in mangrove sediments (Terminos Lagoon, Mexico). Limnol Oceanogr 41(2):284–296

Rodgers AR, Carr A, Beyer H, Smith L, Kie J (2007) HRT: home range tools for ArcGIS. Ontario Ministry of Natural Resources, Centre for Northern Forest Ecosystem Research, Thunder Bay

Rodríguez-Boza L (2015) Relación de los patrones de actividad del jaguar (*Panthera onca*) con los del venado cola blanca (*Odocoileus virginianus*) y el pecarí de collar (*Tayassu tajacu*) en la región de Palizada, Campeche. Dissertation. Universidad Júarez Autónoma de Tabasco

Salek M, Drahníková L, Tkadlec E (2015) Changes in home range sizes and population densities of carnivore species along the natural to urban habitat gradient. Mamm Rev 45(1):1–14

Sánchez-Rojas G, Gallina S, Mandujano S (1997) Área de actividad y uso del hábitat de dos venados cola blanca (*Odocoileus virginianus*) en un bosque tropical de la costa de Jalisco, Mexico. Act Zool Mex (n.s.) 72:39–54

Schwede G, Hendrichs H, McShea W (1993) Social and spatial organization of female white-tailed deer, *Odocoileus virginianus*, during the fawning season. Anim Behav 45(5):1007–1017

Sikes RS, Gannon WL (2011) Society of mammalogists and the animal care and use committee of American. Guidelines of the American Society of Mammalogists for the use of wild mammals in research. J Mamm 92(1):235–253

Sparrowe RD, Springer PF (1970) Seasonal activity patterns of white-tailed deer in eastern South Dakota. J Wildl Manag 34:420–431

Turchin P (1998) Quantitative analysis of movement. Measuring and modeling population redistribution in animals and plants. Sinauer Associates, Sunderland

Van Beest FM, Rivrud IM, Loe LE, Milner JM, Mysterud A (2011) What determines variation in home range size across spatiotemporal scales in a large browsing herbivore? J Anim Ecol 80:771–785

Webb SL, Riffell SK, Gee KL, Demarais S (2009) Using fractal analyses to characterize movement paths of white-tailed deer and response to spatial scale. J Mammal 90(5):1210–1217

Webb SL, Gee KL, Strickland BK, Demarais S, DeYoung RW (2010) Measuring fine-scale white-tailed deer movements and environmental influences using GPS collars. Int J Ecol 459610:1–12 https://doi.org/10.1155/2010/459610

Weber M (2005) Ecology and conservation of sympatric tropical deer populations in the Greater Calakmul Region, south-eastern Mexico. Ph.D dissertation, Durham University

Yáñez-Arancíbia A, Day JW, Twilley RR, Day RH (2014) Manglares: ecosistema centinela frente al cambio climático, Golfo de México. Madera y bosques 20:39–75

Zar JH (1999) Biostatistical analysis, 4th edn. Prentice Hall, Upper Saddle River

Chapter 9
Terrestrial Locomotion and Other Adaptive Behaviors in Howler Monkeys (*Alouatta pigra*) Living in Forest Fragments

Juan Carlos Serio-Silva, Ricarda Ramírez-Julián, Timothy M. Eppley, and Colin A. Chapman

9.1 Introduction

It is estimated that 2.3 million km^2 of forest was lost globally between 2000 and 2012 and tropical forest decreased by 2101 km^2 per year during this time (Hansen et al. 2013). As a result, understanding which species can survive in degraded habitats and how they accomplish this is critically important. One of the most common results of forest degradation is forest fragmentation (Chapman et al. 2007; Harcourt and Doherty 2005; Marsh and Chapman 2013). The genus *Alouatta* is arguably one of the most successful primates in coping with fragmentation because it survive in highly fragmented habitats (Glander 1975; Milton 1980; Estrada and Coates-Estrada 1984; Pozo-Montuy and Serio-Silva 2006), while many other primate species gradually disappear/die out in fragmented landscapes (Nijman 2013 – Indonesia; Aggimarangsee 2013 – Thailand; Chalise 2013 – Nepal). For example, Chapman et al. (2013) studied 20 fragments containing 5 species of primates near Kibale

J. C. Serio-Silva (✉)
Red DE Biología y Conservación de Vertebrados, Instituto de Ecología A.C., Veracruz, Mexico
e-mail: juan.serio@inecol.mx

R. Ramírez-Julián
POSGRADO Instituto de Ecología, A.C., Veracruz, Mexico

T. M. Eppley
San Diego Zoo Global – Institute for Conservation Research, San Diego, CA, USA

C. A. Chapman
Department of Anthropology, McGill University, Montreal, QC H3A 2A7, Canada

School of Life Sciences, University of KwaZulu-Natal, Scottsville, Pietermaritzburg, South Africa

Shaanxi Key Laboratory for Animal Conservation, Northwest University, Xi'an, China
e-mail: colin.chapman@mcgill.ca

R. Reyna-Hurtado, C. A. Chapman (eds.), *Movement Ecology of Neotropical Forest Mammals*, https://doi.org/10.1007/978-3-030-03463-4_9

National Park, Uganda, and documented an 81% decline in fragment occupancy over just 15 years.

Howler monkeys (*Alouatta* spp.) are widely recognized as folivorous primates, who consume fruits when these are available; this mixed diet allows howler monkeys to obtain the majority or all of their water from food sources (Bicca-Marques and Calegaro-Marques 1995; Neville et al. 1988; Milton 1980; Milton et al. 1980; Estrada 1984; Pavelka and Knopff 2004). Although primarily arboreal, occasional terrestrial locomotion has been described in *Alouatta fusca* (da Silva 1981), *A. caraya* (Bicca-Marques and Calegaro-Marques 1995), *A. seniculus* (Racenis 1951; Terborgh 1983), *A. palliata* (Gilbert and Stouffer 1989; Serio-Silva and Rico-Gray 2000), *A. guariba* (Almeida-Silva et al. 2005), and *A. pigra* (Bonilla-Sanchez et al. 2012; Pozo-Montuy and Serio-Silva 2007; Pozo-Montuy et al. 2011). In addition, while they typically obtain water from their diet, *Alouatta* have also been observed drinking water from tree hollows, bromeliads, or groundwater (Gilbert and Stouffer 1989; Bicca-Marques and Calegaro-Marques 1995). The drinking of such water has been observed in *A. palliata* (Glander 1978; Gilbert and Stouffer 1989; Serio-Silva and Rico-Gray 2000), *A. caraya* (Bicca-Marques and Calegaro-Marques 1995), *A. guariba* (Almeida-Silva et al. 2005), and *A. pigra* (Pozo-Montuy and Serio-Silva 2006). Drinking water has been associated with the consumption of mature leaves and median daytime temperature (Glander 1978; Bicca-Marques and Calegaro-Marques 1995; Serio-Silva and Rico-Gray 2000). However, attempts to analyze any association with season have produced contrasting results with some studies reporting more drinking during the dry season (Gilbert and Stouffer 1989; Serio-Silva and Rico-Gray 2000) and at least one study reporting increased drinking in the rainy season (Glander 1978). In the current study, we explore how unusual terrestrial locomotion and direct water intake may be adaptive behaviors facilitating long-term forest fragment use by five groups of *Alouatta pigra* living in forest remnants scattered throughout pastures in Balancán, Mexico. We assess relationships between these two behaviors and the characteristics of the howler monkeys' home ranges, foods eaten, and activity budgets during both rainy and dry seasons.

9.2 Methods

The study was conducted on the Leona Vicario Ranch in Balancán Mexico (17°40′ N. 91°30′ W.; Fig. 9.1) from September to November 2007 and March to May 2008. Mean annual precipitation in the area is 1818 mm. The five small study fragments are composed of secondary vegetation, though these were previously low-flooded forest dominated by bloodwood trees (*Haematoxylum campechianum*). The fragments are surrounded by large expanses of pastures and crop fields that are farmed when the land is not flooded (Pozo-Montuy 2003). In this region of Mexico, the government encouraged colonization and land conversion for grazing cattle and growing crops in the 1960s (Pozo-Montuy et al. 2008; Reyes-Castillo 1978), and the deforestation rate for the middle Usumacinta watershed has been estimated to be

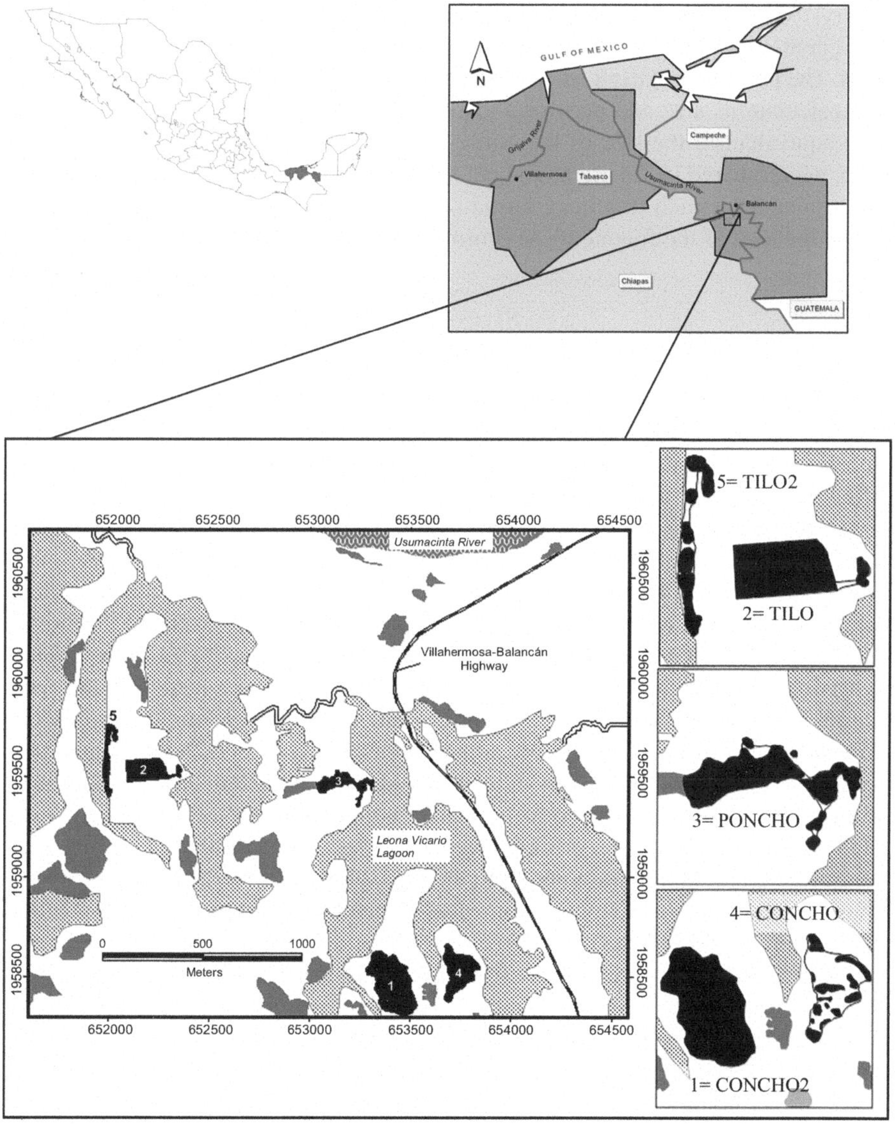

Fig. 9.1 Location of the forest remnants inhabited by five groups of *A. pigra* in Balancán, Tabasco

as high as 1.6% annually (Mendoza and Dirzo 1999); however, the owners of the fragments used by the study groups agreed to not significantly exploit the forest for the duration of the study (Serio-Silva et al. 2015).

To describe the structure and vegetation of each group's home range, ten transects (50×2 m), separated by at least 10 m, were established in each fragment

(Gentry 1982). All trees with a diameter at breast height (DBH) greater than 10 cm were counted, and the species, DBH, and height were recorded (Chapman et al. 1992). The perimeter of each fragment was mapped (Garmin eTrex Vista), allowing us to calculate its size and perimeter (Arc View GIS 3.2). All groups had a home range equivalent to the size of the entire fragment, with the exception of *Poncho* group, which shared its fragment with another group (Table 9.1). A shape index (SI) was estimated for each fragment, based on the relationship between the perimeter (P) and the area of the fragment (A) (Forman and Godron 1986).

$$\mathrm{SI} = \frac{P}{2\sqrt{\pi A}}$$

The five groups differed in composition (Table 9.1). Their behavior was quantified using 2-h focal animal sampling of non-infants for a total of 1117 h of observation. We used a simplified ethogram and recorded all behaviors of interest (i.e., resting, feeding, locomotion, and socializing) if these lasted at least 1 min. Focal animal observations were rotated among group members to obtain similar amounts of behavioral data for each individual during all sample periods (2 days – see below). We used ad libitum observations to record all instances of unusual terrestrial locomotion and water drinking. If these occurred during focal animal follows, observation of the focal animal was stopped and attention transferred to the animal engaged in these behaviors. The order in which monkeys descended to the ground or drank, the amount of time spent drinking, and the source of food or water were all recorded. We flagged which trees group members used to reach the ground and then return to measure the straight-line ground distance traveled. Given that groups were small and all members were typically within sight, we believe all or most instances of terrestriality and drinking were observed. Each group was followed for 2 consecutive days before moving on to the next, until there were 10 days of observation per group for each season (rainy season, September to November 2007; dry season, March to May 2008).

To determine if there were behavioral differences in groups observed, a one-way analysis of variance was used to compare mean daily observation time. This analysis was done for the rainy and dry seasons separately, and no significant differences were found ($P > 0.05$ in all cases). The percentage of time dedicated to the main activities was compared among groups using the nonparametric Kruskal-Wallis analysis and the post hoc Student-Newman-Keuls (SNK) test. We used linear regression to examine the relationship between terrestrial movement (number of events) and the characteristics of the habitat use. Behavioral data were evaluated and summarized using a principal component analysis (PCA; Statistica7). Associations between the first two principal component behavioral axes and terrestrial behavior, drinking water, and the proportion of time spent consuming different plant parts were evaluated using linear regression.

Table 9.1 Characteristics of the groups of *A. pigra* observed in Balancán, Tabasco, Mexico, and of the remnants of vegetation they inhabited during the rainy and dry seasons

Group	Concho2		Tilo		Poncho		Concho		Tilo2	
Variable	Rainy	Dry	Rainy	Dry	Rainy	Dry	Rainy	Dry	Rainy	Dry
No. of monkeys	6	5	4	2	3	2	3	2	2	2
Sex-age composition	2MA 2FA Mjuv Fjuv	2MA FA FA(inf) Mjuv	MA FA 2Mjuv	MA Mjuv	2MA FA(inf)	MA FA(inf)	MA FA FA(inf)	MA FA(inf)	MA FA	MA FA(inf)
Total area (ha)	3.88	3.88	2.9	2.06	1.42	1.15	2.86	1.84	1.2	0.58
Treed area (ha)	3.88	3.88	2.18	2.06	1.27	1.15	1.16	0.78	0.93	0.32
Treed proportion	1	1	0.86	1	0.89	1	0.4	0.42	0.77	0.55
Polygons	1	1	2	1	5	2	13	6	5	3
Shape index (SI)	2.78	2.78	2.48	2.35	4.69	4.5	2.8	2.63	4.48	3.4

M male, *F* female, *A* adult, *juv* juvenile, (*inf*) infant carried by its mother.

9.3 Results

Forest fragments differed in size and vegetation composition among groups and across seasons (Table 9.1). During the rainy season, the growth of shrubs at fragment edges was associated with an expansion in the forested area, creating more usable space for animal movement. These seasonal differences affected the monkeys' area of movement, such that groups reduced the size of their habitat use during the dry season. There were also significant differences in seasonal activity budgets within groups (Kruskal-Wallis $P < 0.008$) and differences among groups within a season in the time spent engaging in these activities (Table 9.2).

The matrix of factorial coefficients from the principal component analysis indicates that rainy and dry season behaviors can be accurately summarized in two components; PC1 includes the proportion of time dedicated to resting (rainy = 1.00; dry = 0.99), feeding (rainy =−0.73; dry = −0.98), and locomotion (rainy = −0.98; dry = −0.94), while PC2 comprises time spent engaging in social activities (rainy = −0.71; dry = 0.99; Fig. 9.2). Behavioral differences between the rainy and dry season, in relation to these two PC axes, varied by group (Fig. 9.2).

9.3.1 Terrestrial Behavior

A total of 63 terrestrial events were observed, with more events occurring during the rainy season ($n = 43$) than dry season ($n = 20$). The *Poncho* ($n = 26$) and *Concho* ($n = 23$) groups had the most events and covered the greatest cumulative distance on the ground (Fig. 9.3). Six of the events (12.2%) involved the groups climbing along barbed wire fences, and four (8.2%) occurred in seasonally flooded areas. The average time for a whole group to climb down from the canopy, walk in single file to another tree, and climb back into the canopy (i.e., the time the first monkey descended to the last animal ascended) varied among groups and seasons (range: 0.5–8.5 min), with the longest ground travel time lasting 18 min (Table 9.3). However, each individual typically spent less than 1 min traveling on the ground.

Table 9.2 Median values of the percentage of time that each group of *A. pigra* spent on their main activities in Balancán, Tabasco, Mexico, during the rainy and dry seasons

Activity	Resting		Feeding		Locomotion		Socializing	
Group	Rainy	Dry	Rainy	Dry	Rainy	Dry	Rainy	Dry
Poncho	63[a]	70.35[a]	31.7[a]	23.6	6.6[a]	4.3[a]	0[a]	0.35[a]
Concho2	59.35[a]	75.35[a]	26.15[a]	21.5	7.75[a]	4.3[a]	4.9	0.15[a]
Tilo	67.55	81.55[b]	25.35[a]	13.5[a]	5.9[a]	3.25[a]	0[a]	1.85[a]
Tilo2	73.6[b]	81.45[b]	23.3[b]	15.65[a]	3.05[b]	2.25	0.1[a]	0.05[a]
Concho	74.1[b]	85.4[b]	22.05[b]	12.45[a]	3[b]	1.35	0[a]	0[a]

Shared letters indicate no significant differences between groups for a given activity and season ($P < 0.05$).

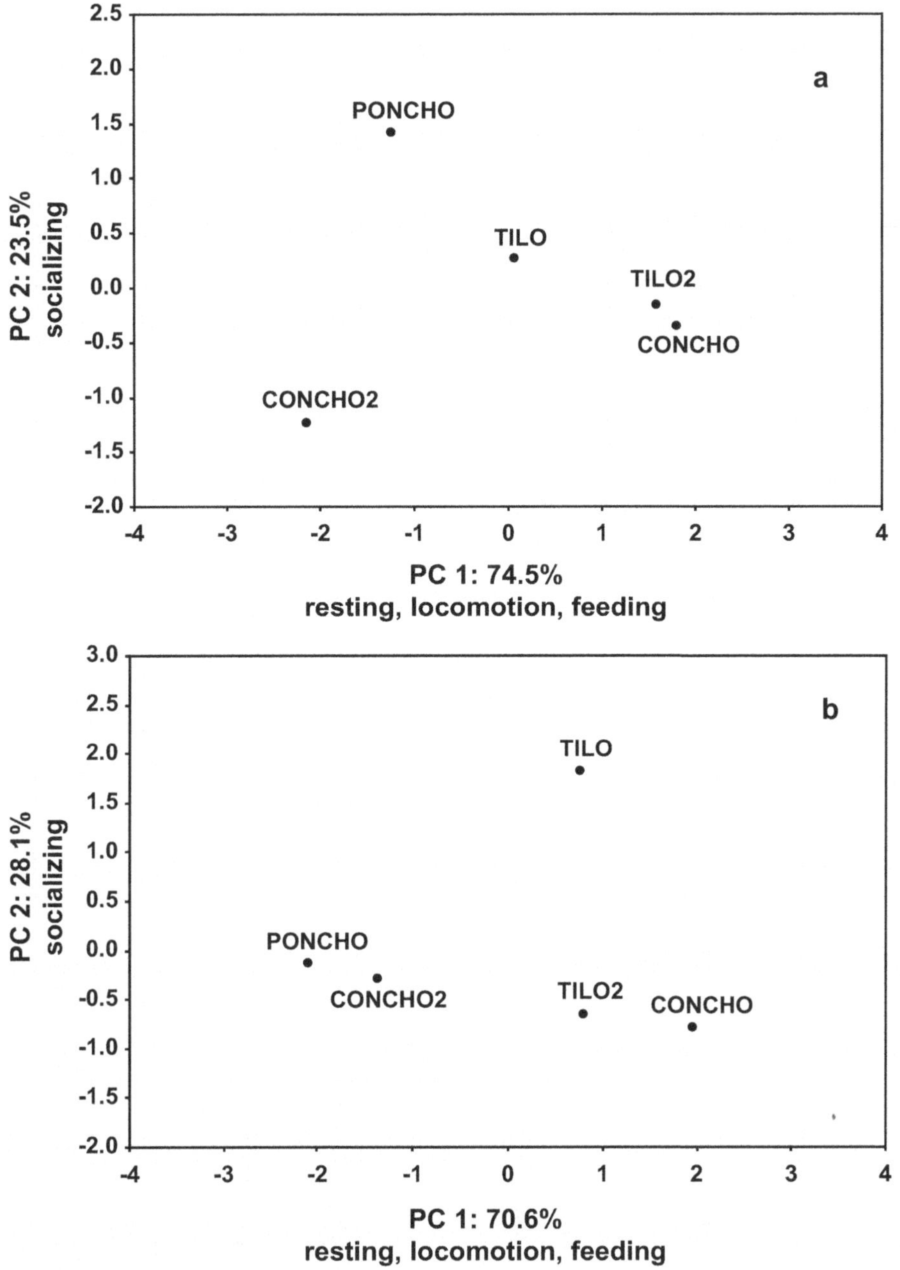

Fig. 9.2 Arrangement of the different groups of *A. pigra* in Balancán, Tabasco, Mexico, according to principal components 1 and 2 characterizing their behavior during the rainy (**a**) and dry (**b**) seasons

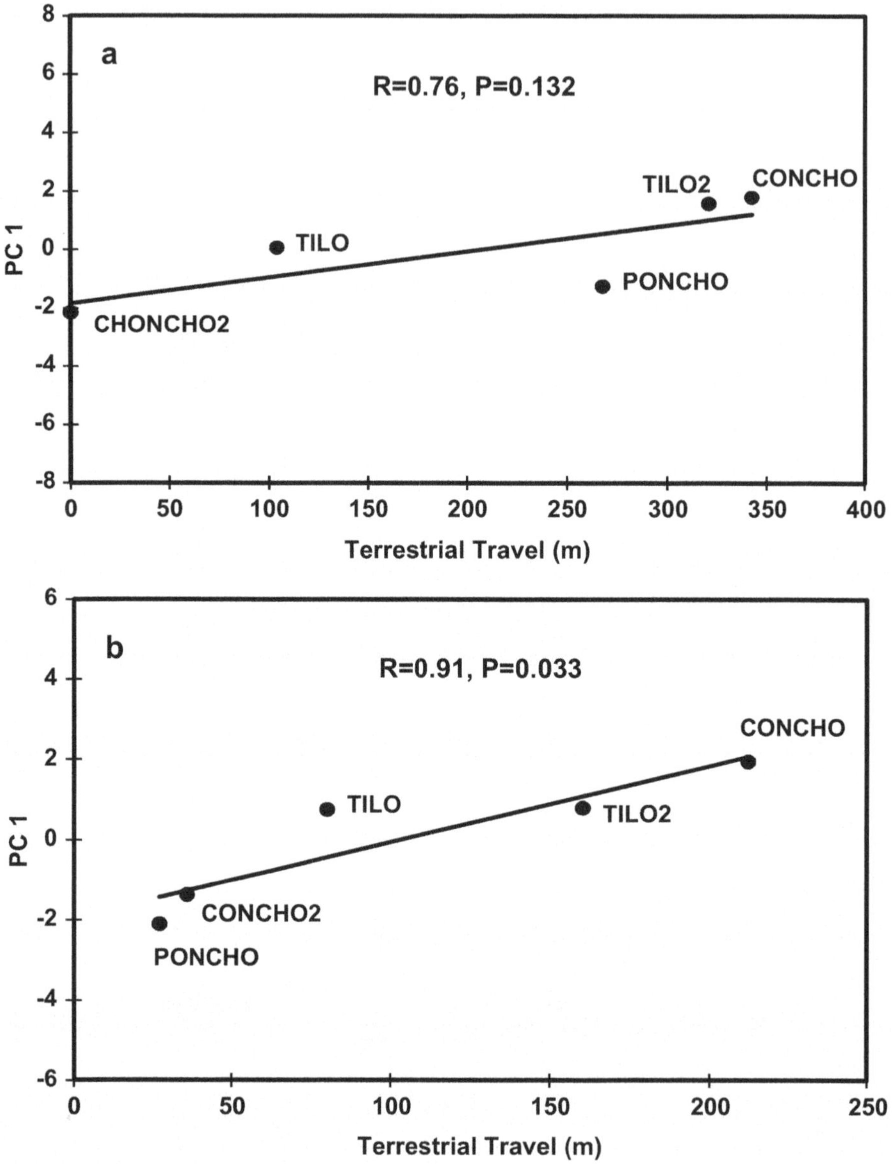

Fig. 9.3 Linear regression for the cumulative distance of terrestrial locomotion of *A. pigra* in Balancán, Tabasco, Mexico, with principal component 1 that summarizes the behavior of the groups during the rainy (**a**) and dry (**b**) seasons

Table 9.3 Characteristics of terrestrial locomotion events observed in the *A. pigra* groups studied in Balancán, Tabasco, Mexico

	Concho2		Tilo		Poncho		Concho		Tilo2	
Group	Rainy	Dry	Rainy	Dry	Rainy	Dry	Rainy	Dry	Rainy	Dry
Events (*n*)	0	2	2	1	20	6	17	6	4	5
Mean duration (interval) minutes	0	0.5	8.5	8	3.7	1	6.2	4	3.1	2
			(3–14)		(0.5–18)	(1–3)	(1–17)	(1–8)	(0.5–9)	(1–4)
Cumulative distance (m)	0	36	104.1	80.3	268.1	27.4	342.7	212.3	97.9	160.6
									(+224)[a]	

[a]In the Tilo2 group, 224 m were added to reflect their movement through the scrub (*Mimosa pigra*) that separates the trees

There was seasonal variation in the time of day during which groups traveled on the ground (Table 9.4), with groups descending to the ground more than twice as often in the rainy season ($n = 44$) than the dry season ($n = 20$). During the rainy season, terrestrial locomotion occurred throughout the day, though never after 18h00, with most events occurred between 14h00 and 15h59. During the dry season, there were fewer terrestrial locomotion bouts, but the majority of those occurred between 08h00 and 09h59 and between 16h00 and 17h59.

Animals in the *Concho* and *Tilo* groups went to the ground to feed on non-canopy plants, such as vines like *Sycidium tamnifolium* and *Syngonium angustatum*. On one occasion an adult female from the *Poncho* group descended to the ground to drink water that had accumulated in a cow's hoofprint, and members of the *Tilo2* group went to the ground in the rainy season to drink from the lagoon (Table 9.5). Once an adult female from the *Concho2* group descended to the ground to retrieve her fallen infant.

Terrestrial excursions were initiated by both males and females but varied by group. In the *Poncho* group, during the rainy season, most terrestrial use was led by the oldest male (75%); however, during the dry season, this male disappeared and terrestrial use was led by another adult male. In the *Concho* group, most terrestrial events occurred during the rainy season, and in this group most (59%) events were

Table 9.4 Time of the day when the member of each *A. pigra* group went down to the ground in Balancán, Tabasco, Mexico, during the rainy (R) and the dry (D) seasons and the number of times they did so

	Concho2		Tilo		Poncho		Concho		Tilo2		Total	
Time	R	D	R	D	R	D	R	D	R	D	R	D
0600–0759 h	–	–	–	–	3	–	–	–	–	–	3	0
0800–0959 h	–	–	1	–	4	3	–	3	–	1	5	7
1000–1159 h	–	–	–	–	5	2	2	–	–	1	7	3
1200–1359 h	–	–	1	–	4	–	1	–	–	1	6	1
1400–1559 h	–	–	–	–	4	–	8	1	3	1	15	2
1600–1759 h	–	2	–	–	1	1	6	3	1	–	8	6
1800–1959	–	–	–	1	–	–	–	–	–	–	0	1

Table 9.5 Reason that *A. pigra* descended to the ground in Balancán, Tabasco, Mexico, during the rainy (R) and the dry (D) seasons

Group	Concho2		Tilo		Poncho		Concho		Tilo2		Total	
Reason	R	D	R	D	R	D	R	D	R	D	R	D
To get to another tree	–	1	1	1	20	6	13	5	3	5	37	18
To feed	–	–	1	–	–	–	3	–	–	–	4	0
To drink water	–	–	–	–	1	–	–	–	–	–	1	0
To get to another tree to feed	–	–	–	–	–	–	1	1	–	–	1	1
To get to another tree to drink	–	–	–	–	–	–	–	–	1	–	1	0
To rescue offspring	–	1	–	–	–	–	–	–	–	–	0	1

led by an adult female carrying an infant. In both the *Tilo* and *Tilo2* groups, bouts of terrestrial locomotion were usually initiated by an adult male, while in the *Concho2* group, such events were rare but were led by an adult of either sex.

There was no relationship between group size and the number of terrestrial events ($r_{sp} = 0.558$, $P = 0.328$), nor between the number of terrestrial events and area used ($r_{sp} = 0.420$, $P = 0.428$) and the treed area ($r_{sp} = 0.639$, $P = 0.246$).

There was a significant relationship between the cumulative distance traveled on the ground and the first principal component (PC 1), which accounts for the behavior of the groups during the dry season ($r = 0.91$, $P = 0.033$). The greater the distance covered, the more time was spent resting and less time dedicated to arboreal locomotion and feeding. This same trend was observed in the rainy season, but the relationship was not significant ($r = 0.76$, $P = 0.132$).

9.3.2 *Water Drinking*

Monkeys drank water on 18 occasions: 11 occurred during the rainy season and 7 in the dry season. The main source of water was tree cavities ($n = 9$), followed by the leaves of the Mexican palmetto (*Sabal mexicana*), which collect water in the center of their starburst-shaped leaves ($n = 4$). On one occasion an animal drank from a depression in the ground, while on another animal drank from a lagoon.

In the rainy season, the groups mainly ate mature leaves, with the exception of the *Concho*2 group, which ate a greater percentage of fruit (Table 9.6). In contrast, during the dry season, the consumption of mature leaves was still highest in the *Concho* and *Tilo2* groups, but the consumption of other plant parts, such as leaf shoots, young leaves, and flowers, increased, while in the *Concho2*, *Tilo*, and *Poncho* groups, ripe fruit was most frequently eaten. There was no relationship between the plant parts eaten and the number of water drinking events ($P > 0.10$ in all cases).

Table 9.6 Percentage of feeding time that the *A. pigra* groups in Balancán, Tabasco, Mexico, spent eating different plant parts during the rainy (R) and the dry (D) seasons

	Concho2		Tilo		Poncho		Concho		Tilo2	
Plant part	R	D	R	D	R	D	R	D	R	D
Flower buds	0	1.09	5.76	29.15	25.83	10.74	0	19.4	0	0
Flowers	0	22.23	0	9.04	8.36	7.66	0.13	5	0.46	10.64
Petioles	0	0.88	4.32	0.87	3.48	0	13.47	3	0	0
Shoots	0	12.58	0	10.69	0	15.57	0	11.4	0	5.22
Young leaves	5.83	11.42	16.8	2.53	16.13	21.79	16.46	17.9	25.46	25.7
Mature leaves	33.67	8.16	48.3	6.03	38.05	10.14	47.58	26.8	74.07	40.16
Unripe fruit	2.18	0	8.16	3.21	0	0	0	0	0	0
Ripe fruit	58.33	43.64	16.7	38.48	8.15	34.1	22.36	16.5	0	18.27

9.4 Discussion

As different regions experience deforestation due to anthropogenic land conversions or forest loss due to climate change, understanding the adaptive behaviors that facilitate the use of fragments becomes critical to predict which species will be most threatened in the future and to construct inform conservation and management plans. All of the fragments we studied can be considered small (1.0–5.0 ha) relative to the size viewed suitable to support primate populations (deVleeschower and Raboy 2013; Harcourt and Doherty 2005; Anderson et al. 2007; Laurance et al. 2000; Tutin et al. 1997; Marsh 1999; Marsh and Chapman 2013).

Our research suggests that the ability to travel on the ground to isolated trees, thereby effectively expanding the group's habitat use, is a trait that facilitates the survival and success of primate populations living in fragmented landscapes. Observations suggest the main reason animals go to the ground is to get to an isolated tree or set of trees. On some occasions they just traveled through the tree they entered from the ground, while in other instances animal fed in the isolated tree (Table 9.5). However, it is possible that larger groups conduct terrestrial locomotion more often simply because there are many individuals or because they used a larger area; also larger groups might be less vulnerable to predation.

For these howlers, animal drank more in the rainy season than in the dry season, which is counter-intuitive if water is a limiting resource for this population. The dry season is the time when water is restrictive and succulent fruit or flowers are typically unavailable or limiting, and one would expect this could impact population success. These observations suggest water never becomes limiting resources that affect population growth for these fragment-living howler monkeys. The nature with which the howlers used terrestrial locomotion, the rarity that they descended to the ground, and the fact that a female and her infant were killed by dogs when on the ground all illustrate that coming to the ground poses a predation risk (see also Brussard 1985; Pozo-Montuy and Serio-Silva 2007; van Belle and Estrada 2006). Thus, we suggest that howler monkey survival in such severely fragmented landscapes depends on conservation and restoration of the native vegetation, so that groups have access to sufficient food sources, food resources that provide their water, and corridors of trees, decreasing the need for terrestrial locomotion and its associated predation risk. The planting of corridors among isolated forest trees or clumps of trees would allow groups to expand their home range to include new food sources, while not necessitating terrestrial locomotion avoiding the elevation risk of predation. In Balancán, feral dogs are not a natural element of the ecosystem, and restricting the movement of dogs to near human housing would be beneficial to howler populations. These conservation recommendations were derived from a study of howler monkeys, a species very capable of prospering in fragmented landscapes; however, since many arboreal primates living in fragments have been described to come to the ground (Campbell et al. 2005; Gebo and Sargis 2005; Onderdonk and Chapman 2000; Chapman et al. 2013), we believe that our conclusions are applicable to many species.

Similar results have come from other research (Marsh and Chapman 2013). The question that logically stems from this research is: "what can be done when these fragments are scattered across a large landscape that is all privately owned?" In such situations, for any conservation initiative to be successful, community involvement becomes critical. From the above research, our team developed a model for howler monkey conservation in the fragmented habitats of the Municipality of Balancán. This involved four interrelated elements. First, we helped to secure the establishment of a protected area for howler monkeys in a private ranch now named *Primatology and Wildlife Research Field Station* where since 2006 we have been studying the population of howlers in the fragmented landscape surrounding the station.

Second, to take conservation to the larger landscape level, the most urgent activity was to promote environmental education and to inform the community of the perspective that wildlife, including primates, and the forest were of value. Third, we worked at the political level, and after several meetings, the municipal authorities of Balancán agreed to develop a formal declaration, where from October 2013 Balancán is identified as the *Sanctuary of the Sacred Black Howler Monkey.* With this declaration Balancán is now the only city in Mexico that officially recognizes the importance of monkeys. The fourth element of our plan was to make this official declaration widely known to the local community, and this involved a celebration. We developed a festival called the "International Black Howler Monkey Week" with the support of local authorities, state and federal government (SERNAPAM and PROCER-CONANP respectively), local schools, and the community itself. During the first years, we estimated that not less than 20,000 people participated.

It is our hope that these activities will connect the results of the research to real conservation that will operate on a landscape level on privately owned land. This outreach and the fate of the monkeys are being monitored so that it will be possible to evaluate the value of this conservation model. The approach will have to be modified for each region in which it is applied, but this represents an attempt to take research data and apply it to conservation action at the appropriate spatial and temporal scale. If we are able to have real impact in perception and identity about primates by local people, surely these habitat fragments will increase their surface area, and then monkeys will not need going to the ground exposing themselves for predation or other not healthy conditions (ectoparasites, diseases risks, among others). But most importantly, understanding how monkeys develop behavioral adaptive responses in fragmented areas shows us their strategies for habitat use and surviving, including items selection for food, preferences for arboreal substrate, and how they decide (as group or individually) cost-benefits for movement on the ground or in not-tall trees seeking better food sources. There is need to develop more studies on this topic and compare results between species and different ecological niches where monkeys survive along tropical fragmented areas; besides this, we consider that only involving ecological-scientific research and explaining to local people all effects of forest fragmentation and isolating distances for monkeys movements, we could contribute to integrative conservation strategies in habitat management and primate conservation.

Acknowledgments We are grateful to Consejo Nacional de Ciencia y Tecnología (CONACyT) for the graduate studies scholarship (204435) awarded to RR-J for her master studies. The Instituto de Ecología AC (INECOL) provided the opportunity for graduate studies and a knowledge-rich environment. NSERC and CRC grants helped fund the writing stage. We thank Gilberto Pozo and Magali Bonilla for their generous support while RR-J was in Balancán. The Pozo Family and, especially, Mr. Paco Pozo and Mr. Rufino Gómez for their help with the fieldwork, and with great affection RR-J thanks Charo Mosqueda for unconditional support. Nelly Jiménez kindly assisted with plant identification and Valerie Schoof provided valuable comments on the project.

References

Aggimarangsee N (2013) Status monitoring of isolated populations of macaques and other non-human primates in Thailand. In: Marsh LK, Chapman CA (eds) Primates in fragments: complexity and resilience. Springer, New York

Almeida-Silva B, Guedes PG, Boubli JP, Strier KB (2005) Deslocamento terrestre e o comportamento de beber em um groupo de barbados (*Alouatta guarida clamitan* Cabrera, 1940) en Ninas Gerais, Brasil. Neotropical Primates 13:1–3

Anderson J, Cowlishaw G, Rowcliffe JM (2007) Effects of forest fragmentation on the abundance of *Colobus angolensis palliatus* in Kenya's coastal forests. Int J Primatol 28:637–655

Bicca-Marques JC, Calegaro-Marques C (1995) Drinking behavior of the black howler monkey (*Alouatta caraya*). Folia Primatol 58:107–111

Bonilla-Sanchez YM, Serio-Silva JC, Pozo-Montuy G, Chapman CA (2012) Does the expansion of Eucalyptus plantations signal hope or despair for black howler monkeys? Int J Primatol 33:233–245

Brussard PF (1985) Minimum viable populations: how many are too few? Restor Manag Notes 3(1):21–25

Campbell CJ, Aureli F, Chapman CA, Ramos-Fernandez G, Matthews K, Russo SE, Suarez S, Vick L (2005) Terrestrial behavior of Ateles spp. Int J Primatol: 26: 1039–1051

Chalise MK (2013) Fragmented primate populations of Nepal. In: Marsh LK, Chapman CA (eds) Primates in fragments: complexity and resilience. Springer, New York, pp 329–356

Chapman CA, Chapman LJ, Wrangham RW, Hunt K, Gebo D, Gardner L (1992) Estimators of fruit abundance of tropical trees. Biotropica 24:527–531

Chapman CA, Naughton-Treves L, Lawes MJ, Wasserman MD, Gillespie TR (2007) The conservation value of forest fragments: explanations for population declines of the colobus of western Uganda. Int J Primatol 23:513–578

Chapman CA, Ghai RR, Jacob AL, Koojo SM, Reyna-Hurtado R, Rothman JM, Twinomugisha D, Wasserman MD, Goldberg TL (2013) Going, going, gone: a 15-year history of the decline of primates in forest fragments near Kibale National Park, Uganda. In: Marsh LK, Chapman CA (eds) Primates in fragments: complexity and resilience. Springer, New York

da Silva EC (1981) A preliminary survey of brown howler monkeys (*Alouatta fusca*) at the Cantareira Reserve (Sao Paulo, Brazil). Rev Bras Biol 41:897–909

deVleeschower KM, Raboy BE (2013) Multi-level and multi-disciplinary approaches to understanding endangered primates in complex landscapes: golden-headed lion tamarins in Southern Bahia, Brazil. In: Marsh LK, Chapman CA (eds) Primates in fragments: complexity and resilience. Springer, New York

Estrada A (1984) Resource use by howling monkeys (*Alouatta palliata*) in the rain forest of Los Tuxtlas, Veracruz, Mexico. Int J Primatol 5:105–131

Estrada A, Coates-Estrada R (1984) Some observations on the present distribution and conservation of *Alouatta* and *Ateles* in Southern Mexico. Am J Primatol 7:133–137

Forman RTT, Godron M (1986) Landscape ecology. Wiley, New York

Gebo DL, Sargis EJ (2005) Terrestrial adaptations in the postcranial skeletons of guenons. Am J Phys Anthropol 93:341–371

Gentry AH (1982) Patterns of neotropical plant species diversity. Evol Biol 15:1–84

Gilbert KA, Stouffer PC (1989) Use of a ground water source by mantled howler monkeys (*Alouatta palliata*). Biotropica 21:380

Glander KE (1975) Habitat description and resource utilization: an ecological view of social organization in mantled howler monkeys. In: Tuttle RH (ed) Socioecology and psychology of primates. Mouton, The Hague, pp 37–57

Glander KE (1978) Drinking from arboreal water sources by mantled howling monkeys (*Alouatta palliata* gray). Folia Primatol 29:206–217

Hansen MC, Potapov PV, Moore R, Hancher M, Turubanova SA, Tyukavina A, Thau D, Stehman SV, Goetz SJ, Loveland TR, Kommareddy A, Egorov A, Chini L, Justice CO, Townshend JRG (2013) High-resolution global maps of 21st-century forest cover change. Science 342:850–853. https://doi.org/10.1126/Science.1244693

Harcourt AH, Doherty DA (2005) Species-area relationships of primates in tropical forest fragments: a global analysis. J Appl Ecol 42:630–637

Laurance WF, Vasconcelos HL, Lovejoy TE (2000) Forst loss and fragmentation in the Amazon: implications for wildlife conservation. Oryx 34:39–45

Marsh LK (1999) Ecological effect of the black howler monkey (*Alouatta pigra*) on fragmented forests in Community Baboon Sanctuary, Belize. Ph.D. Thesis, University of Washington, St. Louis

Marsh LK, Chapman CA (2013) Primates in fragments: complexity and resilience. Springer, New York

Mendoza E, Dirzo R (1999) Deforestation in Lacandonia (southeast Mexico): evidence for the declaration of the northernmost tropical hot-spot. Biodivers Conserv 8:1621–1641

Milton K (1980) The foraging strategies of howler monkeys: a study in primate economics. Columbia University Press, New York

Milton K, Van Soest P, Robertson JB (1980) Digestive efficiencies of wild howler monkeys. Physiol Zool 53:402–409

Neville MK, Glander KE, Braza F, Ryland AB (1988) The howling monkeys, genus *Alouatta*. In: Mittermeier RA, Rylands AB, Coimbra-Filho AF, Da Fonseca GAB (eds) Ecology and behavior of neotropical primates, vol 2. WWF-US, Washington, DC, pp 349–453

Nijman V (2013) One hundred years of solitude: effects of long-term forest fragmentation on the primate community of Java, Indonesia. In: Marsh LK, Chapman CA (eds) Primates in fragments: complexity and resilience. Springer, New York

Onderdonk DA, Chapman CA (2000) Coping with forest fragmentation: the primates of Kibale National Park, Uganda. Int J Primatol 21:587–611

Pavelka MSM, Knopff KH (2004) Diet and activity in black howler monkeys (*Alouatta pigra*) in southern Belize: does degree of frugivory influence activity level? Primates 45:105–111

Pozo-Montuy G (2003) Comportamiento de monos aulladores (*Alouatta pigra*) en habitat fragmentado, en la Rancheria Leona Vicario, Balancan, Tabasco. Universidad Juarez Autonoma de Tabasco, Mexico

Pozo-Montuy G, Serio-Silva JC (2006) Comportamiento alimentario de monos aulladores negros (*Alouatta pigra* Lawrence, Cebidae) en hábitat fragmentado en Balancán, Tabasco, México. Acta Zool Mex 22:53–66

Pozo-Montuy G, Serio-Silva JC (2007) Movement and resource use by a group of *Alouatta pigra* in a forest fragment in Balancan, Mexico. Primates 48:102–107

Pozo-Montuy G, Serio-Silva JC, Bonilla-Sánchez YM, Bynum N, Landgrave R (2008) Current status of the habitat and population of the black howler monkey (*Alouatta pigra*) in Balancán, Tabasco, Mexico. Am J Primatol 70:1–8

Pozo-Montuy G, Serio-Silva JC, Bonilla-Sanchez YM (2011) Influence of the landscape matrix on the abundance of arboreal primates in fragmented landscapes. Primates 52:139–147

Racenis J (1951) Some observations on the red howling monkey (*Alouatta seniculus*) in Venezuela. J Mammology 33:114–115
Reyes-Castillo P (1978) La fauna silvestre en el Plan Balancán Tenosique. INIREB. Biblioteca del Instituto de Ecologia Xalapa, Mexico
Serio-Silva JC, Rico-Gray V (2000) Use of a stream by Mexican howler monkeys. Southwest Nat 45:332–333
Serio-Silva JC, Olguín EJ, Garcia-Feria L, Tapia-Fierro K, Chapman CA (2015) Cascading impacts of anthropogenically driven habitat loss: deforestation, flooding, and possible lead poisoning in howler monkeys (*Alouatta pigra*). Primates 56(1):29–35
Terborgh J (1983) Five new world primates. Princeton University Press, Princeton
Tutin CEG, White LJT, Mackanga-Missandzou A (1997) The use of rainforest mammals of natural forest fragments in an equatorial African savanna. Conserv Biol 11:1190–1203
van Belle S, Estrada A (2006) Demographic features of *Alouatta pigra* populations in extensive and fragmented forest. In: Estrada A, Garber P, Pavelka M, Luecke L (eds) New perspectives in the study of Mesoamerican Primates: distribution, ecology, behavior and conservation. Kluwer Academic/Plenum Publishers, New York, pp 121–142

Chapter 10
Variation in Space Use and Social Cohesion Within and Between Four Groups of Woolly Monkeys (*Lagothrix lagotricha poeppigii*) in Relation to Fruit Availability and Mating Opportunities at the Tiputini Biodiversity Station, Ecuador

Kelsey Ellis and Anthony Di Fiore

10.1 Introduction

For group-living organisms, daily activities such as traveling, foraging, and resting are rarely performed in isolation, and individuals' movement patterns are often dependent on one another. As such, there is a growing understanding among animal ecologists that many movement processes must be considered and quantified within a social context (Westley et al. 2018). By studying the concurrent movements of more than one animal, researchers are better able to elucidate how competitive dynamics both within and between social groups – including attraction to or avoidance of particular individuals or sets of individuals – may influence overall patterns of space use and access to critical resources, such as sleeping sites, food, and mates (Kays et al. 2015).

Animal ecologists and behavioral biologists have also increasingly come to appreciate the high degree of variation and complexity that exists in the spatiotemporal association patterns of group-living animals (e.g., Aureli et al. 2008; Whitehead 2008). In many species, individuals live, travel, and forage together in cohesive units typically referred to as "social groups," where these groups can be identified as the set of animals that regularly are seen in spatial proximity to one another and who interact, at least much of the time, in a tolerant or affiliative manner. In some species, however, socio-spatial associations are more transient and flexible, with individuals associating and interacting at different points in time with varying sets of other animals. Species that exhibit these higher degrees of "fission-fusion" social

K. Ellis (✉) · A. Di Fiore
Department of Anthropology, University of Texas at Austin, Austin, TX, USA
e-mail: kelseyellis@utexas.edu

R. Reyna-Hurtado, C. A. Chapman (eds.), *Movement Ecology of Neotropical Forest Mammals*, https://doi.org/10.1007/978-3-030-03463-4_10

dynamics may form subgroups with stable and predictable membership, such as the single male units in multilevel societies of geladas or hamadryas baboons (Grueter et al. 2012), or they may form subgroups with more individualistic and flexible subgroup membership, as commonly observed in spider monkeys, chimpanzees, dolphins, or hyenas (reviewed in Aureli et al. 2008).

Members of the ateline clade of primates (howler monkeys, spider monkeys, woolly monkeys, and muriquis) show a variety of association patterns with varying degrees of flexibility in size, composition, and spatial cohesion of groups over time, ranging from one single cohesive group (e.g., some species of howler monkeys) to groups showing high degrees of fission-fusion dynamics where individuals or subgroups remain separated from one another for several hours or even days (e.g., spider monkeys; Di Fiore et al. 2011). Such dynamic grouping patterns are thought to reflect an optimization strategy by which individuals actively balance the costs and benefits of group living (Chapman and Chapman 2000; Lehmann and Boesch 2004; Aureli et al. 2008). Costs most often include competition over preferred, ephemeral food items, such as ripe fruits, that occur in discrete patches too small to support all group members simultaneously (Klein and Klein 1977; Wrangham 1980, 1986; Symington 1990), while benefits may include the ability to adaptively respond to perceived risks of predation (Link et al. 2011) or shifting reproductive opportunities (Goodall 1986; Symington 1987; Chapman 1990; Hashimoto et al. 2001; Anderson et al. 2002; Mitani et al. 2002; Shimooka 2003). Because larger groups must commonly travel farther per day to compensate for faster depletion rates than smaller groups, subgrouping behavior and adjustments to subgroup size may further reflect strategies to reduce relative ranging costs (Chapman et al. 1995; Janson and Goldsmith 1995; Chapman and Chapman 2000; Asensio et al. 2009).

In this chapter, we present a detailed description of ranging patterns and subgrouping behavior among four neighboring groups of lowland woolly monkeys (*Lagothrix lagotricha poeppigii*) living in a tropical rainforest in eastern Ecuador. Similar to other large-bodied atelines, woolly monkeys live in large mixed-sexed groups with adult sex ratios that are generally female biased (reviewed in Di Fiore et al. 2011). Males are usually philopatric (Di Fiore et al. 2009) with females dispersing from their natal groups when they reach sexually maturity, at around 6 years of age (Nishimura 2003). Woolly monkeys are highly frugivorous, preferring ripe fruit and supplementing their diet with young leaves, shoots, and flowers (Stevenson 1992; Peres 1994; Stevenson et al. 1994; Defler and Defler 1996; Di Fiore 2004; Dew 2005). However, unlike other atelines – and despite their large body size – woolly monkeys also devote a substantial amount of time searching for and consuming animal prey (Dew 2005; Di Fiore 2004; Di Fiore and Rodman 2001; Stevenson 1992; Stevenson et al. 1994). Similar to spider monkeys and muriquis, woolly monkeys are further characterized as having long periods of infant dependency, and they demonstrate reproductive seasonality in which most mating behavior and reproductive milestones (e.g., conceptions and births) are clustered to a few months out of the year (Nishimura 2003; Ellis et al. in review).

Early reports on patterns of spatial cohesion and grouping dynamics among woolly monkeys were equivocal, and descriptions have ranged from social groups that appear to exhibit high degrees of fission-fusion dynamics (Kavanagh and Dresdale 1975) to social groups that remained relatively cohesive throughout the day, albeit widely spread out (Nishimura 1990, 1994; Peres 1996; Di Fiore 1997). Such varied assessments most likely stemmed from the high spatial dispersion of woolly monkey social groups as they perform their daily activities, with group members often spread over several hundred meters (Di Fiore 1997; Peres 1996), thus limiting the ability of a single observer to visually detect more than a subset of individual group members at any given time.

Although precise quantitative metrics of grouping dynamics and social cohesion are still lacking (Stevenson 2006; Aureli et al. 2008), woolly monkeys are, at present, most commonly characterized as living in relatively cohesive social groups that occasionally fission into tightly grouped feeding parties and/or coordinated subgroups that persist for hours or even days (Defler 1996; Di Fiore 1997). Subgroup composition is generally mixed-sex (Peres 1996), but solitary males and all-male subgroups have also been observed (Di Fiore et al. 2009). Cohesion and coordination among the members of social groups and subgroups is often mediated through the frequent use and reciprocation of low-volume contact calls. In some studies, measures of group cohesion (e.g., group spread and interindividual distance among groupmates) have been shown to be negatively correlated with both fruit availability (Peres 1996) and group size (Zárate and Stevenson 2014), meaning that groups spread out more in times of low fruit availability and that members of larger groups are separated by greater distances, on average, than members of smaller groups, presumably in response to within-group feeding competition. However, to our knowledge, no study has investigated how temporal variation in available mating partners also may affect subgrouping behavior and group cohesion, despite its influence on subgroup size and composition in a number of other primate taxa, including chimpanzees (Goodall 1986; Matsumoto-Oda et al. 1998; Boesch and Boesch-Achermann 2000; Wrangham 2000; Hashimoto et al. 2001; Anderson et al. 2002; Mitani et al. 2002) and spider monkeys (Symington 1987; Chapman 1990; Shimooka 2003).

Adding more complexity, woolly monkeys do not actively defend exclusive areas of their home range (i.e., they are non-territorial Burt 1949), and the home ranges of neighboring groups can overlap extensively (45–100%) (Di Fiore 2003; Di Fiore et al. 2011). In these shared areas, groups occasionally encounter one another, with varying degrees of acceptance or tolerance shown toward members of other groups. Thus, at times, socially cohesive groups will coalesce with others to form temporary "supergroups" that rest, travel, and forage together for up to several hours (Nishimura 1990; Peres 1994; Defler 1996; Di Fiore 1997; Di Fiore et al. 2009). Likewise, subsets of individuals from different neighboring social groups have also been observed to form subgroups that range independently from the rest of the members in each of their respective social groups (Ellis et al. 2014). Individuals have also been noted to occasionally visit other social groups for varying periods of time without conflict (Nishimura 1990, 2003). However, in contrast to these tolerant associations, some intergroup encounters among these same groups are clearly hostile and are accompanied by distressed vocalizations, display behavior (e.g., branch shaking, conspicuous branch crashing), and chasing or lunging at members of the opposing group (Di Fiore 1997). The factors determining why some intergroup encounters in woolly monkeys are peaceful, while others are hostile, remain to be investigated, but may be related to location of encounters, the identity and composition of the groups involved, the specific interaction history of those groups, or the presence of particular individuals, as well as to seasonal variation in food resources and mating opportunities (e.g., Fashing 2001; Crofoot et al. 2008; Koch et al. 2016).

While a number of studies have described ranging behavior and space use patterns of woolly monkeys, particularly in relation to fruit availability (reviewed in Di Fiore et al. 2011), few have focused on multiple social groups at the same site or have considered how variation in spatial cohesion and subgrouping dynamics, both within and between *Lagothrix* groups, may affect observed ranging patterns (but see Peres 1996), nor have they considered how the availability of both food and mating opportunities might concurrently impact ranging and subgrouping behavior. Here we use a combination of location and subgroup composition records collected over 14 months by multiple observers following animals from four neighboring social groups of *Lagothrix lagotricha poeppigii* (1) to describe subgrouping behavior and ranging patterns and (2) to examine variation in spatial cohesion among animals within and between groups, over time and in relation to both food availability and mating opportunities. Our study is unique in that observers worked in tandem and could individually recognize animals from multiple social groups and thus were able to collect precise data on subgroup compositions (including on subgroups comprising animals from different social groups) and on interindividual and intersubgroup distances. Individual identifications were corroborated, post hoc, with genotype data collected for the same animals over time to ensure consistency in identity assignment across time.

We first summarize subgrouping behavior and within social group cohesion for each of the four study groups, including the percent of time that social groups divide into subgroups, variation in subgroup size, and spatial dispersion between focal individuals within and between subgroups. We then present data on ranging and space use for each of the four social groups, including information on daily path length, home range size, core area size, and patterns of range overlap. Rates of intergroup encounters are then compared to indices of overlap among the home ranges of the different social groups to determine if intergroup encounters occur more or less often than expected by chance. Finally, we evaluate how these patterns of group cohesion, subgroup size, and ranging behavior vary with temporal fluctuations in both habitat-wide estimates of ripe fruit availability and the availability of mating opportunities.

10.2 Methods

10.2.1 Data Collection

10.2.1.1 Study Site

This study was conducted at the Tiputini Biodiversity Station (TBS) in lowland Ecuador. The research station is located along the northern bank of the Tiputini River (76°08′W, 0°38′S) and is part of the Yasuní Biosphere Reserve, one of the largest remaining tracts of western Amazonian lowland rainforest. Laid out over a series of rolling hills and drainages, an extensive network of well-marked trails provides access to nearly 650 ha of terra firme and seasonally flooded forest. Annual rainfall at the field site generally exceeds 2800 mm (Di Fiore et al. 2009; Blake et al. 2010), with March through July often marked as the wettest months of the year. Fruit abundance varies throughout the year, with January through March typically showing the greatest abundance of ripe fruit and July and August the smallest abundance of ripe fruit (Snodderly et al. in review) . The field site hosts an intact predator community (Blake et al. 2012) and ten different species of nonhuman primate, including three atelines: lowland woolly monkeys (*Lagothrix lagotricha poeppigii*), red howler monkeys (*Alouatta seniculus*), and white-bellied spider monkeys (*Ateles belzebuth*).

10.2.1.2 Study Population

At least eight socially cohesive woolly monkey groups have been identified to range, entirely or partially, within the TBS trail system. These social groups vary in size from 13 to 40 individuals, with typical composition including 2–5 adult males, 4–10 adult females, and 5 or more immatures. At some point in time between 2006 and the present, at least one individual from each of these eight social groups (referred to as *Lagothrix* A, C, D, G, HP, I, L, and P) was anesthetized via remote intramuscular injection of either ketamine HCl or a reconstituted mixture of tiletamine HCl and zolazepam HCl (Zoletil®) using a CO_2-powered darting rifle. While anesthetized, animals were fitted with a Telonics™ or Holohil™ VHF transmitting radio collar to facilitate tracking via radiotelemetry. Social groups with radio-collared animals were then the subjects of at least occasional contact during successive field seasons for the lifetime of the collar on the animal (i.e., until either the radio-collar battery expired, the animal lost the collar or disappeared from the group, or the animal died and the collar was recovered). The radio collars were useful for confirming and monitoring the presence of, at minimum, the collared individual in a particular social group, even without visual contact with that individual.

Over the years and across different researchers, woolly monkey groups at the site have been studied at varying intensities, with more consistent behavioral and ranging data being collected on groups D, G, and I between 2008 and 2009, on groups C and G in 2013, and on groups C, D, G, and P from 2014 to the present.

10.2.1.3 Behavioral Data

The present study focuses on the behavioral and spatial data collected by KME and seven field assistants between August 2014 and December 2015, excluding January 2015 to March 2015, on four of the known social groups at the site: *Lagothrix* C, D, G, and P. These groups occupied contiguous home ranges within the TBS trail system and were previously well habituated to observer presence. Animals were individually recognized by the presence of radio collars with identification tags, facial scarring, variation in pelage color, broken digits, and/or differences in anogenital size and pigmentation. Following Schmitt and Di Fiore (2014), we estimated each individual's age-sex category based on external morphology and behavior. The number of adults and subadults in these four study groups varied between 8 and 19 individuals, with total group size, including juveniles and infants, ranging from 14 to 30 individuals (Table 10.1).

Behavioral data were recorded using a combination of instantaneous and continuous sampling in the context of daylong focal follows (Altmann 1974). Due to the dispersed nature of woolly monkey social groups, a strict schedule of rotating among predetermined focal animals was not feasible. Instead, observers were assigned a particular social group, and once members of that social group were located, observers opportunistically selected one adult or subadult to follow as their focal subject. When selecting a focal animal, preference was given to individuals with the fewest number of accumulated hours of focal sampling for that given month. In cases where social groups were fissioned into two or more subgroups and when there were fewer eligible focal animals within a subgroup than observers, supernumerary observers were tasked with searching throughout the assigned social

Table 10.1 Composition of four woolly monkey groups between August 2014 and December 2015 at the Tiputini Biodiversity Station, Ecuador

	Group C		Group D		Group G		Group P	
Age-sex	2014	2015	2014	2015	2014	2015	2014	2015
Adult male	2	3	5	4	3	4	4	4
Subadult male	1	–	1	–	2	2	–	1–2
Adult female	5	4–5	6	5–6	10–11	8–9	9	10
Subadult female	1	1	–	–	4	3–4	2	2
Juvenile	5	4	5	5	6	7	7	7
Infant	–	1–2	2	1	4	2	2	3
Adults and subadults	9	8–9	12	9–10	19–20	17–19	15	17–18
Total with dependents	14	13–15	19	15–16	29–30	26–28	24	27–28

group's home range for eligible animals in other subgroups to follow. During focal animal follows, the behavior of the focal individual and the identity and distance category (contact, <1 m, <5 m, <10 m) of each neighbor within a 10 m radius were recorded instantaneously every 5 min. Feeding behavior and social interactions were recorded continuously. Rare events such as mating behaviors (e.g., copulations, mating solicitations) between non-focal individuals were recorded ad libitum.

Every 15 min (on the hour and at 15, 30, and 45 min after the hour), observers evaluated subgroup composition and recorded the identity and/or age-sex class of each subgroup member, the general activity of the majority of subgroup members ("resting," "moving," "soraging," "socializing"), and an estimate of subgroup spread (i.e., distance between the farthest two subgroup members, in meters). We defined a subgroup as a subset of individuals that maintained visual or low intensity vocal contact with one another 15 min before, during, or after the sampling (Chapman et al. 1995; Bezanson et al. 2008) and were located at a distance of <100 m from at least one other subgroup member following a chain rule (e.g., Ramos-Fernández 2005). Social groups were thus considered "subgrouped" when the number of individuals present during consecutive sampling periods was fewer than the total group size and those individuals remained out of visual and vocal contact with other members of their social group for two or more sampling points. Subgroup determinations were often corroborated through telemetry checks in which observers scanned the vicinity for radio-collared animals belonging to the social group being followed.

Simultaneous observers continuously monitored the location (using a data-logging GPS: see below), dimension, and cohesion of subgroups being followed, regularly alerting one another to their focal individual's position within the subgroup, as well as the arrival and departure of other individuals, through a series of "whoop" vocalizations and two-way walkie-talkie radios. Typically, there would be one to three observers in the field simultaneously monitoring different individuals from the same social group, but on rare occasions, up to five observers collected focal data simultaneously on different members of the same social group. Focal animal behavioral data and subgroup composition data collected concurrently from different subjects allowed us to measure the spatial dispersion between pairs of individuals within a subgroup as well as the spatial dispersion between individuals in two or more subgroups. To reduce spatial and temporal autocorrelation in ranging analyses, we pooled together all simultaneously recorded subgroup composition and location records in which multiple observers were following focal animals within the same subgroup. For these cases, subgroup location was assigned as the average location between all observers tracking that subgroup. We followed different social groups from dawn to dusk (circa 06:00–18:00) on a rotating schedule, contacting each group on average four to eight times per month (range, 3–19 days), contingent upon weather and the number of eligible focal individuals in each group.

To aid in the collection of ranging data, each observer was assigned a handheld GPS unit (Garmin 76CSx or GPSMap64) that was programmed to store the observer's location every ~20 s, and observers attempted to always remain underneath members of the subgroup being followed and within ten horizontal meters of their focal subjects. Subgroup locations were subsequently determined by extracting locations from the observer's "tracks" at the same 15-min sampling points where subgroup composition was recorded. Location records were collected in the WGS84 coordinate system and projected to Universal Transverse Mercator (UTM 18S) units. To compensate for the minor positional errors inherent in GPS readings, we averaged the locations of all fixes recorded in the 2-min window surrounding the time of interest when extracting a location from the GPS track. For example, to extract a location for 09:00, we would average the set of ~6 locations recorded between 08:59 and 09:01. When possible, subgroup locations were additionally recorded in geometric vector notation from a previously georeferenced landmarks (i.e., the distance and angle from a marked and mapped tree or trail marker).

In some cases, it was impossible to collect or extract a location for a particular sampling point, either (1) because of observer or GPS error (e.g., failure to record a location), (2) because the observer was not positioned close enough to the subgroup (e.g., because the monkeys crossed over a flooded area that was inaccessible to terrestrial observers), or (3) because of poor GPS coverage. In these cases, if possible, we reconstructed a location for the sampling point based on linear extrapolation from the surrounding records. We reconstructed up to a maximum of two successive location points in this manner.

10.2.1.4 Intergroup Encounters

Intergroup encounters (IGEs) were recorded opportunistically any time an individual from another social group was visually or acoustically observed to be within 100 m of an observer's focal subgroup. During IGEs, observers recorded the identities and/or age-sex class of as many members from the other social group as possible, as well as the location of the encounter and the general activity of both groups prior to, during, and after the IGE. The IGE was considered to have ended when individuals from each social group were once again separated by >100 m.

10.2.1.5 Fruit Availability

Biweekly estimates of fruit availability at TBS have been collected regularly since September 2006 (Link et al. in preparation). Following methods outlined in Stevenson (2004), we surveyed nearly 9 km of narrow belt transects that intersect a large portion of the TBS trail system. During each survey we recorded the species (or morphospecies) and diameter at breast height (DBH) of all trees whose crowns overlapped the phenological transect and were bearing ripe, fleshy fruit (for lianas, the DBH of the main supporting trunk was measured). The basal area of the trunk, $\pi\left(\frac{\text{DBH}}{2}\right)^2$, was then calculated as a proxy for ripe fruit production (Chapman et al. 1992; Stevenson 2004), and the effective width of the phenological transects sampled was calculated as the average perpendicular distance from all monitored trees to the center of the transects. The total area sampled each monitoring period, calculated as the effective transect width x total transect length, was 5.25 ha. For trees and lianas that produced fruits in more than one biweekly period, we assigned a proportion of the tree's basal area to each fruiting period following the coefficients of Pascal's triangle, again following Stevenson (2004). For example, if a tree had a basal area of 100 cm^2 and produced fruit during four successive monitoring periods, the basal area index of ripe fruit production would be distributed among those four periods using Pascal's coefficients of 1:3:3:1, resulting in proxy scores of 12.5 cm^2, 37.5 cm^2, 37.5 cm^2, and 12.5 cm^2, for the same respective periods. To then index habitat-wide fruit availability for a given monitoring period, we summed these portions of the basal area of each tree and liana producing ripe fruits and divided by the total area sampled to obtain the total basal area of plants bearing ripe, fleshy fruits per hectare that period.

10.2.1.6 Mating Opportunities

Because we were not monitoring female reproductive hormone profiles, precise measures of female estrous cycles and receptivity are not available for the time period in question. Therefore, to characterize the availability of potentially receptive females (i.e., mating opportunities) across the landscape during this study, we used a behavioral indicator: the total number of different individual females who were observed copulating across all four study groups in a given month. To correct for differences in behavioral sampling intensity across months, we divided this value by the number of cumulative observer hours for that same month to yield a mating opportunity index, which we present as the number of different females seen copulating per 100 observation hours. Mating behavior in woolly monkeys is a fairly conspicuous activity, with solicitations and copulations occurring in front of other group members. Furthermore, the frequency of observed mating behavior and rates of conception have been found to be highly correlated (Ellis et al. in review).

10.2.2 Analyses

10.2.2.1 Home Range Size and Overlap

A total of 11,853 15-min subgroup location records were collected during subgroup follows, with an average of 2,963 ± 646.7 location records collected per social group. From these records, we computed annual and monthly kernel density estimates of habitat utilization for each social group. Home ranges constructed from kernel density estimates (KDE) are based on use distributions and correspond to the minimum area for which the probability of locating a given social group is equal to the probability specified. Following Laver and Kelly's (2008) recommendation, we used the 95% KDE isopleth to define each social group's home range, as it is one of the most commonly used contour values and allows for more consistent comparisons across studies. Each social groups' "core area" was defined as the 50% KDE isopleth.

A key parameter in constructing biologically meaningful home ranges is the bandwidth or smoothing parameter (h). Small values of h tend to undersmooth data, creating fragmented home range contours, whereas large values of h tend to oversmooth the data, leading to an overestimation of home range size and incorporating excessive space around the outermost locations (Kie et al. 2010). We chose to use an ad hoc ($h_{\text{ad hoc}}$) smoothing parameter as this method has provided better results compared to using the reference (h_{ref}) or least-squared cross-validation (h_{lscv}) smoothing factors, which tend to oversmooth and undersmooth the data, respectively (Kie 2013). Starting with an initial smoothing parameter of h = 50 m and iteratively adding 5 m up to h = 100 m, we found that an ad hoc smoothing parameter of h = 85 m was sufficient for yielding a contiguous 95% KDE isopleth for each study group without adding excessive space around the outermost locations. Prior to constructing home range contours, we employed an incremental area analysis (Kenward and Hodder 1996) to ensure that home range areas reached an asymptote and were thus a good estimate of home range size. Home range contours were constructed in R version 3.3 (*R Core Team* 2016) using the "adehabitatHR" package (Calenge 2006) and imported into ArcGIS 10 (Environmental Systems Research Institute, Redlands, CA) for visualization.

We assessed the monthly variation in shared space use between social groups using two measures of overlap. The first measure investigates how far a neighboring group penetrates into a focal group's home range and is taken as the ratio between the area of overlap between the two groups and the total area of the focal group's home range, resulting in two values per dyad. However, this measure does not account for the time each group spent in the area of overlap (e.g., two groups may have high degrees of overlap but spend relatively little time in those overlap zones). Therefore, as a second measure of shared space use, we calculated the utilization distribution overlap index (UDOI). This index is similar to Hurlbert's (1978) index of niche overlap and has been suggested to be the most appropriate index when trying to quantify overlap in terms of common space use within a particular period of time (Fieberg and Kochanny 2005). Values of this index generally range between 0 and 1, with 0 indicating that two home ranges do not overlap at all and 1 indicating that home ranges are uniformly distributed and overlap perfectly. In some cases, if the utilization distributions being compared are not uniformly distributed and there is a high degree of overlap, UDOI may be >1.

10.2.2.2 Daily Path Length

We calculated daily path lengths (DPL) as the Euclidean distance displacement of successive ranging locations recorded during subgroup follows. We performed such calculations for all follows lasting 10.5 h or more (N = 114), as there was no significant difference between follows lasting 10.5–11.75 h (N = 88) compared to follows lasting 12.0–12.25 h (N = 26) (t = −0.256, df = 112, p-value = 0.798). Because members of a given social group may fission from and coalesce into different subgroups throughout the day, some individuals from the same social group may travel farther than others. To account for this variation, we calculated DPLs for all unique focal individuals followed on a given day when they were in different subgroups and averaged the length of these unique paths to provide a single DPL estimate for the social group that day.

10.2.2.3 Factors Influencing Subgrouping, Group Cohesion, Ranging Patterns, and Intergroup Encounters

To investigate the effects of social group size, fruit availability, and mating opportunities on observed patterns of subgrouping, group cohesion, and ranging, we used a combination of mixed effects models. First, we evaluated the probability of a social group being divided into one or more subgroups using a generalized linear mixed effects model (GLMM) for proportions fit with a binomial error distribution and logit link. The response variable was the proportion of subgroup composition (SGC) records collected for which a social group was observed "subgrouped" in relation to total number of records collected for that group during the month, with social group size, the monthly estimate of fruit availability, and the mating opportunity index set as fixed effects. Variation in average monthly subgroup size and interindividual distance were evaluated with linear mixed effects models (LMM), again with group size, the monthly estimate of fruit availability, and the mating opportunity index set as fixed effects. Variation in DPL was also examined using LMM, and fixed effects included the social group's average subgroup size on the day the length of the daily path was collected, as well as the monthly estimate of fruit availability and the mating opportunity index. Variation in home range size, core area size, and their corresponding overlap was also evaluated with LMMs. In these models, group size, the monthly estimate of fruit availability, and the mating opportunity index were set as fixed effects. In all models, group sizes and subgroup sizes only included adult and subadult individuals, as the movement decisions of juveniles and infants are largely dependent on those of their mothers.

Prior to inclusion in models, all predictor variables were converted into unitless Z scores to remove the undue influence of using variables measured in different units (e.g., group size measured as an integer number versus fruit basal area measured as cm^2/ha). In LMMs, all response variables associated with size or distance (e.g., subgroup size, interindividual distance, DPL, home range size, and core area size) were log transformed, while estimates of overlap were square-root transformed using $\sqrt{Y+0.5}$. Given that our data included repeated measures on the same social group and uneven sampling of groups across months, we included social group ID, and the combination of month and year data was collected as random effects. All models were fit with the *lmer* and *glmer* functions found in the R package "lme4"

(Bates et al. 2015), and corresponding *p* values were calculated using *lmerTest* (Kuznetsova et al. 2014).

To select the combination of predictor variables best fitting each dataset, we used the Akaike's information criterion corrected for small sample size (AICc) (Burnham and Anderson 2002). The model with the lowest AICc score was considered the best-fitting model. However, in cases where the next best-fitting model was within two AICc points, indicating a plausible alternative, we preferentially selected the model that was most parsimonious (i.e., the model with the fewest fixed effects) (Burnham and Anderson 2002). We also calculated the relative importance of each fixed effect from the sum of AICc weights across candidate models. All AICc-associated metrics were performed in R using the *dredge* function of the "MuMin" package (Barton 2016). Finally, we report the significance of the final model compared to the null model using log-likelihood ratio tests calculated with the *anova* function in R.

10.3 Results

10.3.1 Group Cohesion

Of the location records collected between August 2014 and December 2015, 10,859 had accompanying subgroup composition (SGC) records, and of these records, 10,743 included 1 or more individuals from our 4 focal study groups (*Lagothrix* C, D, G, and P). Overall, social groups C and D were subgrouped in 62% and 59% of SGC records, respectively, while groups G and P were subgrouped in 85% and 83% of SGC records, respectively. Monthly variation in the proportion of 15-min SGC samples in which social groups were "subgrouped" was best predicted by social group size and the mating opportunity index, but not fruit availability (Tables 10.2 and 10.5), with larger groups tending to be subgrouped more often and all groups subgrouping less often when more mating females were present in the population. For each standard deviation increase in social group size, the chances of being found "subgrouped" increased 3.5-fold, and for each standard deviation increase in the mating opportunity index, the chances of being "subgrouped" decreased by a factor of 0.42.

Excluding subgroup composition records collected during IGEs (n = 829 of 10,743 records), subgroup sizes ranged from 1 to 17 individuals (Fig. 10.1). Despite a twofold difference in total social group size, the median subgroup size across the four study groups was similar, ranging between 7 and 8 individuals. Contrary to our expectations, our measure of monthly habitat-wide ripe fruit availability was not a significant predictor of average subgroup size. We also found that total social group size was not a significant predictor of average subgroup size, suggesting that social groups, regardless of size, nonetheless break into similarly sized subgroups. Instead, monthly variation in subgroup size was best predicted by the mating opportunity index, with subgroup size increasing as the number of different females seen copulating

Table 10.2 Results of the final generalized linear mixed model and linear mixed models predicting the effects of group size, fruit abundance, and mating opportunity index on observed patterns of subgrouping and within-group cohesion. In all models, group ID and month-year were set as random effects

Test	Response variable	Fixed effects	Estimate	Std. error	z or *t* value	*p*
Probability of subgrouping GLMM with proportions (family: binomial link: logit)	# SGC records where social group is "subgrouped"/# total SGC records	(Intercept)	1.161	0.347	3.343	<0.001
		Social group size	1.262	0.166	7.614	<0.001
		Mating opportunity index	−0.842	0.266	−3.169	<0.002
		Model fit: $\chi2 = 101.73$; $p < 0.001$				
Subgroup size LMM	Average monthly SG size	(Intercept)	2.009	0.055	36.541	<0.001
		Mating opportunity index	0.140	0.024	5.914	<0.001
		Model fit: $\chi2 = 16.664$; $p < 0.001$				
Interindividual distance LMM	Average monthly interindividual distance (m)	(Intercept)	4.25	0.1909	22.264	<0.001
		Social group size	0.3623	0.1228	2.952	0.041
		Model fit: $\chi2 = 7.392$; $p < 0.007$				

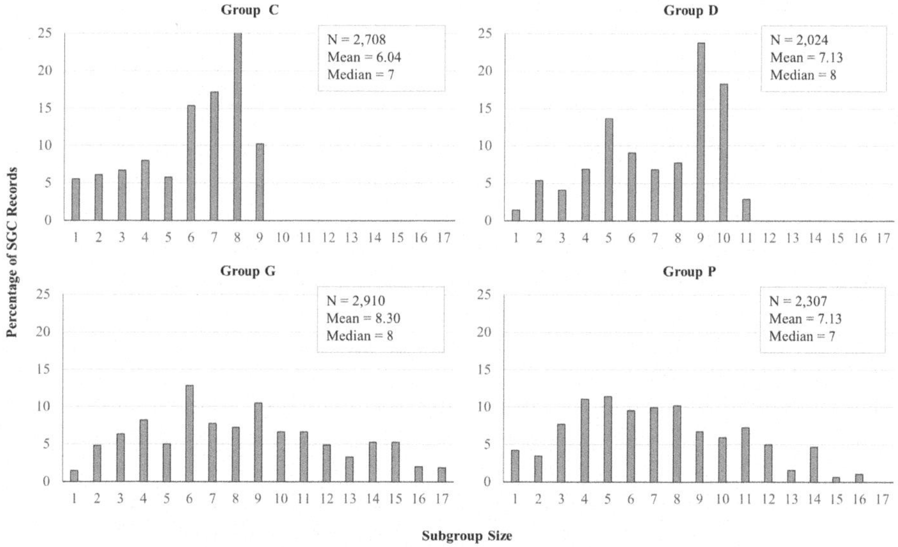

Fig. 10.1 Percent distribution of woolly monkey subgroup sizes among four focal groups from 15-min subgroup composition (SGC) records

across all four study groups increased (Tables 10.2 and 10.5). In all social groups, subgroups were predominantly mixed-sex (88.6%), although 8.2% and 3.2% of subgroup composition records consisted of single- or all-female subgroups and single- or all-male subgroups, respectively.

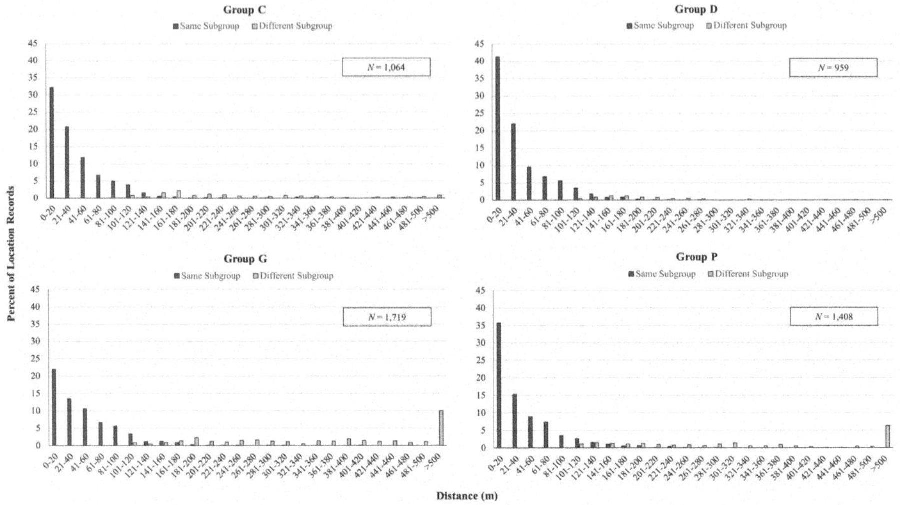

Fig. 10.2 Percent distribution of distances between simultaneously followed animals from four social groups (*N* = 5,150 15-min SGC records). Dark bars represent distances between animals in the same subgroup; light bars indicate distances between animals in separate subgroups

Interindividual distances between members of the same social group were estimated from 5,150 15-min SGC samples where 2 observers were simultaneous with more than 1 member of the same social group and collected location records. The average distance between individuals from the same social group who were associating in the same subgroup (*N* = 4,011 of these simultaneous location records) was ~42 m and ranged between 0 and 493 m, while distances between individuals from the same social group who were simultaneously followed while in different subgroups (*N* = 1,139) ranged from 112 to 1,252 m (Fig. 10.2). As expected, social group size had a significant and positive effect on the average distance between simultaneously followed group members each month (Tables 10.2 and 10.5). This is not surprising, as larger groups were more often divided into subgroups than were smaller groups, and the subgroups into which larger groups were split frequently traveled farther apart from one another (*N* = 906; average, 412 ± 244 m) than the subgroups of smaller groups (*N* = 233; average, 244 ± 112 m). However, contrary to our expectations, fruit availability had no effect on the average distance between simultaneously followed group members each month, nor did the mating opportunity index.

10.3.2 Daily Path Length

Combining daily path length data for the four focal study groups, the average distance traveled each day was 1,887 m (*N* = 114; range, 9223695 m; Table 10.3). Although group size varied substantially across groups, daily path lengths were not

Table 10.3 Average daily path length (m) among four neighboring groups of woolly monkeys at the Tiputini Biodiversity Station

Group	Average	SD	Min	Max	N
Lagothrix C	1,857.81	657.94	922.20	3,695.43	36
Lagothrix D	2,087.18	571.82	1137.98	3,296.22	21
Lagothrix G	1,769.20	464.59	1031.68	3,130.27	36
Lagothrix P	1,941.12	548.34	1193.10	2,844.23	21

Table 10.4 Results of the final linear mixed models predicting the effects of group size (or subgroup size), fruit abundance, and mating opportunity index on ranging patterns. In all models, group ID and month-year were set as random effects

Test	Response variable	Fixed effects	Estimate	Std. error	*t* value	*p*
Daily path length LMM	Daily path length (m)	(Intercept)	7.488	0.040	188.900	<0.001
		Mating opportunity index	0.146	0.039	3.728	<0.004
		Model fit: $\chi 2 = 8.249$; $p < 0.005$				
Home range size LMM	Monthly HR size (ha)	(Intercept)	4.419	0.055	80.393	<0.001
		Social group size	0.148	0.052	2.866	0.044
		Model fit: $\chi 2 = 7.375$; $p < 0.007$				
Core area size LMM	Monthly CA size (ha)	(Intercept)	3.099	0.043	71.808	<0.001
		Social group size	0.123	0.043	2.881	0.042
		Model fit: $\chi 2 = 6.764$; $p < 0.01$				

significantly different between groups (Kruskal-Wallis test: $H = 5.405$ df = 3, $p = 0.14$). Eight of the 114 daily path lengths calculated were missing complementary subgroup composition records and subsequently excluded from the LMM exploring the effects of subgroup size, fruit availability, and mating opportunities on daily path length. After accounting for repeated measures and uneven sampling across groups over time, neither subgroup size nor fruit availability had a significant effect on daily path length (Table 10.5). However, a group's daily path length was significantly and positively related to the mating opportunity index, with DPLs increasing by 15% for every standard deviation increase in the number of copulating females observed that month (Table 10.4).

10.3.3 Home Range Size and Overlap

Home ranges for each social group across the entire study period – defined as the 95% isopleth of kernel density estimates – ranged from 93 ha (group C) to 204 ha (group D) (Fig. 10.3). Overall, groups C, G, and P used, on average, 60–70% of their home range when followed each month, whereas group D only used ~45% of their home range when followed each month. Core areas for the entire study period,

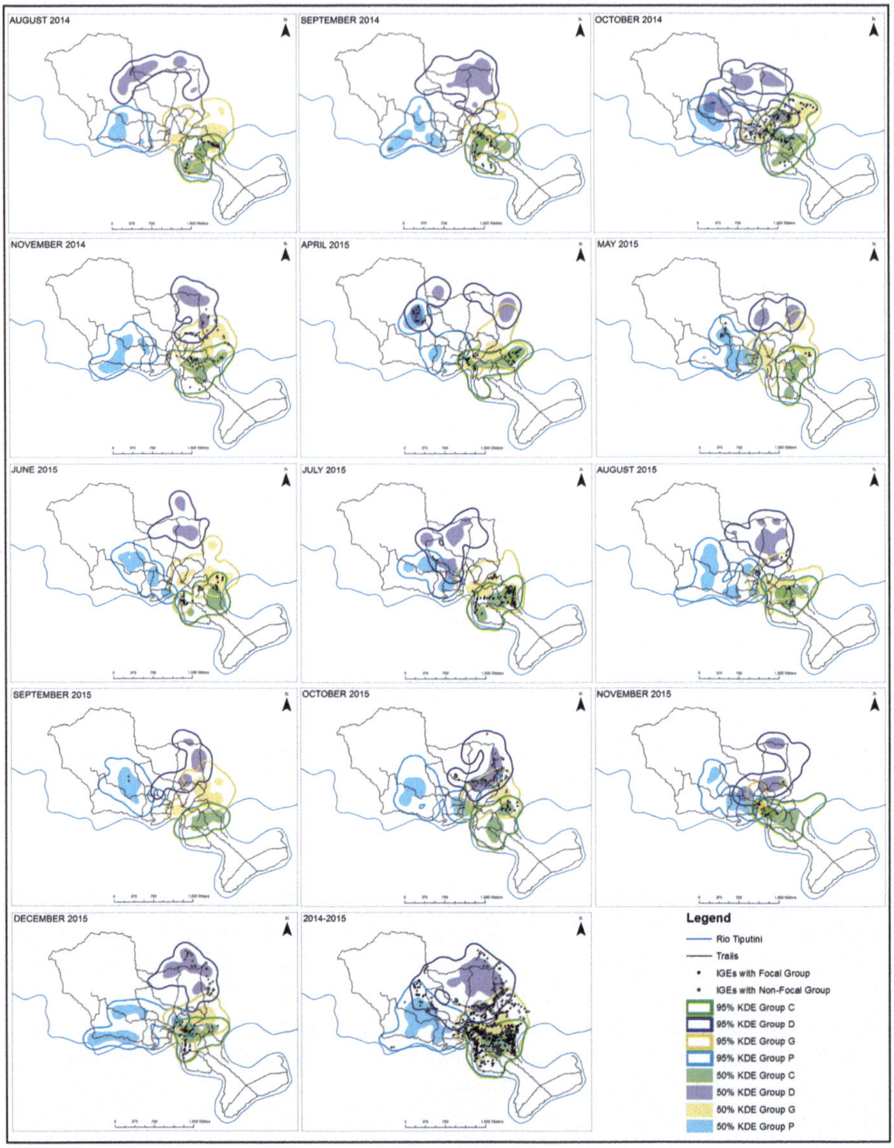

Fig. 10.3 Location of home ranges (95% KDE contours), core areas (50% KDE contours), and intergroup encounters (IGEs) of four neighboring groups of woolly monkeys at the Tiputini Biodiversity Station from August 2014 to November 2014 and May 2015 to December 2015

defined as the 50% isopleth of kernel density estimates, ranged in size from 25 ha (group C) to 53 ha (group D). On average, the monthly core areas of groups C and G represented ~20% of their overall home range, while the core areas of groups D and P only represented 12% and 16% of their respective home ranges. Group size had a significant effect on monthly home range size, with larger groups having larger home ranges (Table 10.4, Table 10.5). However, the significance of this effect is just below the 0.05 probability threshold and should be taken with caution. Neither monthly fruit availability nor our index of mating opportunities was significant predictors of monthly home range size. Similarly, when evaluating monthly variation in core area size, we found no effect of group size, fruit availability, or the mating opportunity index, with AICc values and likelihood ratio tests indicating that none of the alternative models fit significantly better than the null model (Table 10.5).

Home range and core area overlap, taken as the percent of home range or core area a focal group shares with a neighboring group, varied extensively over time and between pairs of groups. Home ranges calculated for the entire study showed that focal groups shared between 2.7% (C and P) and 67.8% (C and G) of their home range with a neighboring focal group, with an average of 22.6 ± 20.8%. Core areas calculated for the entire study period revealed that only groups C and G had overlapping portions of their core areas at this time scale, with group C sharing 28.4% of its total core area with group G and group G sharing 19.8% of its total core area with group C. At a monthly time scale, multiple pairs of groups were found to share portions of their core areas with one another. On average, groups shared 4.8 ± 6.4% of their monthly core area with a neighboring focal group, but this value was highly variable across months (range, 0–50%). Despite monthly variation in average percent home range and core area overlap between groups, we found no relationship between these measures and either fruit availability or mating opportunities (Table 10.5). It should be noted, however, that these overlap values only include the four focal groups, and considerable overlap may exist between the focal groups and neighboring groups not included in this study (e.g., the home ranges of at least two other social groups are known to overlap that of group D to the east and northwest of its home range).

Shared space use on a monthly scale, as measured by UDOI, also varied between pairs of groups and across months. Groups C and G exhibited the highest monthly UDOI values (average, 0.235 ± 0.145), spending more time in their shared home range than other pairs. Considering only months in which the 95% KDE contours overlapped, the UDOI values for other pairs of groups were relatively low in comparison (average, 0.084 ± 0.124), indicating that despite home range overlap, these other pairs used less of their overlap area in the same month than groups C and G. As observed for monthly estimates of home range and core area overlap, neither fruit availability nor mating opportunities had an effect on the time groups spent in shared areas (Table 10.5).

10.3.4 Intergroup Encounters

Over the entire study period, we recorded 102 IGEs, 61 of which occurred directly between 2 or more of our focal study groups and 41 of which occurred between a focal group and a neighboring but non-focal social group. Nearly 60% of these IGEs were considered "tolerant" associations, where affiliative behaviors (e.g., grooming bouts and copulations) were sometimes observed between members of different groups, as well as more neutral interactions in which members from different groups could be seen traveling or feeding less than 10 m from one another with no accompanying agonistic behavior. Distressed vocalizations (e.g., "clucking," which is characteristic of agitation or uneasiness), display behaviors (such as branch shaking), and/or chasing and lunging at members of opposing groups occurred in 33% of IGEs. Avoidance behavior (i.e., a group's leaving an area immediately following the detection of non-group members prior to confrontation) was seen in 6% of IGEs, although this number is likely to be underestimated as observers may often have failed to detect the presence of members from another social group during non-agonistic IGEs. Excluding time spent in overnight associations, IGEs lasted an average of 122 mins and ranged from between several minutes to entire days. The locations of intergroup encounters were not limited to the periphery of a group's home range. In fact, 88 of the 102 IGEs had 1 or more location records sampled during the span of the IGE that fell within the monthly core area of at least 1 of the interacting groups.

A strong majority (73.8%) of the IGEs recorded between focal study groups occurred between groups C and G. Overall, these two groups had a rate of 0.30 encounters per 12 h of observation time with either group or slightly less than one IGE every 3 days. In comparison, the pair of groups with the next highest encounter rate (groups D and G) had only 0.05 encounters per 12 h or roughly one IGE every 20 days. As expected, the observed rate of intergroup encounters among pairs of groups each month was highly and significantly correlated to their monthly UDOIs ($R^2 = 0.762$, df = 45, $p < 0.001$).

Finally, on 18 occasions we observed subgroups that were formed by a subset of individuals belonging to separate social groups. Individuals from groups C and G formed a majority (72.2%) of these mixed-group subgroups, followed by members of groups D and G (11.1%). Mixed-group subgroups almost always included both males and females, the exception being 45 min in which a female and her juvenile from group C traveled with several females and their dependents from group G before being joined by an adult male, also from group G. Mixed-group subgroups would often travel, feed, socialize, and rest as a cohesive unit. These associations lasted from 60 min up to full days and in some cases persisted overnight ($N = 5$). The average subgroup size for mixed-group subgroups was eight individuals (range, 4–12), similar to that of subgroups comprising members of only one social group. We should note that we do not include in this overview of mixed-group subgroups the occasional visits that some unknown, extragroup individuals – males and females alike – would make to our focal social groups. Although visiting animals could stay in association with group members for several hours up to several days, we did not know what social group they were from, nor if they were in the act of potentially immigrating or emigrating.

10.4 Discussion

Together, our results demonstrate that lowland woolly monkeys at the Tiputini Biodiversity Station, show both flexible and fluid association patterns, with subgroup size, composition, and cohesion changing throughout the day and across months. These results also reinforce earlier speculation about the dynamic and variable nature of woolly monkey association patterns (Kavanagh and Dresdale 1975; Peres 1996; Di Fiore 1997; Di Fiore and Strier 2004). However, these dynamic grouping patterns vary considerably from those observed in classically described "fission-fusion" societies. Unlike the individualistic form of fission-fusion dynamics exhibited by spider monkeys and chimpanzees, or the predictable subgrouping patterns observed in geladas or hamadryas baboons, woolly monkeys appear intermediary between these two extremes (Kummer 1968, 1971; Di Fiore and Strier 2004; Aureli et al. 2008). While social groups could be found ranging as a cohesive unit (i.e., where all group members were observed within 100 m from at least one other group member), finding them separated into two or more subgroups was much more common, particularly among the larger groups.

In species showing a high degree of fission-fusion dynamics, changes in subgroup size and cohesion are often thought to reflect an adaptive response to within-group feeding competition over patchily distributed resources (e.g., ripe fruit), and, as such, larger more cohesive subgroups are expected during periods of higher food availability (Wrangham 1977, 1980, 1986; Symington 1990; Chapman et al. 1995; Boesch 1996; Chapman and Chapman 2000; Newton-Fisher et al. 2000; Anderson et al. 2002; Mitani et al. 2002; Shimooka 2003; Asensio et al. 2009). In the present study, however, we found no relationship between habitat-wide estimates of fruit availability and measures of either subgrouping (i.e., the monthly proportion of subgroup composition records in which groups were divided into subgroups and average subgroup size) or intragroup cohesion (i.e., interindividual distance of simultaneously followed group members) among the four social groups. This result may not be that surprising given that previous studies on the grouping dynamics and feeding ecology of spider monkeys and woolly monkeys within the Yasuní region found the same lack of relationship between estimates of fruit availability and measures of within-group feeding competition (Di Fiore and Rodman 2001; Di Fiore 2003, 2004; Link 2011). For example, for woolly monkeys at a nearby field site (the Proyecto Primates Research Area, located about 35 km from the TBS site), the proportion of the monthly diet comprising ripe fruit (a preferred resource) was not correlated with habitat-wide ripe fruit availability, nor were feeding party sizes particularly well predicted by feeding patch size (as measured by diameter at breast height; Di Fiore 2004). Moreover, habitat-wide estimates of ripe fruit availability accounted for only 23% of the observed variation in subgroup size among sympatric white-bellied spider monkeys (*Ateles belzebuth belzebuth*) at TBS (Link 2011).

Such weak relationships may stem from the relatively persistent availability of fruit resources and the less markedly seasonal variation in fruit production at Yasuní compared to other Amazonian sites, but the dispersed spatial organization of woolly monkeys while engaged in subsistence activities and their devotion to searching for animal prey likely also play important roles (Di Fiore 2004; Di Fiore and Rodman 2001).

Subgrouping behavior may also reduce the relative ranging costs individuals incur at larger group and subgroup sizes due to scramble competition (Chapman et al. 1995; Janson and Goldsmith 1995; Chapman and Chapman 2000; Asensio et al. 2009). However, despite variation in both overall group sizes and subgrouping behavior among the four groups, we found no relationship between daily path length and total group size, nor did we find that larger subgroups had longer average daily paths. Again, this lack of relationship may be due to the greater spatial dispersion and less synchronized behavior of individual woolly monkeys as they perform their daily activities, leading to a relaxation of scramble competition among groupmates. Indeed, it was rare to see all subgroup members occupy feeding trees simultaneously. Instead, typically only a small subset of subgroup members would enter a single feeding tree at a time, while the remaining members either rested, foraged on food items found in other tree crowns and lianas, or searched for arthropods and other animals (Di Fiore 2004; Peres 1996). Stemming from this observation is the consideration that subgrouping and resultant ranging patterns may not be reflective of intragroup feeding competition, but rather of an inability to coordinate activities at larger group and subgroup sizes (Dias and Strier 2003). This consideration may help explain why we see larger groups splitting into subgroups more often than smaller groups, yet find no differences in daily path lengths. Speculatively, the fact that all four social groups exhibited median subgroup sizes of seven to eight individuals may indicate that there is an upper threshold to the number of animals that can successfully coordinate their activities at any given time.

Interestingly, while subgrouping and ranging behavior were not demonstrably related to fruit availability, both fission-fusion dynamics and daily path length were strongly predicted by our behavioral index of mating opportunities. We found that as the number of observed sexually active females across the landscape increased, social groups were more likely to stay cohesive (i.e., less likely to fission) and to be seen in larger, mixed-sex subgroups. This result is consistent with observations of a number of other primates that are also characterized by high fission-fusion dynamics and polygynandrous mating systems (e.g., chimpanzees Goodall 1986; Boesch 1996; Matsumoto-Oda et al. 1998; Wrangham 2000; Hashimoto et al. 2001; Mitani et al. 2002 and spider monkeys Symington 1987; Chapman 1990; Shimooka 2003), although mating opportunities are seldom formally discussed as a potential driver of fission-fusion dynamics in these taxa. Taken together with the fact that woolly monkeys demonstrate strong reproductive seasonality, where mating behavior and major reproductive events (i.e., conceptions and births) are each limited to several months out of the year (Nishimura 2003; Ellis et al. in review), the subgrouping dynamics of woolly monkeys at TBS may best be explained by variation in reproductive opportunities, rather than foraging costs and intragroup feeding competition.

The ranging patterns (DPLs and home range sizes) of the four social groups at TBS were, overall, very similar to those reported for two social groups followed by Di Fiore (2003) at the nearby Proyecto Primates Research Area in the Yasuní National Park, with daily path lengths from both sites typically falling between 1700 and 2100 m and annual home ranges between 93 and 204 ha. Although we found a significant and positive relationship between group size and home range size at the monthly scale, the largest overall home range was occupied by group D, one of our smallest groups. The large home range of group D is enigmatic but may be a reflection of poorer habitat quality or the uneven distribution of key resources within their home range (Isbell et al. 1998; Ganas and Robbins 2005), especially given that location records for that group were generally concentrated in the periphery and relatively sparse near the center of their KDE-defined home range (Fig. 10.3). Subjectively, the "center" of group D's home range was not habitually used for resting or foraging, but rather was simply traversed on the way to more productive areas. Another potential reason why the long-term home range of group D was larger than that of other groups may be that they were undergoing a home range shift throughout the course of the study, as evidenced by increased and more repeated use of the northwest portion of their home range as the study progressed. In chimpanzees and gorillas, such shifts have been observed after the disappearance of neighboring groups (Nakamura et al. 2013), as the direct result of mating competition (Watts 1998), and from the splitting and formation of new groups (Watts 1998; Caillaud et al. 2014). What may have prompted a home range shift in group D is not entirely clear, but may include the loss of several adult males between 2014 and 2015 and the potential immigration of new adult males from a neighboring group, HP, whose home range overlapped with that of group D to the northwest. Molecular analyses from noninvasively sampled individuals in both social groups are currently underway to shed light on this possibility.

Long-term home range overlap among neighboring woolly monkey groups at TBS was extensive and comparable to that reported at other sites (reviewed in Di Fiore et al. 2011, but see Shanee 2014), although at finer time scales (i.e., monthly values), the extent of home range overlap between neighboring groups was often reduced. Core area overlap, on the other hand, was always rare, with only groups C and G sharing significant portions of their long-term core areas, although some pairs of groups did share core areas at the monthly time scale. Given that most pairs of neighboring groups had minimal core area overlap – and given that the percent of overlap experienced by groups was often reduced at finer time scales – there may be support for the idea that woolly monkeys retain exclusive access to at least some portions of their home ranges (predominantly their core areas) over short temporal scales. This may be surprising given that woolly monkeys are considered non-territorial and do not actively defend their home range from conspecifics. Still, recent evidence has shown that mountain gorillas – also considered non-territorial – may nonetheless maintain exclusive access to a majority of their core areas through active avoidance of neighbors (Seiler et al. 2017). Similarly, non-territorial baboon troops with substantial degrees of home range overlap demonstrate an avoidance-based spacing pattern, with less spatial overlap and less time spent in areas of overlap at finer time scales (Markham et al. 2013). Given that measures of concurrent space use by neighboring groups of woolly monkeys (UDOI), approached 0 for almost all pairs of groups examined (groups C and G were an exception), temporal partitioning of space is potentially one such mechanism by which woolly monkey groups may reduce direct competition with their neighbors.

Rates of intergroup encounters were highly correlated with monthly measures of shared space use. As such, they varied substantially between neighboring groups, with groups C and G meeting each other on average once every 3 days, while other neighboring pairs of groups only encountered each other once every 20 days, on average. During intergroup encounters, interactions observed between members of different groups ranged from affiliative (e.g., copulation) and tolerant (e.g., co-feeding in food trees) to moderately hostile (e.g., display behavior, chasing), with more than half of IGEs classified as the former. The duration of intergroup encounters varied from as little as several minutes to several hours and even overnight associations, indicating that although woolly monkeys may attempt to avoid their neighbors in some contexts, they may also seek out and maintain associations with members from other social groups in others. This impression is highlighted by the occasional formation of mixed-group subgroups that also lasted from several hours to overnight. Individuals may opt to maintain spatial proximity to extragroup members in order to minimize the risk of predation, glean information about the availability or location of food resources, gain access to reproductive opportunities, or investigate groups into which they may transfer to in the future (e.g., macaques Majolo et al. 2005; Zhao 1997). Furthermore, tolerant intergroup encounters may be facilitated by high levels of genetic relatedness between individuals residing in neighboring groups, as has been suggested for western gorillas (Bermejo 2004; Bradley et al. 2004). Given the genetic evidence that both male and female woolly monkeys potentially disperse from their natal groups (Di Fiore et al. 2009) and the identification of at least one mother-son dyad residing in separate, but neighboring social groups (Ellis and Di Fiore, unpublished data), we find this an attractive explanation and will be investigating its plausibility in further work.

The movement ecology paradigm, at its core, attempts to understand how and why animals move the way they do (Nathan et al. 2008). For group-living animals, such as woolly monkeys, the movements of any given individual are often influenced by social cues from and interactions with both habitual groupmates and extra-group individuals, some of whom may be familiar and perhaps even related animals. As such, attempts to analyze movement of individuals without considering social interactions and group dynamics may be misleading (Langrock et al. 2014). In this chapter, we explicitly explored broad scale patterns of ranging and space use among four groups of woolly monkeys living in lowland Amazonia as they relate to fission-fusion dynamics and temporal variation in fruit availability and mating opportunities. The woolly monkeys of the Tiputini Biodiversity Station exhibited highly flexible and fluid association patterns, both within and between social groups. Contrary to our expectations, these dynamic grouping patterns were not related to intragroup feeding competition, but rather reflected variation in potential mating opportunities across time. For animals inhabiting rich environments – where food resources may be relatively predictable in time and space – the location and number of potential mates should also be included in models predicting movement and grouping patterns. The ability of woolly monkeys to maintain social cohesion and coordinate movements at larger group and subgroup sizes may also be a constraint on association patterns, resulting in larger groups dividing into subgroups more often than smaller groups.

The ability of group-living animals to coordinate their ranging behavior has long fascinated researchers, and recent advances in both technology and computational methods have encouraged the integration of collective behavior research with that of the movement ecology paradigm (Nathan et al. 2008; Westley et al. 2018), providing promising new insights into how animals use social cues to make decisions about travel. Although woolly monkey social groups may use avoidance-based spacing mechanisms to reduce direct competition with neighboring groups, some individuals may also seek out affiliative interactions with conspecifics from other groups, as evidenced by the occurrence of the tolerant intergroup encounters and affinitive interactions we recorded among the members of mixed-group subgroups. Together, our data on woolly monkey ranging behavior highlight the necessity to include social dynamics and social context into models of animal movements and emergent patterns of space use.

Acknowledgments We would like to thank the Ecuadorian government and the Ministerio de Ambiente for permission to work in the Yasuní region and the wonderful directors and staff of the Universidad San Francisco de Quito and Tiputini Biodiversity Station for facilitating our primatological research at the site. Special thanks are due to David and Consuelo Romo, Kelly Swing, Diego Mosquera, Gaby Vinueza, Tomi Sugahara, Majo Rendón, and all of the station's "tigres" who have provided immeasurable logistical support and friendship to our research team in the field. We are grateful for the hard work and dedication of many of our fellow Proyecto Primates team members (Laura Abondano, Rebekah Ellis, Miguel García, Nausica de Gibert, Janel Mayo, Pearson McGovern, Lucy Millington, Evelyn Pain, Sebastián Ramírez, Robyn Reeder, Kelly Sampeck) who spent extended periods in the field collecting the behavioral and phenological data on which the study relies. We further thank our colleagues Becca Lewis, Mariah Hopkins, Andrés Link, and Denné Reed who assisted with the design of this study and Maryjka Blaszczyk and Amanda Perofsky who provided helpful editorial comments in the early stages of writing. This research was funded by NSF BCS-1062540, NSF BCS-1540403, NSF BCS-1638822; the Wenner-Gren, L.S.B. Leakey, and Nacey Maggioncalda Foundations; the National Geographic Society; the Explorers Club; Idea Wild; New York University; and the University of Texas at Austin.

Appendix

Table 10.5 Candidate models and subsequent selection (in bold) examining the influence of group size (GS), subgroup size (SGS), fruit availability (FA), and mating opportunity index (MO) on estimates of subgrouping behavior, within-group cohesion, ranging patterns, and range overlap. Random effects include group ID (ID) and the combination of month and year (MY) data was collected. Models are ranked from lowest AICc to highest

	Test	Response variable	Models	df	AICc	ΔAICc	w_i	Relative importance			
								GS	SGS	FA	MO
Subgrouping and cohesion	**Probability of subgrouping** GLMM with proportions (family: binomial link: logit)	# SGC records where social group is "subgrouped"/total # of SGC records	**GS + MO + (1|ID) + (1|MY)**	**5**	**1552.40**	**0.00**	**0.71**				
			GS + FA + MO + (1|ID) + (1|MY)	6	155.84	2.45	0.21				
			MO + (1|ID) + (1|MY)	4	1557.37	4.97	0.06				
			FA + MO + (1|ID) + (1|MY)	5	1559.77	7.38	0.02	1.00	NA	0.23	0.92
			GS + (1|ID) + (1|MY)	4	1645.10	92.71	0.00				
			GS + FA + (1|ID) + (1|MY)	5	1647.46	95.07	0.00				
			1 + (1|ID) + (1|MY) [NULL]	3	1649.32	96.92	0.00				
	Subgroup size LMM	Average SG size	**MO + (1|ID) + (1|MY)**	**5**	**−25.01**	**0.00**	**0.61**				
			GS + MO + (1|ID) + (1|MY)	6	−22.45	2.57	0.17				
			FA + MO + (1|ID) + (1|MY)	6	−22.44	2.57	0.17				
			GS + FA + MO + (1|ID) + (1|MY)	7	−19.75	5.26	0.04	0.21	NA	0.21	1.00
			1 + (1|ID) + (1|MY) [NULL]	4	−10.81	14.20	0.00				
			FA + (1|ID) + (1|MY)	5	−8.74	16.27	0.00				
			GS + (1|ID) + (1|MY)	5	−8.51	16.51	0.00				
	Interindividual distance LMM	Monthly average interindividual distance (m)	**GS + (1|ID) + (1|MY)**	**5**	**86.44**	**0.00**	**0.51**				
			GS + MO + (1|ID) + (1|MY)	6	88.15	1.71	0.22				
			GS + FA + (1|ID) + (1|MY)	6	89.08	2.64	0.14				
			GS + FA + MO + (1|ID) + (1|MY)	7	91.04	4.60	0.05	0.92	NA	0.20	0.29
			1 + (1|ID) + (1|MY) [NULL]	4	91.25	4.81	0.05				
			MO + (1|ID) + (1|MY)	5	93.09	6.65	0.02				
			FA + (1|ID) + (1|MY)	5	93.80	7.36	0.01				

Ranging and overlap	**Daily path length** LMM	Daily path length (m)	FA + MO + (1|ID) + (1|MY)	6	−12.10	0.00	0.38	NA	0.33	0.57	0.97
			MO + (1|ID) + (1|MY)	**5**	**−11.47**	**0.63**	**0.28**				
			SGS + FA + MO + (1|ID) + (1|MY)	7	−10.70	1.40	0.19				
			SGS + MO + (1|ID) + (1|MY)	6	−9.95	2.15	0.13				
			1 + (1|ID) + (1|MY) [NULL]	4	−5.43	6.68	0.01				
			SGS + (1|ID) + (1|MY)	5	−4.29	7.81	0.01				
			FA + (1|ID) + (1|MY)	5	−3.43	8.67	0.00				
	Home range size LMM	Monthly HR size (ha)	GS + (1|ID) + (1|MY)	5	−10.80	0.00	0.55	0.92	NA	0.24	0.21
			GS + FA + (1|ID) + (1|MY)	6	1.16	2.25	0.18				
			GS + MO + (1|ID) + (1|MY)	6	1.49	2.58	0.15				
			1 + (1|ID) + (1|MY) [NULL]	4	3.83	4.91	0.05				
			GS + FA + MO + (1|ID) + (1|MY)	7	3.85	4.93	0.05				
			FA + (1|ID) + (1|MY)	5	6.10	7.18	0.02				
			MO + (1|ID) + (1|MY)	5	6.26	7.34	0.01				
	Core area size LMM	Monthly CA size (ha)	**GS + (1|ID) + (1|MY)**	**5**	**8.93**	**0.00**	**0.50**	0.90	NA	0.28	0.21
			GS + FA + (1|ID) + (1|MY)	6	10.75	1.82	0.20				
			GS + MO + (1|ID) + (1|MY)	6	11.48	2.55	0.14				
			1 + (1|ID) + (1|MY) [NULL]	4	13.23	4.30	0.06				
			GS + FA + MO + (1|ID) + (1|MY)	7	13.33	4.40	0.06				
			FA + (1|ID) + (1|MY)	5	15.10	6.17	0.02				
			MO + (1|ID) + (1|MY)	5	15.61	6.68	0.02				

(continued)

Table 10.5 (continued)

<table>
<tr><th rowspan="2"></th><th rowspan="2">Test</th><th rowspan="2">Response variable</th><th rowspan="2">Models</th><th rowspan="2">df</th><th rowspan="2">AICc</th><th rowspan="2">ΔAICc</th><th rowspan="2">w_i</th><th colspan="4">Relative importance</th></tr>
<tr><th>GS</th><th>SGS</th><th>FA</th><th>MO</th></tr>
<tr><td rowspan="12">Ranging and overlap</td><td rowspan="4">Home range overlap
LMM</td><td rowspan="4">Monthly HR overlap (%)</td><td>1 + (1|ID) + (1|MY) [NULL]</td><td>4</td><td>591.37</td><td>0.00</td><td>0.53</td><td rowspan="4">NA</td><td rowspan="4">NA</td><td rowspan="4">0.28</td><td rowspan="4">0.27</td></tr>
<tr><td>FA + (1|ID) + (1|MY)</td><td>5</td><td>593.28</td><td>1.91</td><td>0.20</td></tr>
<tr><td>MO + (1|ID) + (1|MY)</td><td>5</td><td>593.34</td><td>1.97</td><td>0.20</td></tr>
<tr><td>FA + MO + (1|ID) + (1|MY)</td><td>6</td><td>595.33</td><td>3.96</td><td>0.07</td></tr>
<tr><td rowspan="4">Core area overlap
LMM</td><td rowspan="4">Monthly CA overlap (%)</td><td>1 + (1|ID) + (1|MY) [NULL]</td><td>4</td><td>494.81</td><td>0.00</td><td>0.50</td><td rowspan="4">NA</td><td rowspan="4">NA</td><td rowspan="4">0.26</td><td rowspan="4">0.32</td></tr>
<tr><td>MO + (1|ID) + (1|MY)</td><td>5</td><td>496.29</td><td>1.48</td><td>0.24</td></tr>
<tr><td>FA + (1|ID) + (1|MY)</td><td>5</td><td>196.94</td><td>2.13</td><td>0.17</td></tr>
<tr><td>FA + MO + (1|ID) + (1|MY)</td><td>6</td><td>498.45</td><td>3.64</td><td>0.08</td></tr>
<tr><td rowspan="4">Utilization distribution overlap index
LMM</td><td rowspan="4">Monthly UDOI of 95% KDE contours</td><td>1 + (1|ID) + (1|MY) [NULL]</td><td>4</td><td>−76.87</td><td>0.00</td><td>0.47</td><td rowspan="4">NA</td><td rowspan="4">NA</td><td rowspan="4">0.22</td><td rowspan="4">0.40</td></tr>
<tr><td>MO + (1|ID) + (1|MY)</td><td>5</td><td>−76.05</td><td>0.82</td><td>0.31</td></tr>
<tr><td>FA + (1|ID) + (1|MY)</td><td>5</td><td>−74.36</td><td>2.51</td><td>0.13</td></tr>
<tr><td>FA + MO + (1|ID) + (1|MY)</td><td>6</td><td>−73.52</td><td>3.35</td><td>0.09</td></tr>
</table>

AICc Akaike's information criterion, Δ*AICc* difference in AICc compared to the best model, w_j Akaike weight

References

Altmann J (1974) Observational study of behavior: sampling methods. Behaviour 49:227–266

Anderson DP, Nordheim EV, Boesch C, Moermond TC (2002) Factors influencing fission-fusion grouping in chimpanzees. In: Boesch C, Hohmann G, Marchant L (eds) Behavioral diversity in chimpanzees and bonobos. Cambridge University Press, Cambridge, UK, pp 90–101

Asensio N, Korstjens AH, Aureli F (2009) Fissioning minimizes ranging costs in spider monkeys: a multiple-level approach. Behav Ecol Sociobiol 63:649–659

Aureli F, Schaffner CM, Boesch C et al (2008) Fission-fusion dynamics: new research frameworks. Curr Anthropol 49:627–654

Barton K (2016) MuMIn: multi-model inference. R package version 1.15.6. http://CRAN.R-project.org/package=MuMIn

Bates D, Mächler M, Bolker B, Walker S (2015) Fitting linear mixed-effects models using lme4. J Stat Softw 67(1):1–48. https://doi.org/10.18637/jss.v067.i01

Bermejo M (2004) Home-range use and intergroup encounters in western gorillas (*Gorilla g. gorilla*) at Lossi forest, North Congo. Am J Primatol 64:223–232. https://doi.org/10.1002/ajp.20073

Bezanson M, Garber PA, Murphy JT, Premo LS (2008) Patterns of subgrouping and spatial affiliation in a community of mantled howling monkeys (*Alouatta palliata*). Am J Primatol 70:282–293. https://doi.org/10.1002/ajp.20486

Blake JG, Guerra J, Mosquera D et al (2010) Use of mineral licks by white-bellied spider monkeys (*Ateles belzebuth*) and red howler monkeys (*Alouatta seniculus*) in Eastern Ecuador. Int J Primatol 31:471–483. https://doi.org/10.1007/s10764-010-9407-5

Blake JG, Mosquera D, Loiselle BA et al (2012) Temporal activity patterns of terrestrial mammals in lowland rainforest of eastern Ecuador. Ecotropica 18:137–146

Boesch C (1996) Social grouping in Tai chimpanzees. In: McGrew W, Marchant L, Nishida T (eds) Great ape societies. Cambridge University Press, Cambridge, UK, pp 101–113

Boesch C, Boesch-Achermann H (2000) The chimpanzees of the Taï forest: behavioural ecology and evolution. Oxford University Press, New York

Bradley BJ, Doran-Sheehy DM, Lukas D et al (2004) Dispersed male networks in Western gorillas. Curr Biol 14:510–513. https://doi.org/10.1016/j.cub.2004.02.062

Burnham KP, Anderson DR (2002) Model selection and multimodel inference: a practical information-theoretic approach. Springer Science & Business Media, New York

Burt WH (1949) Territoriality. J Mammal 30:25–27. https://doi.org/10.2307/1375192

Caillaud D, Ndagijimana F, Giarrusso AJ et al (2014) Mountain gorilla ranging patterns: influence of group size and group dynamics. Am J Primatol 76:730–746. https://doi.org/10.1002/ajp.22265

Calenge C (2006) The package "adehabitat" for the R software: a tool for the analysis of space and habitat use by animals. Ecol Model 197:516–519. https://doi.org/10.1016/j.ecolmodel.2006.03.017

Chapman CA (1990) Association patterns of spider monkeys: the influence of ecology and sex on social organization. Behav Ecol Sociobiol 26:409–414

Chapman CA, Chapman LJ (2000) Determinants of group size in social primates: the importance of travel costs. In: Boinski S, Garber PA (eds) On the move: how and why animals travel in groups. University of Chicago Press, Chicago, pp 24–42

Chapman CA, Chapman LJ, Wrangham R et al (1992) Estimators of fruit abundance of tropical trees. Biotropica 24:527–531

Chapman CA, Chapman LJ, Wrangham RW (1995) Ecological constraints on group size: an analysis of spider monkey and chimpanzee subgroups. Behav Ecol Sociobiol 36:59–70

Crofoot MC, Gilby IC, Wikelski MC, Kays RW (2008) Interaction location outweighs the competitive advantage of numerical superiority in *Cebus capucinus* intergroup contests. Proc Natl Acad Sci 105:577–581. https://doi.org/10.1073/pnas.0707749105

Defler TR (1996) Aspects of the ranging pattern in a group of wild woolly monkeys (*Lagothrix lagothricha*). Am J Primatol 38:289–302

Defler TR, Defler SB (1996) Diet of a group of *Lagothrix lagothricha lagothricha* in southeastern Colombia. Int J Primatol 17:161–190

R Core Team (2016) R: a language and environment for statistical computing. R Foundation for Statistical Computing, Vienna

Dew JL (2005) Foraging, food choice, and food processing by sympatric ripe-fruit specialists: *Lagothrix lagotricha poeppigii* and *Ateles belzebuth belzebuth*. Int J Primatol 26:1107–1135. https://doi.org/10.1007/s10764-005-6461-5

Di Fiore A (1997) Ecology and behavior of lowland woolly monkeys (*Lagothrix lagotricha poeppigii*, Atelinae) in eastern Ecuador. University of California, Davis

Di Fiore A (2003) Ranging behavior and foraging ecology of lowland woolly monkeys (*Lagothrix lagotricha poeppigii*) in Yasuní National Park, Ecuador. Am J Primatol 59:47–66. https://doi.org/10.1002/ajp.10065

Di Fiore A (2004) Diet and feeding ecology of woolly monkeys in a western Amazonian rain forest. Int J Primatol 25:767–801

Di Fiore A, Rodman PS (2001) Time allocation patterns of lowland woolly monkeys (*Lagothrix lagotricha poeppigii*) in a neotropical terra firma forest. Int J Primatol 22:449–480

Di Fiore A, Strier KB (2004) Flexibility in social organization in atelin primates. In: XXth Congress of the International Primatological Society. Folia Primatologica, Turin, Italy, p 140

Di Fiore A, Link A, Schmitt CA, Spehar SN (2009) Dispersal patterns in sympatric woolly and spider monkeys: integrating molecular and observational data. Behaviour 146:437–470

Di Fiore A, Link A, Campbell C (2011) The atelines: behavioral an socioecological diversity in a New World monkey radiation. In: Primates in perspective. Oxford University Press, New York, pp 155–188

Dias LG, Strier KB (2003) Effects of group size on ranging patterns in *Brachyteles arachnoides hypoxanthus*. Int J Primatol 24:209–221. https://doi.org/10.1023/A:1023054513354

Ellis K, Pain E, Schmitt CA, Di Fiore A (2014) Ranging and association patterns of woolly monkeys (*Lagothrix poeppigii*) in lowland Ecuador. Am J Phys Anthropol 153:119

Ellis K, Abondano L, Montes-Rojas A et al (in review) Reproductive seasonality in two sympatric primates (*Ateles belzebuth* and *Lagothrix lagotricha poeppigii*) from Amazonian Ecuador. Am J Primatol

Fashing PJ (2001) Male and female strategies during intergroup encounters in guerezas *Colobus guereza*: evidence for resource defense mediated through males and a comparison with other primates. Behav Ecol Sociobiol 50:219–230. https://doi.org/10.1007/s002650100358

Fieberg J, Kochanny CO (2005) Quantifying home-range overlap: the importance of the utilization distribution. J Wildl Manag 69:1346–1359. https://doi.org/10.2193/0022-541X(2005)69[1346:QHOTIO]2.0.CO;2

Ganas J, Robbins MM (2005) Ranging behavior of the mountain gorillas (*Gorilla beringei beringei*) in Bwindi Impenetrable National Park, Uganda: a test of the ecological constraints model. Behav Ecol Sociobiol 58:277–288. https://doi.org/10.1007/s00265-005-0920-z

Goodall J (1986) The chimpanzees of Gombe. Belknap Press, Cambridge

Grueter CC, Chapais B, Zinner D (2012) Evolution of multilevel social systems in nonhuman primates and humans. Int J Primatol 33:1002–1037. https://doi.org/10.1007/s10764-012-9618-z

Hashimoto C, Furuichi T, Tashiro Y (2001) What factors affect the size of chimpanzee parties in the Kalinzu Forest, Uganda? Examination of fruit abundance and number of estrous females. Int J Primatol 22:947–959

Hurlbert SH (1978) The measurement of niche overlap and some relatives. Ecology 59:67–77. https://doi.org/10.2307/1936632

Isbell LA, Pruetz JD, Young TP (1998) Movements of vervets (*Cercopithecus aethiops*) and patas monkeys (*Erythrocebus patas*) as estimators of food resource size, density, and distribution. Behav Ecol Sociobiol 42:123–133. https://doi.org/10.1007/s002650050420

Janson CH, Goldsmith ML (1995) Predicting group size in primates: foraging costs and predation risks. Behav Ecol 6:326–336

Kavanagh M, Dresdale L (1975) Observations on the woolly monkey (*Lagothrix lagothricha*) in northern Colombia. Primates 16:285–294. https://doi.org/10.1007/BF02381556

Kays R, Crofoot MC, Jetz W, Wikelski M (2015) Terrestrial animal tracking as an eye on life and planet. Science 348:aaa2478. https://doi.org/10.1126/science.aaa2478

Kenward RE, Hodder KH (1996) Ranges V. An analysis system for biological location data. Institute of Terrestrial Ecology, Dorset

Kie JG (2013) A rule-based ad hoc method for selecting a bandwidth in kernel home-range analyses. Anim Biotelemetry 1:13. https://doi.org/10.1186/2050-3385-1-13

Kie JG, Matthiopoulos J, Fieberg J et al (2010) The home-range concept: are traditional estimators still relevant with modern telemetry technology? Philos Trans R Soc Lond B Biol Sci 365:2221–2231. https://doi.org/10.1098/rstb.2010.0093

Klein LL, Klein DJ (1977) Feeding behaviour of the Colombian spider monkey, *Ateles belzebuth*. In: Clutton-Brock TH (ed) Primate ecology: studies in feeding and ranging behavior in lemurs, monkeys, and apes. Academic, London, pp 153–181

Koch F, Signer J, Kappeler PM, Fichtel C (2016) The role of the residence-effect on the outcome of intergroup encounters in Verreaux's sifakas. Sci Rep 6:28457. https://doi.org/10.1038/srep28457

Kummer H (1968) Social organization of hamadryas baboons. Chicago University Press, Chicago

Kummer H (1971) Primate societies: group techniques of ecological adaptation. Aldine, Chicago

Kuznetsova A, Brockhoff PB, Christensen RHB (2014) lmerTest: tests for random and fixed effects for linear mixed effect models. R. Package version 2.0-20. http://cran.r-project.org/web/packages/lmerTest/index.html

Langrock R, Hopcraft JGC, Blackwell PG et al (2014) Modelling group dynamic animal movement. Methods Ecol Evol 5:190–199. https://doi.org/10.1111/2041-210X.12155

Laver PN, Kelly MJ (2008) A critical review of home range studies. J Wildl Manag 72:290–298. https://doi.org/10.2193/2005-589

Lehmann J, Boesch C (2004) To fission or to fusion: effects of community size on wild chimpanzee (*Pan troglodytes verus*) social organisation. Behav Ecol Sociobiol 56:207–216. https://doi.org/10.1007/s00265-004-0781-x

Link A (2011) Social and ecological determinants of "fission-fusion" sociality and grouping strategies in the white bellied spider monkey (*Ateles belzebuth belzebuth*) in a lowland rainforest in Western Amazonia. New York University, New York

Link A, Galvis N, Fleming E, Di Fiore A (2011) Patterns of mineral lick visitation by spider monkeys and howler monkeys in Amazonia: are licks perceived as risky areas? Am J Primatol 73:386–396. https://doi.org/10.1002/ajp.20910

Majolo B, Ventura R, Koyama NF (2005) Sex, rank and age differences in the Japanese macaque (*Macaca fuscata yakui*) participation in inter-group encounters. Ethology 111:455–468. https://doi.org/10.1111/j.1439-0310.2005.01087.x

Markham AC, Guttal V, Alberts SC, Altmann J (2013) When good neighbors don't need fences: temporal landscape partitioning among baboon social groups. Behav Ecol Sociobiol 67:875–884. https://doi.org/10.1007/s00265-013-1510-0

Matsumoto-Oda A, Hosaka K, Huffman MA, Kawanaka K (1998) Factors affecting party size in chimpanzees of the Mahale Mountains. Int J Primatol 19:999–1011. https://doi.org/10.1023/A:1020322203166

Mitani JC, Watts DP, Lwanga JS (2002) In: Boesch C, Hohmann G, Marchant L (eds) Ecological and social correlates of chimpanzee party size and composition. Cambrige University Press, Cambridge, UK, pp 102–111

Nakamura M, Corp N, Fujimoto M et al (2013) Ranging behavior of Mahale chimpanzees: a 16 year study. Primates 54:171–182. https://doi.org/10.1007/s10329-012-0337-z

Nathan R, Getz WM, Revilla E et al (2008) A movement ecology paradigm for unifying organismal movement research. PNAS pnas.0800375105. https://doi.org/10.1073/pnas.0800375105

Newton-Fisher NE, Reynolds V, Plumptre AJ (2000) Food supply and chimpanzee (*Pan troglodytes schweinfurthii*) party size in the Budongo forest reserve, Uganda. Int J Primatol 21:613–628. https://doi.org/10.1023/A:1005561203763

Nishimura A (1990) A sociological and behavioral study of woolly monkeys, *Lagothrix lagotricha*, in the upper Amazon. Sci Eng Rev Doshisha Univ 31:87–121

Nishimura A (1994) Social interaction patterns of woolly monkeys (*Lagothrix lagotricha*): a comparison among the atelines. Sci Eng Rev Doshisha Univ 35:91–110

Nishimura A (2003) Reproductive parameters of wild female *Lagothrix lagotricha*. Int J Primatol 24:707–722

Peres CA (1994) Diet and feeding ecology of gray woolly monkeys (*Lagothrix lagotricha cana*) in central Amazonia: comparisons with other atelines. Int J Primatol 15:333–372

Peres CA (1996) Use of space, spatial group structure, and foraging group size of gray woolly monkeys (*Lagothrix lagotricha cana*) at Urucu, Brazil. In: Adaptive radiations of neotropical primates. Springer, New York, pp 467–488

Ramos-Fernández G (2005) Vocal communication in a fission-fusion society: do spider monkeys stay in touch with close associates? Int J Primatol 26:1077–1092. https://doi.org/10.1007/s10764-005-6459-z

Schmitt CA, Di Fiore A (2014) Life history, behavior, and development of wild immature lowland woolly monkeys (*Lagothrix poeppigii*) in Amazonian Ecuador. In: The woolly monkey. Springer, New York, pp 113–146

Seiler N, Boesch C, Mundry R et al (2017) Space partitioning in wild, non-territorial mountain gorillas: the impact of food and neighbours. R Soc Open Sci 4:170720. https://doi.org/10.1098/rsos.170720

Shanee S (2014) Ranging behaviour, daily path lengths, diet and habitat use of yellow-tailed woolly monkeys (*Lagothrix flavicauda*) at La Esperanza, Peru. In: Defler TR (ed) The woolly monkey: behavior, ecology, systematics, and captive research. Springer, New York, pp 167–185

Shimooka Y (2003) Seasonal variation in association patterns of wild spider monkeys (*Ateles belzebuth belzebuth*) at La Macarena, Colombia. Primates 44:83–90. https://doi.org/10.1007/s10329-002-0028-2

Snodderly, D.M., Ellis, K.M., Lieberman, S.R., Link., A., Fernandez-Duque, E., and Di Fiore, A. [in review]. Initiation of feeding by four sympatric Neotropical primates (Ateles belzebuth, Lagothrix lagotricha poeppigii, Plecturocebus (Callicebus) discolor, and Pithecia aequatorialis) in Amazonian Ecuador: Relationships to photic and ecological factors. PLoS ONE

Stevenson PR (1992) Diet of woolly monkeys (*Lagothrix lagotricha*) at La Macarena, Colombia. Field Stud New World Monkeys Macarena Colomb 6:3–14

Stevenson PR (2004) Phenological patterns of woody vegetation at Tinigua Park, Colombia: methodological comparisons with emphasis on fruit production/Patrones fenológicos de vegetación leñosa en el Parque Tinigua, Colombia: comparaciones metodológicas con énfasis en la producción de frutos. Caldasia 26:125–150

Stevenson PR (2006) Activity and ranging patterns of Colombian woolly monkeys in North-Western Amazonia. Primates 47:239–247

Stevenson PR, Quinones MJ, Ahumada JA (1994) Ecological strategies of woolly monkeys (*Lagothrix lagotricha*) at Tinigua National Park, Colombia. Am J Primatol 32:123–140

Symington MM (1987) Sex ratio and maternal rank in wild spider monkeys: when daughters disperse. Behav Ecol Sociobiol 20:421–425

Symington MM (1990) Fission-fusion social organization in *Ateles* and *Pan*. Int J Primatol 11:47–61. https://doi.org/10.1007/BF02193695

Watts DP (1998) Long-term habitat use by mountain gorillas (*Gorilla gorilla beringei*). 1. Consistency, variation, and home range size and stability. Int J Primatol 19:651–680. https://doi.org/10.1023/A:1020324909101

Westley PAH, Berdahl AM, Torney CJ, Biro D (2018) Collective movement in ecology: from emerging technologies to conservation and management. Philos Trans R Soc B 373:20170004. https://doi.org/10.1098/rstb.2017.0004

Whitehead H (2008) Analyzing animal societies: quantitative methods for vertebrate social analysis. University of Chicago Press, Chicago

Wrangham RW (1977) Feeding behavior of chimpanzees in Gombe National Park, Tanzania. In: Clutton-Brock TH (ed) Primate ecology. Academic, London

Wrangham RW (1980) An ecological model of female-bonded primate groups. Behaviour 75:262–300

Wrangham RW (1986) Ecology and social relationships in two species of chimpanzee. In: Rubenstein DI, Wrangham RW (eds) Ecology and social evolution: birds and mammals. Princeton University Press, Princeton, pp 352–378

Wrangham RW (2000) Why are male chimpanzees more gregarious than mothers? A scramble competition hypothesis. In: Kappeler PM (ed) Studies of feeding and ranging behavior in lemurs, monkeys, and apes. Cambridge University Press, Cambridge, UK, pp 248–258

Zárate DA, Stevenson PR (2014) Behavioral ecology and interindividual distance of woolly monkeys (*Lagothrix lagothricha*) in a rainforest fragment in Colombia. In: The woolly monkey. Springer, New York, pp 227–245

Zhao Q-K (1997) Intergroup interactions in Tibetan macaques at Mt. Emei, China. Am J Phys Anthropol 104:459–470. https://doi.org/10.1002/(SICI)1096-8644(199712)104:4<459::AID-AJPA3>3.0.CO;2-N

Chapter 11
Home Range and Daily Traveled Distances of Highland Colombian Woolly Monkeys (*Lagothrix lagothricha lugens*): Comparing Spatial Data from GPS Collars and Direct Follows

Leidy Carolina García-Toro, Andrés Link, Elsy Johanna Páez-Crespo, and Pablo R. Stevenson

11.1 Introduction

Movement ecology integrates four mechanistic components, which are necessary to describe movements and the use of an organism's space (Nathan et al. 2008). The *internal state* is related to the individual abilities and defines the intrinsic motivation to move. An organism's movement is controlled by the *motion capacity*, either under its own locomotion or by being carried by physical means (e.g., winds, water, etc.) or other organisms. The navigation capacity is the ability of organisms to orient, what allows them to obtain a realized movement path (Holyoak et al. 2008). The last component is the *external state*; this represents the biotic and abiotic aspects that influence individual's movement (Nathan et al. 2008). In this context, animals do not use all the environment in a uniform or random way. In contrast, they usually concentrate their activity in those areas that provide the necessary conditions and resources to satisfy their energetic requirements and optimize their survival and reproductive success (internal state) (Manly et al. 1993; Nathan et al. 2008). In other words, the use of space by animals is influenced by the availability and spatiotemporal distribution of resources, such as food, sleeping sites, and refuge (external state) (Ramos-Fernandez et al. 2013; Nathan et al. 2008) as well as by inter- and intraspecific interactions (Stevenson and Castellanos 2000). The movement is associated with a sequence of steps and stops, called by Nathan and colleagues in 2008 as "movement phases," which allow the fulfillment of a particular goal (e.g., searching for food or escaping a predator). For example, several studies

L. C. García-Toro (✉) · A. Link · E. J. Páez-Crespo · P. R. Stevenson
Laboratorio de Ecología de Bosques Tropicales y Primatología, Departamento de Ciencias Biológicas, Universidad de Los Andes, Bogotá, Colombia
e-mail: lc.garcia13@uniandes.edu.co; a.link74@uniandes.edu.co; ej.paez10@uniandes.edu.co; pstevens@uniandes.edu.co

R. Reyna-Hurtado, C. A. Chapman (eds.), *Movement Ecology of Neotropical Forest Mammals*, https://doi.org/10.1007/978-3-030-03463-4_11

have evidenced that animals concentrate their activity and guide their movements to specific areas within their habitat and use them with a relatively high intensity (Manly et al. 1993). These core areas or central areas tend to change little over time and often indicate the location of fundamental resources, which are frequently used due to the knowledge of the habitat by the individuals (Ramos-Fernandez et al. 2013). Ultimately, the sequence of all movement phases set the lifetime track of an individual (Holyoak et al. 2008).

The frequency and intensity of habitat use by animals can be quantified using different sampling methods, which can be evaluated from accumulation curves as sampling increases (González and Stevenson 2010). Some methods allow studying different aspects of movement ecology, and others provide excellent tools to address specifically spatial data. These methods include direct tracking on an individual or focal group (González and Stevenson 2010; Ramos-Fernández et al. 2013), sampling through linear transects based on the auditory detection of the study animals (Buckland et al. 2010) and use of radio telemetry and GPS collars. However, studies carried out using GPS collars methods to estimate home range sizes and DTD are still scarce in groups or individual primates (Walter et al. 2011; Markham and Altmann 2008; Parga 2011; Ren et al. 2008), and in primatology, the use of GPS collars is particularly useful for both semiterrestrial and arboreal species (Stark et al. 2017). Observational studies are highly valued because they allow direct behavioral assessments but may be by subject to observational bias (Altmann 1974).

It has been suggested that home range analyses require independence of observations in order to guarantee unbiased estimations (Harris et al. 1990; White and Garrott 1990; Cresswell and Smith 1992; Kenward 1992) given that autocorrelation increases the probability of type I error by increasing the degrees of freedom (Legendre 1993), thus interfering with statistical testing hypotheses (Solla et al. 1999). However, animals generally move in a nonrandom or transiently independent manner, and data on their movements are strongly autocorrelated, especially when these observations are recorded at short time intervals. These data are particularly useful to accurately describe animal's movements and space use (Swihart and Slade 1985). In contrast, Solla et al. (1999) explained that nonindependent observations are valid if these data are grouped to estimate a single value of an experimental unit (e.g., home range size). Therefore, the assumption of independence between observations of sequential spatial data to determine home range sizes and DTD is not a constraint. However, when the data does not have spatial structure and is necessary to fulfill the assumption of independence or to perform statistical models that allow the autocorrelation of the data to be contemplated, so, the statistical inference does not lose its foundation (Camus and Lima 1995).

Woolly monkeys (*Lagothrix* spp.) are Ateline primates, which are the largest Neotropical primates (Fleagle 1999). Some studies suggested that woolly monkeys are characterized by fission-fusion social groups, while others recognized them as cohesive units (Defler 2010; Di Fiore and Rodman 2001). However, it seems that the degree of fission-fusion dynamics in this species is influenced by local environmental factors (Stevenson et al. 2015). Frequently females and occasionally males migrate between groups, although males exhibit philopatry to a higher extent (Soini 1986; Nishimura 2003; Di Fiore et al. 2009). Woolly monkeys live in multimalemultifemale social groups, which vary between 10 and 49 individuals (Stevenson et al. 1994; Peres 1996). A clear dominance between males is usually difficult to establish in this species (Nishimura 1990). However, observational studies established the dominance of males over females (Di Fiore and Fleischer 2005; Stevenson 1998; Nishimura 1994). Furthermore, larger males tend to be dominant over the smallest ones (Schmitt and Di Fiore 2014). Their home range size varies between species and study sites (Shanee and Shanee 2014) from approximately 200 ha at Duda river, Colombia (Stevenson 2000), up to 1000 ha at Urucu river, Amazonas, Brazil (Peres 1996). Average DTD range between 1633 and 2880 m (Defler 1996; Di Fiore 2003; Stevenson et al. 1994). Currently, most wild populations of woolly monkey are considered as Critically Endangered (CR) by the IUCN mainly due to deforestation, habitat fragmentation, and hunting pressure (Stevenson and Link 2011). These factors diminish population size and compromise their long-term viability. Woolly monkeys also have a high vulnerability to habitat and demographic perturbations given their feeding habits (e.g., frugivores) and low reproductive rates, which constrains the recovery of its populations (Nishimura 2003; Lizcano et al. 2014).

The aim of this study was to estimate the home range sizes and DTD of two Colombian woolly monkey groups at Cueva de Los Guácharos National Natural Park using two different sampling methods: direct observations of focal animals using manual recording of location data with a handheld GPS unit and through GPS collars located in animals from both study groups. A comparison of sampling methods has not been previously addressed and contributes to a better understanding of the potential limitation of spatial studies conducted by direct focal follows. Our main research questions were:

1. When we standardize for sampling effort, are there differences in the estimates of home range area and DTD between methods?
2. In case of differences, are topography and trail locations associated with sampling bias?
3. For statistical purposes, what is the appropriate sampling period to obtain independent data?

We expected that in steep topographies, which characterize the study area, researchers might have difficulties following monkeys. As a consequence, data provided by GPS collars would cover broader geographical areas, which would include inaccessible areas to researchers.

11.2 Methods

11.2.1 Study Area

Cueva de Los Guácharos National Natural Park is located in the southern part of the Colombian Andes in Huila and Caquetá departments (01°33′N, 76°08′W). The predominant habitats are Sub-Andean primary and secondary forests (Prada and

Stevenson 2016). All these forests are characterized by a rugged topography, represented by mountains with an average slope of 23.3% and a maximum inclination of 51.3%. The study area covers an average altitudinal elevation of 2000 m.a.s.l. Average annual rainfall is 2284 mm; the rainy season covers most of the year. The average annual temperature is 18.8 °C (Vargas et al. 2014).

11.2.2 Individuals Captured

In April and July 2012, January 2013, and October 2016, four woolly monkeys from two different study groups were captured. The group size of the first group called Colombia (C) was 32–34 individuals (approximately 5–7 adult males, 10 adult females, 2 subadult males, 2 subadult females, 8 juveniles, and 5 offspring), and Brazil's (B) group size was 29–35 individuals (approximately 9–11 adult males, 10–12 adult females, 3 subadult males, 1–2 subadult females, 4 juveniles, and 3 offspring). We followed darting procedures according to the suggestions made by Cunningham et al. (2015), Fedigan (2010), and Fernandez-Duque et al. (2017). Four monkeys were temporally anesthetized and captured using Zoletil (dosage of 15 mg/kg body weight). Zoletil was supplied with a CO2-powered rifle and Pneu-Dart darts (type U). Three individuals from C group were captured: an adult male called "Pibe," two adult females "Maya," and "Leyca" (adult female with a young). Additionally, one adult male from B group was captured "Toninho." Field assistants caught immobilized monkeys in capture net and all individuals survived uninjured. During processing, we examined the monkeys' health; each individual was weighed, measured, photographed, and fitted with a GPS collar with very high frequency (VHF) telemetry signal (Telemetry Solutions). The VHF has an independent battery of the GPS settings, and these batteries are made of lithium thionyl chloride 3.6 V. Each collar was set up to take location data every half an hour with 60 s GPS on time to acquire satellite signal, and 45 s GPS additional time for calibration, during the first 4 days of biweekly periods (days 1–4 and 15–18 of each month), from 06:00 to 18:00 h. Later the animals were released and were monitored to establish their health state (Fig. 11.1).

Most guidelines suggest that collar's weight should be <5% of the animal's total body weight (Fedigan 2010; Cuthill 1991). Collars with GPS devices used in this study weighed 150 g which represents approximately 3% of animal's weight. These GPS collars were based on the reception of signals sent by a satellite system with the same name. The GPS collars collected data about the geographical position of the animal in Universal Transverse Mercator (UTM) coordinates with datum WGS84. The "fix" parameter and the horizontal dilution of precision (HDOP) were stored in the device memory (Sub 1 GHz RF) with a capacity to store 65.000 data points. The "fix" parameter indicates the set of satellite signals used to calculate the geographical position. Data from 2D dimension are considered less precise because they only obtain the signs of three satellites; in contrast, data from three dimensions (3D) are more exact because those use four or more satellites (Moen et al. 1996; Di

Fig. 11.1 Adult male woolly monkeys with GPS collars. "Toninho" from group B (left and center). "Pibe" from group C (right)

Orio et al. 2003; Lewis et al. 2007). The HDOP is a value between 0 and 9 that estimates the precision level associated with the X and Y position of the GPS collars, values lower than 1 represent the highest possible confidence level, 1–2 is excellent, 2–5 is good, and so on. The stored information was recovered through a package of data (txt.) that was transmitted from the GPS collar memory placed in the animal to a modem in a laptop with a software provided by Telemetry Solutions Quantum GPS collar SW v 0.278. After the data screening, only 6.7% of localizations obtained were removed since they presented HDOP values greater than five, and also some data was acquired using only two dimensions (2D). Compared to other studies, which have had to remove 8–26% of their points (Stark et al. 2017; Pebsworth et al. 2012; Markham and Altmann 2008), this study only eliminated a small percentage of localizations and is therefore considered a high-quality data set.

11.2.3 Sampling Protocol

Individuals of both groups C and B have been studied almost permanently between July 2009 and January 2017, which supposes they are sufficiently habituated to observation. To determine the direct tracking, we located the study groups using the telemetry signal. Later, we focused on following the individuals that carried GPS collars using focal animal sampling method (Altmann 1974). However, when we lost the focal animal, we selected randomly another focal adult or subadult until we reencounter the focal individual. Woolly monkey groups were followed during periods of variable length throughout the day (average 8.1 h SD ± 2.5), focusing on one individual as long as possible. To estimate home range sizes, DTD, and areas of use, each researcher carried a handheld GPS device (Garmin GPSmap 64s).Those GPS devices have an approximately precision of <10 m 95% of the data. The position of the groups was recorded every half an hour through the option "mark a waypoint" in the GPS. During direct follows we attempted to follow the target monkey for at least 20 days a month, including at least 5 days of 12 h. We record behavioral data, but this information is being analyzed for another publication.

11.2.4 Data Analysis

11.2.4.1 Home Range and Core Area

A sampling effort curve was calculated to represent the cumulative number of hectares as a function of time for both methods: GPS collars and direct tracking (Defler 1996; Di Fiore 2003; Zárate and Stevenson 2014). Home ranges for each individual were evaluated by using Spatial Analyst extensions for ArcGIS (Version 10.3) and the "adehabitatHR" package for the statistical R software. Due to the variability of methods that have been used to estimate home range sizes, and the debate on which one is the most accurate (Powell 2000; Boyle et al. 2009), in this study several methods were used and compared. Minimum convex polygon (MCP) is the simplest and most common method used in spatial analyses in wild primates (Boyle et al. 2009); on the other hand, the grid cell method (GCM) establishes the number of locations per unit of space. For this purpose, the area where woolly monkeys moved was divided into a grid of 100 × 100 m (1 ha) (Di Fiore 2003). Finally, to identify the most common area used by woolly monkeys, we estimated Kernel density using 50, 75, and 95% Kernel density estimators of the data collected (Kernohan et al. 2001).

11.2.4.2 Potential Explanations for Estimation Bias: Topography and Trail Accessibility

To determine if there was an association between broken topography and a deficiency of samples by observers in steep topography, we used a digital elevation model (DEM) of the forest, based on radar images (shuttle radar topography mission: SRTM). The images were taken with a resolution of 30 m, the most accurate for areas with high elevations (Pierce et al. 2006). The generated image showed the height values and it was imported into ArcGIS (Version 10.3). We used the maximum height difference for each 1 ha quadrat as a proxy to the terrain steepness. To test whether researchers used to sample more in areas closer to trail system than areas away from it, we calculated the distance in meters to the nearest trail from the center point of each quadrat used by monkeys (both those carrying collars and followed by researchers).

Finally, we evaluated the independence of sequential locations using both GPS collars data and direct tracking data. Randomizations of subsequent records were made at different time intervals (0.5, 1.0, 1.5 h, and so on) until autocorrelation of the observations was eliminated. Without the previous procedure, it is impossible to perform a test with robust statistics. Therefore, randomizations were done to fulfill the assumption of independence and to be able to evaluate the comparison between methods under a nonparametric Mann-Whitney U test (Camus and Lima 1995).

11.2.4.3 Daily Traveled Distance

DTDs were determined based on the point-to-point displacements between consecutive location records (30 min apart), gathered between 6:00 and 18:00 h. Then, sums of these distances throughout the day were calculated to obtain the daily travel estimate. A Kruskal-Wallis H analysis and Mann-Whitney U test were made to compare the DTD between individuals who carried GPS collars and the direct tracking by the researchers with handheld GPS devices. Subsequently, a post hoc Dunn test with Bonferroni correction was performed to evaluate the equality between the values of the means of the DTD. Given the practical difficulties for researchers to consistently complete 12 h of continuous sampling, a linear regression model was used to estimate the DTD when the sampling was shorter than 12 h (Appendix, Fig. 11.8). Thus, DTD was estimated by extrapolating the regression line to the value of 12 h.

11.3 Results

11.3.1 Home Range and Core Area

The home range of each one of the three individuals with GPS collar from group C overlapped almost entirely (Fig. 11.3, and Appendix, Fig. 11.7). Therefore, the home range size estimated through the counting GCM, which considered all hectares used by monkeys for group C, was close to 580 ha (Table 11.1). The home range size estimated through the MCP was 474 ha, estimated from 95% of the data. The estimated home range size using direct follows by researchers with handheld GPS devices for group C was 242 ha (Figs. 11.2a, b, 11.3, 11.4b, and Appendix, Fig. 11.7). In addition, all estimations of home ranges were incomplete, even for the most highly sampled individuals. For instance, "Pibe's" accumulation curve did not reach an asymptote, even though we had a representative sampling of more than 1800 h (Fig. 11.2a). Nevertheless, home range sizes found for the three individuals with GPS collars of group C were similar (Fig. 11.2a). Estimating the home range size of "Toninho" of group B using data collected every 30 min for 37 days over a year was also incomplete (Fig. 11.2c). This suggests that the home range size is much larger than the 160 ha (GCM) recorded with the GPS collar. Based on direct follows by researchers, the home range size reported for this individual was 152 ha with more than 80 days of sampling. Both results supported the idea that home range size of group B is still incomplete (Fig. 11.2c).

Kernel method analysis showed that group C did not use all areas of their home range uniformly (Fig. 11.4). "Toninho" of group B used a core area size of 74 ha, and for group C, it was 99 ha (based on 50% of the data). The accumulation curves showed that core area sizes were more stable than overall home ranges and that core

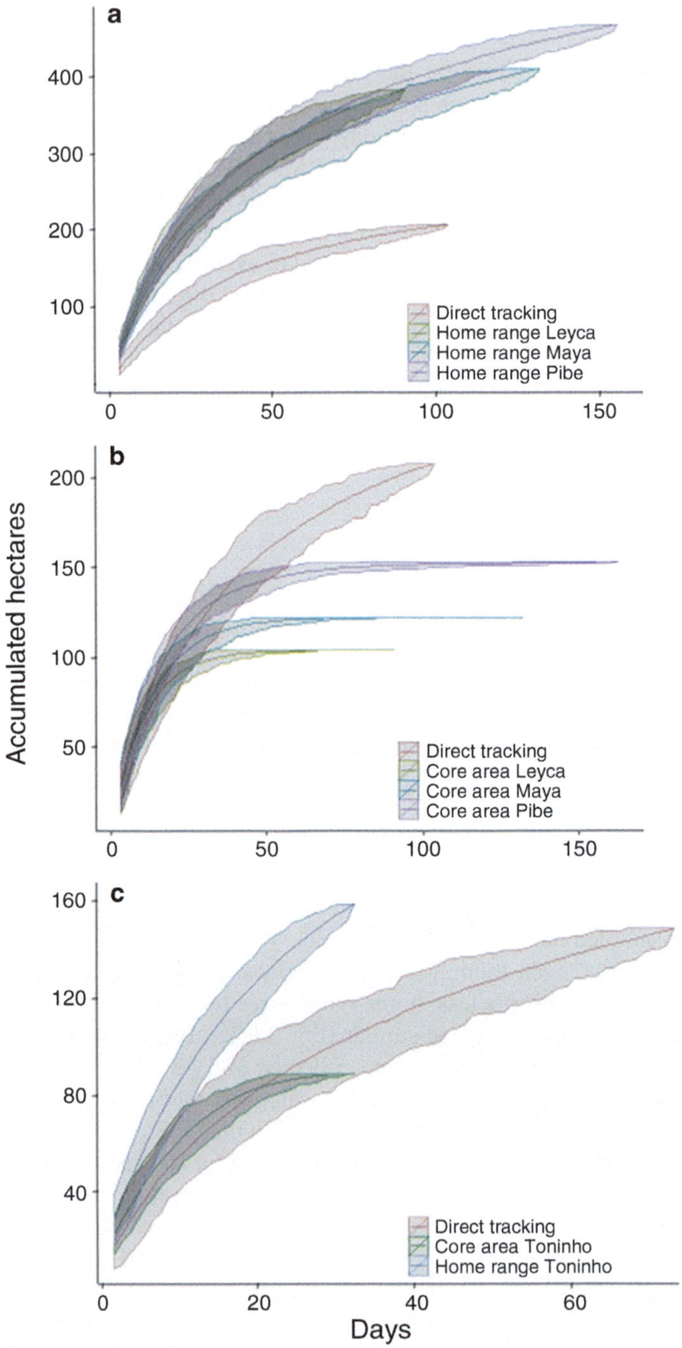

Fig. 11.2 Estimation of the area used by woolly monkeys as a function of cumulative sampling effort. (**a**) Home range used by the three individuals of group C using both GPS collars and direct tracking by researchers with handheld GPS devices. (**b**) Core area (50%) registered for group C and direct tracking. (**c**) Home range and core area used by "Toninho" in group B and direct tracking by researchers with handheld GPS devices. Shaded regions show a 95% confidence interval

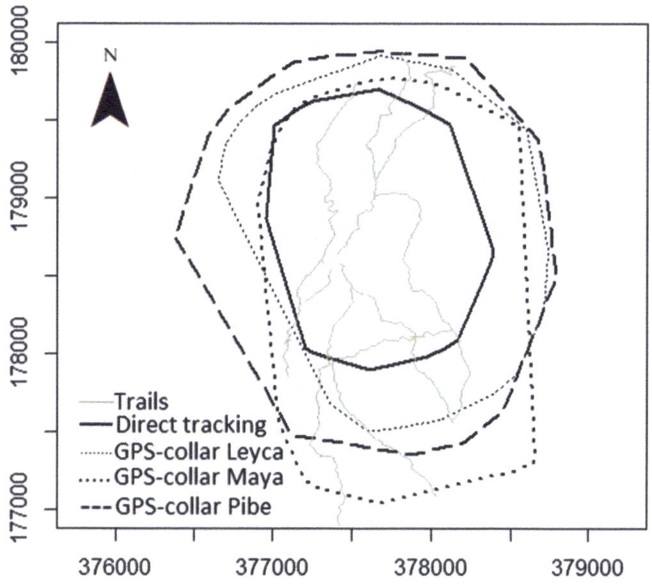

Fig. 11.3 Home range of each GPS-collared individuals of group C using the MCP compared to direct tracking. Axes (*x*, *y*) indicate the UTM coordinates

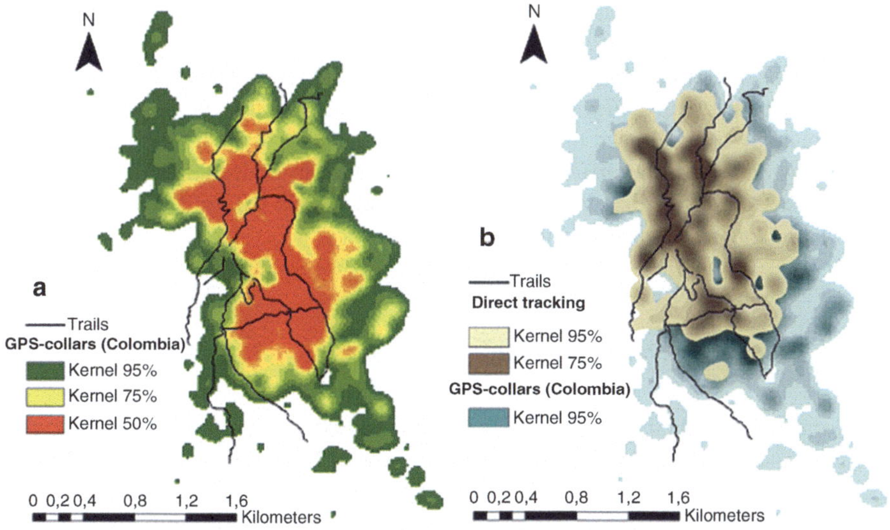

Fig. 11.4 Estimation of the density of use and home range size for group C through GPS collar and area size obtained by direct tracking. (**a**) Estimation of the density using 50, 75, and 95% of the data for group C. (**b**) Blue color represents the distribution of GPS collars data of group C (438 ha), and brown color denotes the distribution of the direct tracking data obtained through handheld GPS devices (242 ha). Axes (*x*, *y*) indicate the UTM coordinates

Table 11.1 Estimated home range sizes and core areas values used by woolly monkeys at PNN Cueva de Los Guácharos for each monkey that carried GPS collars

Individual name	Group	N^{a}	Kernel (ha) Core area (50%)	Kernel (ha) Home range	Minimum convex polygon	Grid cell
Leyca	Colombia	2221	98	385	363	384
Maya	Colombia	3530	82	412	403	410
Pibe	Colombia	3986	113	460	472	467
Total	Colombia	9737	99	438	474	580
Tonihno	Brazil	950	74	243	194	160

[a]Number of records

areas were well sampled, reaching an asymptote over sampling time (Fig. 11.2b, c). This pattern was similar for the three individuals that belong to group C.

11.3.2 Comparison of Sampling Methods

Independence of observations to perform the comparison of the data from both sampling methods was reached after a 5-h period. Therefore, only the data recorded from this interval were analyzed. When comparing the data (GPS collars vs. direct tracking), we recognized that the GPS collars method registered areas with higher unevenness by quadrat and location data that were far away from the trail system in comparison with direct tracking method ($U = 2507.5$, $P = 0.026$ and $U = 2610$, $P = 0.009$, respectively, Fig. 11.5). In this analysis, we only used data recorded simultaneously from the two methods. When all the data obtained through the GPS collars was analyzed, the difference and the variation in both the average unevenness (SD = 15) and the distance to the nearest trail system (SD = 296) were higher than when using simultaneous follows.

11.3.3 Daily Traveled Distance

The average DTD obtained by the GPS collars method was 2064 m and 1256 m for groups C and B, respectively. According to the direct tracking method by researchers with handheld GPS devices, the average DTD was 917 m and 997 m, respectively (Fig. 11.6). The greatest daily distance was recorded by the adult male "Pibe" (4795 m), and the minimum daily distance was registered by "Toninho" representing 452 m daily traveling. On the other hand, the maximum recorded height was recorded by the GPS collars (group C 2376 m and group B 2747 m). DTD showed significant differences between the records obtained from both groups C and B vs. the direct tracking with handheld GPS devices ($X^2(3) = 208.87$, $P = 0.001$ and $U = 768.5$, $P = 0.004$, Fig. 11.6).

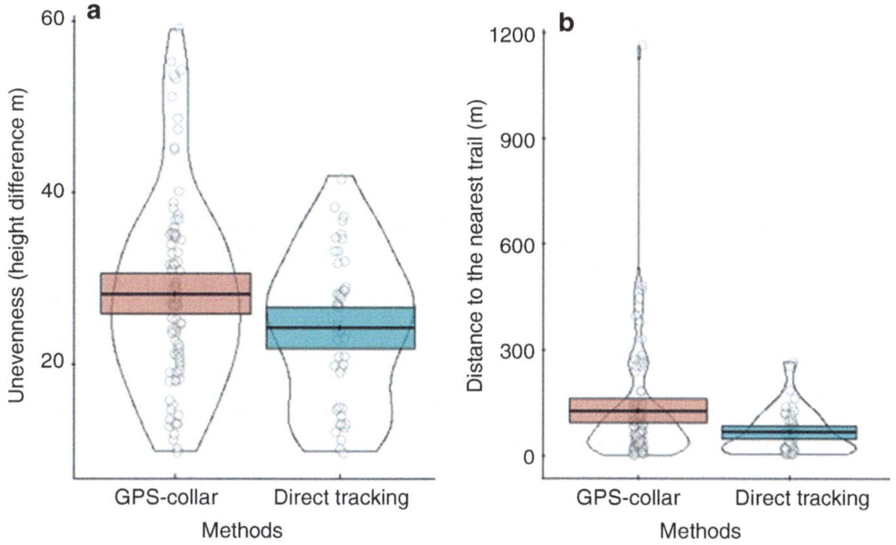

Fig. 11.5 (**a**) Unevenness (m) per 1 ha plot and (**b**) distance to the nearest trail system (m) of the study area where group C moved in relation to the two sampling methods used

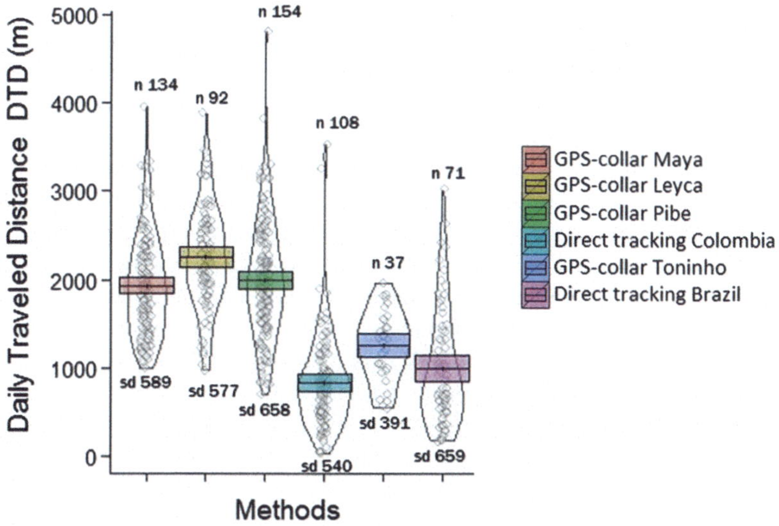

Fig. 11.6 Daily traveled distances of two woolly monkey groups (C and B) using both methods: GPS collars (placed in four individuals) and direct tracking with handheld GPS devices

11.4 Discussion

11.4.1 Differences in Home Ranges and Sampling Methods

Home range size estimated for group C was 580 ha, and this value is within the reported ranges in the literature, although home range size varies considerably compared to other studies of woolly monkeys (Table 11.2). Differences in the home range size in the previous studies could be related to the use of different sampling rules (e.g., focal vs. scan sampling: González and Stevenson 2010). For instance, observers unwilling or uncapable of transiting in flooded forest or too far from trail systems (Peres 1994) may report incomplete and biased home range estimations. It is important to clarify that for purposes of comparison with the other studies, we used the home range size estimated through the GCM, although Kernel UD is one the most accurate methods to analyze home range data (Kernohan et al. 2001).

The home range size estimated in this study for both groups (C and B) was incomplete (Fig. 11.2). Furthermore, the home range of group C obtained through direct tracking was almost half of the home range size obtained by the GPS collars method (Table 11.2; Fig. 11.2a). These results are due to the fact that monkeys did not maintain a stable home range throughout the study period. Is essential to recall that woolly monkeys exhibited fission-fusion dynamics (data in preparation), which can be influencing notable changes in the use of space. For example, during this study, some individuals visited large areas of high forest, where researchers were not able to access, and also some monkeys were absent over long periods.

Likewise, these differences demonstrated that there are limitations in using the direct tracking method in wild primate groups, which may be accentuated when researchers are working with animals that inhabit mountainous areas or zones where located or follow primate groups are complicated (e.g., solitary and/or nocturnal species or species that live in areas of difficult access) (Fedigan et al. 1988). Using GPS collars and the analyses carried out in this study, it was possible to collect a significant amount of data from continuous movements of woolly monkeys. GPS collars data were collected at frequent and generally programmed time intervals in order that the woolly monkey's movement routes are more reliable (Nathan et al. 2008), with a high degree of quality and precision and covering broad geographic areas (D'Eon et al. 2002; Rodgers 2001). This is supported because the location data of woolly monkeys did not coincide with the data collected by the GPS collars to determine the home range size (Figs. 11.3 and 11.4b, Appendix, Fig. 11.7), even though direct tracking was carried out for long periods of time (800 h).

The GPS collars have revolutionized the study of home ranges, habitat use, and behavior of the animals (Hebblewhite and Haydon 2010). This technology has provided us the opportunity to simultaneously track several woolly monkeys, with improved locational accuracy and increased precision without the limitations associated with the direct human observation and biases due to bad weather, topographic difficulties, and the spatial scales used (Moorcroft and Lewis 2006; Hebblewhite and Haydon 2010).

On the other hand, the accumulation curves did not reach an asymptote to determine the total home range size, despite having a representative sample of more than 1800 h (more than 150 days of sampling) with the GPS collars method (Fig. 11.2a). Although habitat use is a dynamic process that can vary depending on external factors, for example, it is dangerous for animals to move within new areas where predation risk increases, and there is no information about the availability and resource location (Janmaat et al. 2009). A study conducted by Di Fiore and Suárez (2007) on the displacement routes of spider monkeys (*Ateles belzebuth belzebuth*) and woolly monkeys (*Lagothrix lagotricha poeppigii*) in the Ecuadorian Amazon showed that these primates use stable routes through the years. It has been suggested that these species remember through mental maps different topographical features, nodes, or intersections that are connected by tree routes constituting reference points. Mental maps allow primates to maintain a stable habitat and therefore optimize their energy expenditure (Di Fiore and Suárez 2007). Particularly, based on 60 and 45 days of sampling, Di Fiore (2003) registered the total home range size for two of the three groups of *Lagothrix lagotricha poeppigii* located in the Yasuní National Park, where the accumulation curves of hectares approached to a clear asymptote and the group of *Lagothrix Lagotricha* studied by Defler (1996) in the biological station Caparú reached an asymptote with 160-day sampling (1500 h). On the other hand, in the study carried out by Zárate and Stevenson (2014) in the experimental station El Trueno with *Lagothrix lagotricha*, a subsampling was observed with 20 days of tracking for one of the groups; their data is very similar to the one registered in this study for "Toninho" of group B; nonetheless, they register almost the total home range size with 60 days of tracking. Therefore, it is clear that there are benefits associated with the repetitive and frequent use of a particular area or territory (Janmaat et al. 2009), and therefore, it is possible to quantify the total home range size for woolly monkeys (Di Fiore 2003). Although our study was carried out for long periods of time with both methods (direct tracking and GPS collars), we did not manage to quantify the total home range size for the highland woolly monkeys.

Core area size obtained through the two methods was very similar, which was supported by the evident overlapping. Mostly, the core area was located in zones closer to the trail system generated by the investigators (Fig. 11.4 and 11.5 and Appendix, Fig. 11.7). That could be explained because trails were built in places were woolly monkey observations were common (although some of the paths were made previously to this investigation, in areas of easy access in the park). When monkeys begin to move away from their core areas, tracking them became more complicated since many of the areas that were used as home range are rugged and away from the trail system (Fig. 11.5). Therefore, the use of the GPS collars method is particularly important when researchers are carrying out studies in remote areas, which is recurrent in primates' studies (Fedigan et al. 1988).

On the other hand, core areas where group C concentrated its activity (Fig. 11.4a seen in red) are classified by Nathan et al. (2008) as one of the movement phases called foraging. It has been recorded that these areas correspond to continuous feeding patches. These core areas generally contain high proportions of mature forest followed by secondary forest, with continuous canopy and large trees (Prada and

Stevenson 2016) with a high production of fruits of several species that woolly monkeys consume frequently (e.g., *Tapirira guianensis subandina*, *Guettarda hirsuta*, *Miconia* sp., and lianas like *Cissus trianae*) (Cifuentes et al. 2013; Vargas et al. 2014). Woolly monkeys may be forced to use large areas when the mature forest does not provide enough food. However, analyzes made in this investigation do not allow us to establish a causal relationship between these two variables.

The GPS collars method has allowed essential advances in studies of animal behavior, ecology, dispersion patterns, and conservation biology (Fernandez-Duque and Rotundo 2003; Fedigan et al. 1988; Fedigan 2010). However, it is important to mention potential problems. In the study conducted by Hopkins and Milton (2016), adverse effects of ball-chain collars on female howler monkeys were evaluated. They reported that several females presented lacerations on their necks. In other studies, animals presented infections, behavioral changes and/or isolation (Hilpert and Jones 2005; Juarez et al. 2011; Müller and Schildger 1994). However, most studies report the absence of negative behavior or physical impacts in the use of collars in primates (Fedigan 2010; Trayford and Farmer 2012). Mainly, in our study, the woolly monkeys to whom GPS collars were placed did not show any evident physical or behavioral changes. However, we emphasize the importance of not placing GPS collars on juveniles neither on old individuals, since the former have not finished their growth stage and the latter may experience motor and energy restrictions. The most prominent difficulties that we found in monitoring with the GPS collars for woolly monkeys at Cueva de Los Guácharos National Natural Park were the period of immobilization and the procedure to place the collar on the animal and download the data. Woolly monkeys are arboreal primates, and the land where they are found is broken, which makes these processes a challenging experience.

11.4.2 Daily Traveled Distances

Daily trajectory length showed significant differences between the sampling methods for the two groups of monkeys (Fig. 11.6). This result was expected, due to the dispersion and natural fluctuation of each study groups, which makes difficult to track them continuously. Although the handheld GPS tracking is significant, there are many limitations in monitoring monkeys due to factors such as difficulties in the mountainous terrain and climate conditions (heavy and constant rains), hindering the optimal tracking of the troop or group. Accordingly, the GPS collars method is more useful to register the DTD in groups of highland woolly monkeys than with the data registered by the direct tracking using handheld GPS devices (Fig. 11.5). Is important to mention that we only measured the horizontal distances and is possible that in an ecosystem with mountains and hills the real traveled distance would have been higher.

Additionally, the DTD estimated through the direct tracking on focal individuals of group C with handheld GPS against the time of sampling, referring to a continuous tracking of 12 h, was 1535 m (Appendix, Fig. 11.8). This number is inferior to the one reported on the average for DTD for the same group (C) using the GPS collars method (2064 m), which indicates that continuous tracks during 12 h with handheld GPS are slightly closer but not similar to those registered by the individuals with GPS collars. For the above, this study demonstrated that the use of GPS collars could be very useful in obtaining fundamental data to improve the knowledge about ecology and behavior of primate's societies (Fernandez-Duque and Rotundo 2003; Fedigan et al. 1988; Walter et al. 2011). In particular, some analyzes may indicate that DTD is closely related to the home range size. In this study, the home range size estimated through GPS collars method for group C was 438 ha with an average DTD of 2064 m, a value slightly higher than values recorded by Stevenson in 2006 in Tinigua Park and much lower compared to the data reported by González and Stevenson (2010) in the Caparú Biological Station or the one registered by Defler (1996) in the same study site (Table 11.2). The variation in the values obtained in this study and those reported by other researchers indicates that the DTD does not depend directly on the home range size, and those can vary significantly depending on the degree of competition, group size, quality of the habitat, or the availability and temporal distribution of fruits (Stevenson and Castellanos 2000). In this case, since woolly monkeys are mainly frugivorous, they need to travel long distances to meet their energy needs (González and Stevenson 2010; Strier 1992). Preliminary analyses of woolly monkeys located in Cueva de Los Guácharos National Natural Park show that the DTD is not as strongly associated with fruit availability in the forest but to subgroup size.

In summary, this study compared the location of simultaneous data using two methods of field sampling, direct tracking with handheld GPS units and GPS collars. We found that GPS collars provided reliable data regarding spatial information to groups of highland primates. In contrast, the direct tracking method provides unique information for behavioral and ecological analysis. However, this method is biased to collect spatial data mainly when the topography is abrupt and difficult to access. In our study, the direct tracking was carried out for long periods of time, which required a great physical effort. Nevertheless, data collected through handheld devices was half of what was reported by the use of GPS collars, since it was often impossible to reach zones where woolly monkeys were, and therefore those areas were not registered in the handheld GPS devices. In the case of group B, we only collared one individual; therefore, the values obtained do not correspond to the whole group, but to a subgroup (data in preparation). Although collars are expensive, we believe that, for future research, it is essential that the number of monkeys that wear GPS collars would be higher, to study not only their home range and DTD but also their social systems.

In conclusion, results and analysis showed that the use of GPS collars provided higher-quality data compared to direct observations using handheld GPS devices on highland woolly monkey home ranges and DTD. The GPS collars method can contribute to a better understanding of the spatial ecology and moving strategies of taxa living in broken terrains, swampy areas, and other habitats that may prove challenging for conducting direct observations. The independence of observations for statistical purposes from the previous analysis was reached after a 5-h period. The home

Table 11.2 Home range sizes and daily traveled distances (in meters) from ten different studies of woolly monkeys

Researches	Home range (ha)	Sampling method	Group size	Mean DTD (m)	Minimal distance (m)	Maximum distance (m)
Brazil, Urucu River[a]	>860	Direct tracking	39–41	–	–	–
Caparú (1987)[b]	760	Direct tracking	24	2880	2016	3582
Duda River (1994)[c]	169	Direct tracking	17–23	1633	–	–
Peru, Pacaya-Samiria[d]	350	Direct tracking	23	–	–	–
Peneya River[e]	370	Direct tracking	30–40	–	–	–
Peneya River[f]	450	Direct tracking	30–40	–	–	–
Caparú (2004–2005)[g]	>440	Direct tracking	~41	2539	1125	4375
Tinigua[h]	200	Direct tracking	14–30	2001		
PNNCG[i]	300	Direct tracking	20	1580	522	2993
*PNNCG (2012–2013)[j]	580/205	GPS collar and direct tracking	32–34	2064/917	896	3888
PNNCG (2016–2017)[k]	160/152	GPS collar and direct tracking	29–35	1256/997	542	1968

[a]Peres (1994); [b]Defler (1996); [c]Stevenson et al. (1994); [d]Soini (1986); [e]Nishimura and Izawa (1975); [f]Nishimura (1990); [g]González and Stevenson (2010); [h]Stevenson (2006); [i]Vargas et al. (2014); [j]This study for group C; [k]This study for group B

*Home range estimated with the Kernel density method was 438/242.

range size for both males and females in high mountain woolly monkeys is similar. The data recorded from direct tracking by observers are representative of the core areas but not represent the entire home range, and perhaps estimating the complete home range of a group highland monkeys is almost impossible logistically. This is due to the fact that woolly monkeys frequently fission into several subgroups and presents marked changes in their use of space. To estimate home range size in highland monkeys' longtime sampling is required, and probably most approaches using focal follows would describe the core areas.

Appendix

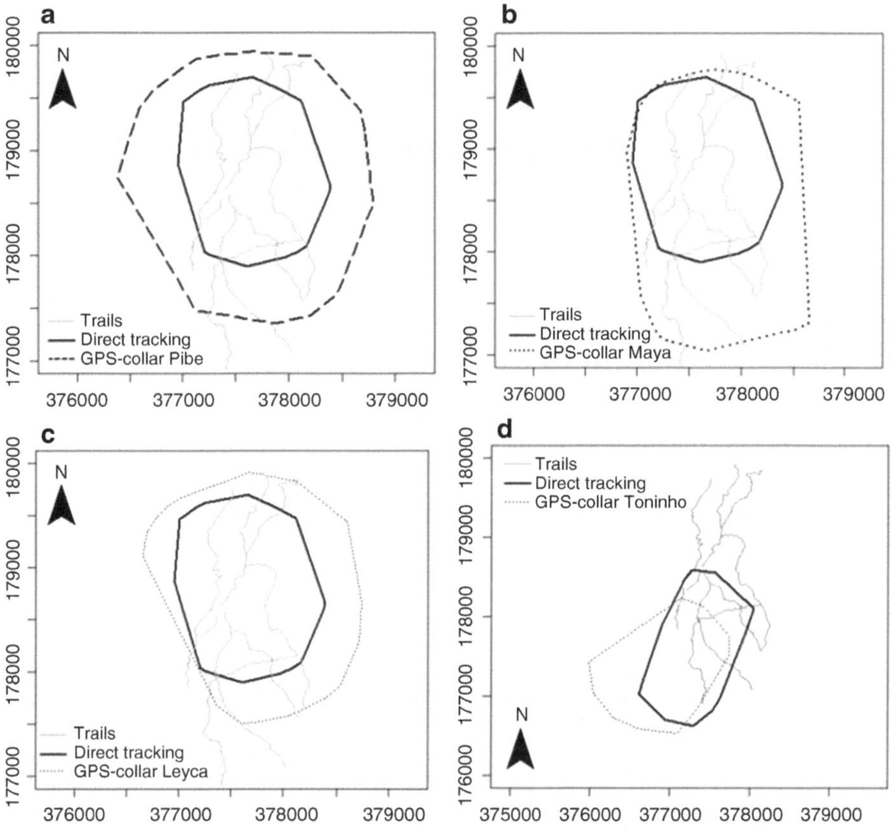

Fig. 11.7 Home range size estimate for each one of the GPS collar individuals with the minimum convex polygon method compared with the size of the area from direct tracking with GPS devices. The external polygon represents the home range size of the GPS collar individuals (a) Pibe. (b) Maya. (c) Leyca. (d) Toninho, while the internal polygon represents the daily monitoring with direct tracking (95% of the data). Axes (*x*, *y*) indicate the UTM coordinates

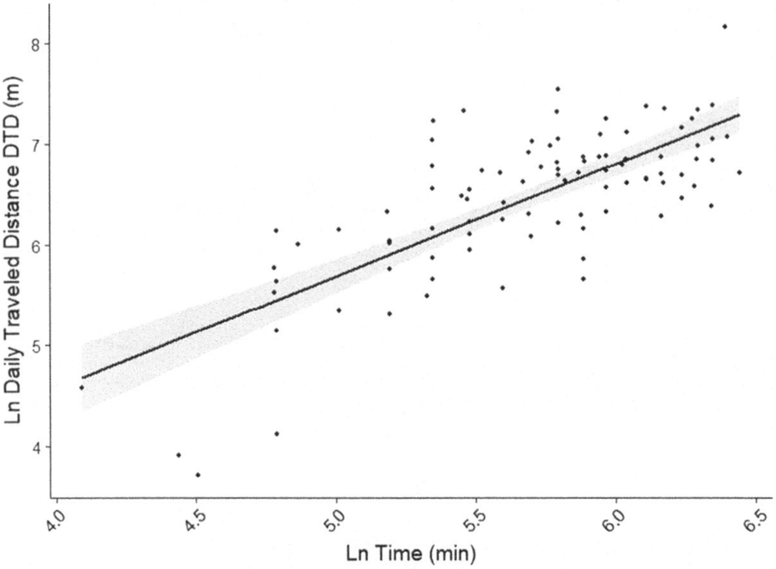

Fig. 11.8 Sampling effort calculated from DTD estimations for direct tracking method for group C ($R^2 = 0.56$)

References

Altmann J (1974) Observational study of behavior: sampling methods. Behaviour 49:227–267

Boyle SA, Lourenço WC, Da Silva LR, Smith AT (2009) Travel and spatial patterns change when *Chiropotes satanas* chiropotes inhabit forest fragments. Int J Primatol 30:515–531

Buckland ST, Plumptre AJ, Thomas L, Rexstad EA (2010) Design and analysis of line transect surveys for primates. Int J Primatol 31:833–847

Camus PA, Lima M (1995) El uso de la experimentación en ecología: supuestos, limitaciones, fuentes de error y su status como herramienta explicativa. Rev Chil Hist Nat 68:19–42

Cifuentes E, Ramírez M, León J, Galvis N, Vargas S, Stevenson P (2013) Dieta de los churucos colombianos (Lagothrix lagothricha lugens) en relación con la productividad de frutos en el Parque nacional natural Cueva de Los Guácharos. In: Defler TR, Stevenson PR, Bueno ML, Guzmán-Caro DC (eds) Primates colombianos en peligro de extinción. Asociación primatológica colombiana Bogotá, pp 294–312

Cresswell WJ, Smith GC (1992) The effects of temporally autocorrelated data on methods of home range analysis. In: Priede IG, Swift SM (eds) Wildlife telemetry: remote monitoring and tracking of animals. Ellis Horwood, Chichester, pp 272-284

Cunningham EP, Unwin S, Setchell JM (2015) Darting primates in the field: a review of reporting trends and a survey of practices and their effect on the primates involved. Int J Primatol 36(5):894–915

Cuthill IC (1991) Field experiments in animal behavior: methods and ethics. Anim Behav 42:1007–1101

D'Eon RG, Serrouya R, Smith G, Kochanny CO (2002) GPS radiotelemetry error and bias in mountainous terrain. Wildl Soc Bull 30:430–439

Defler TR (1996) Aspects of the ranging pattern in a group of wild woolly monkeys (*Lagothrix lagothricha*). Am J Primatol 38(4):289–302

Defler TR (2010) Historia natural de los primates colombianos. Universidad Nacional de Colombia, Bogotá

Di Fiore A (2003) Ranging behavior and foraging ecology of lowland woolly monkeys (*Lagothrix lagotricha poeppigii*) in Yasuni National Park, Ecuador. Am J Primatol 59:47–66

Di Fiore A, Fleischer RC (2005) Social behavior, reproductive strategies, and population genetic structure of *Lagothrix poeppigii*. Int J Primatol 26(5):1137–1173

Di Fiore A, Rodman PS (2001) Time allocation patterns of lowland woolly monkeys (*Lagothrix lagotricha poeppigii*) in a neotropical terra firma forest. Int J Primatol 22(3):449–480

Di Fiore A, Suarez SA (2007) Route-based travel and shared routes in sympatric spider and woolly monkeys: cognitive and evolutionary implications. Anim Cogn 10:317–329

Di Fiore A, Link A, Schmitt CA, Spehar SN (2009) Dispersal patterns in sympatric woolly and spider monkeys: integrating molecular and observational data. Behaviour 146(4):437–470

Di Orio AP, Callas R, Schaefer RJ (2003) Performance of two GPS telemetry collars under different habitat conditions. Wildl Soc Bull 31:372–379

Fedigan LM (2010) Ethical issues faced by field primatologists: asking the relevant questions. Am J Primatol 72(9):754–771

Fedigan LM, Fedigan L, Chapman C, Glander KE (1988) Spider monkey home ranges: a comparison of radio telemetry and direct observation. Am J Primatol 16(1):19–29

Fernandez-Duque M, Chapman CA, Glander KE, Fernandez-Duque E (2017) Darting primates: steps toward procedural and reporting standards. Int J Primatol 39:1–8

Fernandez-Duque E, Rotundo M (2003) Field methods for capturing and marking Azarai night monkeys. Int J Primatol 24(5):1113–1120

Fleagle JG (1999) Primate adaptation and evolution. Academic, San Diego

González M, Stevenson PR (2010) Comparación de los patrones de movimiento diario, actividad y dieta, reportados para los micos churucos (*Lagothrix lagothricha*): diferencias producto de la oferta de frutos y de la metodología usada. In: Pereira-Bengoa V, Stevenson PR, Bueno ML, Nassar-Montoya F (eds) Primatología en Colombia: avances al principio del milenio. Fundación Universitaria San Martín, Bogotá, pp 3–18

Harris S, Cresswell WJ, Forde PG, Trewhella WJ, Woollard T, Wray S (1990) Home-range analysis using radio-tracking data–a review of problems and techniques particularly as applied to the study of mammals. Mammal Rev 20(2–3):97–123

Hebblewhite M, Haydon DT (2010) Distinguishing technology from biology: a critical review of the use of GPS telemetry data in ecology. Philos Trans R Soc B 365(1550):2303–2312

Hilpert AL, Jones CB (2005) Possible costs of radio-tracking a young adult female mantled howler monkey (*Alouatta palliata*) in deciduous habitat of Costa Rican tropical dry forest. J Appl Anim Welf Sci 8(3):227–232

Holyoak E, Casagrandi R, Nathan R, Revilla E, Spiegel O (2008) Trends and missing parts in the study of movement ecology. Proc Natl Acad Sci 105(49):1060–19065

Hopkins ME, Milton K (2016) Adverse effects of ball-chain radio-collars on female mantled howlers (*Alouatta palliata*) in Panama. Int J Primatol 37(2):213–224

Janmaat KR, Olupot W, Chancellor RL, Arlet ME, Waser PM (2009) Long-term site fidelity and individual home range shifts in *Lophocebus albigena*. Int J Primatol 30:443–466

Juarez CP, Rotundo MA, Berg W, Fernández-Duque E (2011) Costs and benefits of radio-collaring on the behavior, demography, and conservation of owl monkeys (*Aotus azarai*) in Formosa, Argentina. Int J Primatol 32(1):69–82

Kenward RE (1992) Quantity versus quality: programmed collection and analysis of radio-tracking data. In: Priede IG, Swift SM (eds) Wildlife telemetry: remote monitoring and tracking of animals. Ellis Horwood, New York, pp 231–245

Kernohan BJ, Gitzen RA, Millspaugh JJ (2001) Analysis of animal space use and movements. In: Millspaugh JJ, Marzluff JM (eds) Radio tracking and animal populations. Academic, New York, pp 15–166

Legendre P (1993) Spatial autocorrelation: trouble or new paradigm? Ecology 74(6):1659–1673

Lewis JS, Rachlow JL, Garton EO, Vierling LA (2007) Effects of habitat on GPS collar performance: using data screening to reduce location error. J Appl Ecol 44:663–671

Lizcano DJ, Ahumanda JA, Nishimura A, Stevenson PR (2014) Population viability analysis of woolly monkeys in western amazonia; the woolly monkey developments in primatology. Prog Prospects 39:267–282

Manly BFJ, McDonald LL, Thomas DL (1993) Resource selection by animals: statistical desing and analysis for field studies. Chapman and Hall, London, p 177

Markham AC, Altmann J (2008) Remote monitoring of primates using automated GPS technology in open habitats. Am J Primatol 70:495–499

Moen R, Pastor J, Cohen Y, Schwartz CC (1996) Effects of moose movement and habitat use on GPS collar performance. J Wildl Manag 60:659–668

Moorcroft PR, Lewis MA (2006) Mechanistic home range analysis. Monographs in population biology. Princeton University Press, Princeton

Müller KH, Schildger J (1994) Capture and radio-telemetry of masked titi monkeys, *Callicebus personatus* melanochir. Neotropical Primates 2(4):7–8

Nathan R, Getz WM, Revilla E, Holyoak M, Kadmon R, Saltz D, Smouse PE (2008) A movement ecology paradigm for unifying organismal movement research. Proc Natl Acad Sci 105(49):19052–19059

Nishimura A (1990) A sociological and behavioral study of woolly monkeys, *Lagothrix lagothricha* in the upper Amazon. Sci Eng Rev Doshiosha Univ 31:1–121

Nishimura A (1994) Social interaction patterns of woolly monkeys (*Lagothrix lagotricha*): a comparison among the atelines. Sci Eng Rev Doshisha Univ 35:236–254

Nishimura A (2003) Reproductive parameters of wild female *Lagothrix lagotricha*. Int J Primatol 24(4):707–772

Nishimura A, Izawa K (1975) The group characteristics of woolly monkeys (*Lagothrix lagothricha*) in the upper Amazonian basin. Contemporary primatol. Karger, Basel, pp 351–357

Parga JA (2011) Nocturnal ranging by a diurnal primate: are ring-tailed lemurs (*Lemur catta*) cathemeral? Primates 52:201–205

Pebsworth PA, Morgan HR, Huffman MA (2012) Evaluating home range techniques: use of global positioning system (GPS) collar data from chacma baboons. Primates 53(4):345–355

Peres CA (1994) Diet and feeding ecology of gray woolly monkeys (*Lagothrix lagotricha*) in central Amazonia; comparisons with other Atelines. Int J Primatol 15(3):333–372

Peres CA (1996) Use of space, spatial group structure, and foraging group size of gray woolly monkeys (*Lagothrix lagotricha cana*) at Urucu, Brazil. In: Marilyn A, Norconk AL, Rosenberger, Paul AG (eds) Adaptive radiations of neotropical primates. Plenum Press, New York, pp 467–488

Pierce L, Kellndorf J, Walker W, Barros O (2006) Evaluation of the horizontal resolution of SRTM elevation data. Photogramm Eng Remote Sens 72(11):1235–1124

Powell RA (2000) Animal home ranges and territories and home range estimators. In: Boitani L, Fuller TK (eds) Research techniques in animal ecology: controversies and consequences. Columbia University Press, New York, pp 65-110

Prada CM, Stevenson PR (2016) Plant composition associated with environmental gradients in tropical montane forest (Cueva de Los Guácharos National Park, Huila, Colombia). Biotropica 48(5):568–576

Ramos-Fernández G, Aguilar SES, Schaffner CM, Vick LG, Aureli F (2013) Site fidelity in space use by spider monkeys (*Ateles geoffroyi*) in the Yucatan Peninsula, Mexico. PLoS One 8(5):e62813

Ren B, Li M, Long Y, Grueter CC, Wei F (2008) Measuring daily ranging distances of *Rhinopithecus bieti* via a global positioning system collar at Jinsichang, China: a methodological consideration. Int J Primatol 29:783–794

Rodgers AR (2001) Tracking animals with GPS: the first 10 years. Tracking animals with GPS. The Macaulay Institute, Aberdeen, pp 1–10

Schmitt CA, Di Fiore A (2014) Life history, behavior, and development of wild immature lowland woolly monkeys (*Lagothrix poeppigii*) in Amazonian Ecuador. In: Defler T, Stevenson PR (eds) The woolly monkey: behavior, ecology, systematics, and captive research. Springer, New York, pp 113–146

Shanee N, Shanee S (2014) Yellow-tailed woolly monkey (Lagothrix favicauda): conservation status, anthropogenic threats, and conservation initiatives. In: Defler TR, Stevenson PR (eds) The woolly monkey: Behavior, ecology, systematics and captive research. Springer, New York, pp 283–299

Soini P (1986) A synecological study of a primate community in the Pacaya-Samiria National. Reserve, Peru. Primate Conserv 7:63–71

Solla DE, Shane R, Bonduriansky R, Brooks RJ (1999) Eliminating autocorrelation reduces biological relevance of home range estimates. J Anim Ecol 68(2):221–234

Stark DJ, Vaughan IP, Saldivar DAR, Nathan SK, Goossens B (2017) Evaluating methods for estimating home ranges using GPS collars: a comparison using proboscis monkeys (*Nasalis larvatus*). PLoS One 12(3):e0174891

Stevenson PR (1998) Proximal spacing between individuals in a group of woolly monkeys (*Lagothrix lagothricha*) in Tinigua National Park, Colombia. Int J Primatol 19(2):299–311

Stevenson PR (2000) Seed dispersal by woolly monkeys (*Lagothrix lagotricha*) at Tinigua National Park, Colombia: dispersal distance, germination rates, and dispersal quantity. Am J Primatol 50:275–289

Stevenson PR (2006) Activity and ranging patterns of Colombian woolly monkeys in noth-western Amazonia. Primates 47(3):239–247

Stevenson PR, Castellanos MC (2000) Feeding rates and daily path range of the colombian woolly monkeys as evidence for between-and within-group competition. Folia Primatol 71:399–408

Stevenson PR, Link A (2011) Lagothrix lugens. IUCN Red list of threatened species. Version

Stevenson PR, Quiñones MJ, Ahumada JA (1994) Ecological strategies of woolly monkeys (*Lagothrix lagotricha*) at Tinigua National Park, Colombia. Am J Primatol 32(2):123–140

Stevenson PR, Zárate DA, Ramírez MA, Henao-Díaz F (2015) Social interactions and proximal spacing in woolly monkeys: lonely females looking for male friends. In: Furuichi T, Yamagiwa J, Aureli F (eds) Dispersing primate females. Springer, Tokyo, pp 45–71

Strier KB (1992) Atelinae adaptations: behavioral strategies and ecological constraints. Am J Phys Anthropol 88:515–524

Swihart RK, Slade NA (1985) Testing for independence of observations in animal movements. Ecology 66(4):1176–1184

Trayford HR, Farmer KH (2012) An assessment of the use of telemetry for primate reintroductions. J Nat Conserv 20(6):311–325

Vargas SA, León J, Ramírez M, Galvis N, Cifuentes E, Stevenson PR (2014) Population density and ecological traits of highland woolly monkeys at Cueva de los Guácharos National Park, Colombia. High Altitude Primates, Springer New York, pp 85–102

Walter WD, Fischer JW, Baruch-Mordo S, VerCauteren KC (2011) What is the proper method to delineate home range of an animal using today's advanced GPS telemetry systems: the initial step. In: Krejcar O (ed) Modern telemetry. In Tech, Rijeka, pp 249–268

White GC, Garrott RA (1990) Analysis of wildlife radio-tracking data. Academic, Toronto, p 383

Zárate D, Stevenson PR (2014) Behavioral ecology and Interindividual distance of woolly monkeys (*Lagothrix lagothricha*) in a rainforest fragment in Colombia. In: Defler T, Stevenson PR (eds) The woolly monkey. Developments in primatology: progress and prospects. Springer, New York, pp 227–245

Chapter 12
Ranging Responses to Fruit and Arthropod Availability by a Tufted Capuchin Group (*Sapajus apella*) in the Colombian Amazon

Carolina Gómez-Posada, Jennifer Rey-Goyeneche, and Elkin A. Tenorio

12.1 Introduction

Patterns of movement and space use in animals are fundamental dynamic responses for a species' ecological and evolutionary processes, as they regulate access to resources (Nathan et al. 2008). Understanding *why*, *where*, *when*, and *how* animals move (Nathan et al. 2008) is essential to forecast wild populations' responses to changes and disturbances in the quality and distribution of habitats (Chapman et al. 2018; Winkler et al. 2014). In relation to the spatiotemporal heterogeneity of the environment, the abundance and distribution of food resources have been proposed as the main drivers of primates' ranging behavior (Reyna-Hurtado et al. 2018; Wallace 2008). As a result, seasonal changes in resource supply enable animals to adopt different strategies to maximize food acquisition. Most studies on the space use patterns of primates that include insects in their diet overlook the variety of consumed food items, limiting predictor variables to fruits or leaves and excluding other important resources, such as invertebrate prey (Gursky 2000). In tropical forests, not only fruit availability changes seasonally for these species; arthropod abundance and composition also suffer drastic changes, generally associated with water availability (Boinski and Fowler 1989; Vasconcellos et al. 2010). Therefore, the

C. Gómez-Posada (✉)
Department of Biology, Universidad del Valle, Cali, Colombia

Instituto de Investigación de Recursos Biológicos Alexander von Humboldt, Bogotá, Colombia
e-mail: cgomez@humboldt.org.co

J. Rey-Goyeneche
Asociación Primatológica Colombiana, Bogotá, Colombia

E. A. Tenorio
Instituto de Investigación de Recursos Biológicos Alexander von Humboldt, Bogotá, Colombia

R. Reyna-Hurtado, C. A. Chapman (eds.), *Movement Ecology of Neotropical Forest Mammals*, https://doi.org/10.1007/978-3-030-03463-4_12

abundance and distribution of both fruits and arthropods can influence consumers' foraging strategies (Rímoli et al. 2008).

The tufted capuchin monkey (*Sapajus apella,* formerly *Cebus apella*) is a medium-sized (~3.7 kg) Neotropical primate that inhabits different types of forest, displaying a generalist behavior in habitat use (Defler 2010). Its omnivorous diet consists of mainly fruits and arthropods, and in less proportion plant parts and small vertebrates, among others (Fragaszy et al. 2004). Due to its highly insectivorous diet, it is valid to consider that the availability of arthropods has a strong influence on its decisions of foraging, activity budget, diet, and space use patterns. Spatial responses of capuchin species to the variation in fruit availability in the forest have been widely studied (Chapman 1988; Peres 1994; Robinson 1986; Stevenson et al. 2000; Terborgh 1983; Zhang 1995). Capuchin species can show contrasting foraging patterns in conditions of fluctuating fruit offer. When fruit abundance is low, capuchins may move extensively throughout its home range in search for food (Terborgh 1983) or rely on a key resource (Brown and Zunino 1990; Peres 1994) and stay near its location (Spironello 2001). In periods of high fruit availability, it may save energy and remain in a portion of its home range (Di Bitetti 2001; Terborgh 1983; Zhang 1995) or increase the travel routes in order to access the wide offer of food (Gómez-Posada 2003). Although fruit abundance and distribution have been proposed as the most important factors affecting the capuchins' ranging behaviors, few studies have taken into account the capuchins' spatial responses to arthropod availability (Terborgh 1983), despite the fact that this resource represents more than 50% of their diet (Fragaszy et al. 2004). Then, little is known about the influence of arthropods supply on the capuchins' space requirements and spatial decisions.

To address the lack of integrated approaches, we simultaneously studied the influence of fruit and arthropod availability on the ranging behavioral responses of a group of wild tufted capuchins. We assessed six ranging variables that are good indicators of a primate's spatial decisions: daily distance traveled, home range size, backtracking, turning angles, travel speed, and arthropod-foraging speed; some of which have been widely studied (e.g., daily path lengths, home range size), and some others need more research (e.g., turning angles or movement tortuosity, recrossing a path in a day or backtracking) (Janson 2000; Robinson 1986; Wallace 2008). To describe the group's responses to the variation of their two main food resources, two approaches were considered: first, we compared the ranging patterns between periods in which ripe fruits and arthropods were abundant and periods of scarcity of both resources; second, we built models to test which predictor variables indicating fruit and arthropod availability could explain each one of the ranging behavior variables.

We hypothesized that during periods of combined scarcity of resources, the group would increase its ranging behavior; in order to find enough arthropods and fruits to maintain the quality of the diet, all the response variables would be larger when both resources are scarce, as capuchins would move extensively throughout their home range in search for food (Di Bitetti 2001; Peres 1994; Terborgh 1983). We also hypothesized that response variables would be affected differently by specific indicators of food resource availability: some would be influenced by fruit offer, others by arthropod availability, and others by the combined supply of resources. To test this hypothesis, we proposed the following predictions: (a) an increase in fruit and arthropod abundance would lead to shorter daily path lengths, as both resources would be highly available in space (Janson 2000); (b) home range size would increase during periods of combined scarcity, due to the low density of fruit resources and its random distribution and the low arthropod encounter rate (Robinson 1986; Terborgh 1983; Zhang 1995); (c) backtracking would increase when fruit resources are scattered and at low densities, as capuchins would prefer to use intensively the few productive feeding sites and revisit the closest feeding trees, instead of farther ones (Garber 2000; Janson 2000); (d) travel speed between feeding sites would be low when fruits are abundant and aggregated, as high productive feeding patches would be proximate, resulting in shorter distances to travel and less pressure to search for food resources (Janson 2000); (e) arthropod-foraging speed would be higher during scarcity periods, due to the low success in prey capture influenced by low arthropod abundance (Terborgh 1983); and (f) during the scarcity period, capuchins would have more sinuous paths (increase in turning angles) due to their arthropod-foraging strategy, which requires a detailed scan of the area (Calengea et al. 2009; Phiphatsuwannachai et al. 2018).

To test our predictions, we used ranging data collected during 1-year study of a *S. apella* group in the Colombian Amazon, which was aimed at evaluating the combined effect of arthropod and fruit availability on the capuchins' diet and activity patterns (Gómez-Posada 2009, 2012). In those studies, it was found that this group was principally insectivorous-frugivorous (48.8–42.9%, respectively) (Gómez-Posada 2012), with a high consumption of fruits and arthropods during periods of high abundance of both food resources (Gómez-Posada 2012). During the scarcity period of both resources, capuchins increased the consumption of vegetable parts and other items (Gómez-Posada 2012) and spent more time foraging and traveling (Gómez-Posada 2009).

12.2 Methods

12.2.1 Study Site and Study Group

This study was conducted between January and December 1999 at the Mosiro Itajura-Caparú Biological Station (MICBS) in the Colombian Amazon (1° 5.55′ S, 69° 30.8′ W). The study site consists of a continuous lowland (200 m a.s.l.) humid tropical forest within the National Natural Park Yaigojé-Apaporis (1,060,603 ha) in the Southeast of the Department of Vaupés (Fig. 12.1). Mean annual temperature at MICBS is 25.1 °C, and the average annual precipitation is 3836 mm, following a bimodal seasonal regime, but without a classical dry season (Defler and Defler 1996). The rainy season comprised from April to June, with a minor peak in January, and the two low rainy periods occurred from February to March and from July to

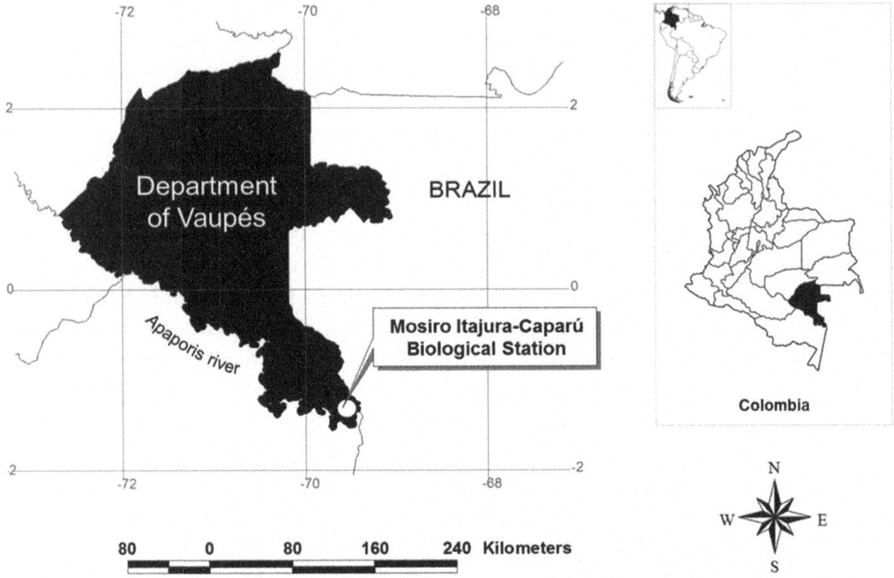

Fig. 12.1 Location of the Mosiro Itajura-Caparú Biological Station in the Colombian Amazon

December. In 1999, total precipitation was 4089 mm (data collected at the study site).

The forest has an upper 25–30 m tall canopy with emergent trees of up to 45 m. In the area, there are six main forested habitats: hills, non-floodable river terraces, seasonally flooded blackwater forests, floodable lowlands along streams, and transitional forest between terraces and flooded areas. The dominant plant families, in both number and species, are Leguminosae, Lecythidaceae, Moraceae, Myristicaceae, and Palmae (for details of the plant species of each forest type, see Defler and Defler 1996). In the MICBS, eight species of primates coexist sympatrically: *Sapajus apella*, *Cebus albifrons*, *Saimiri sciureus*, *Callicebus torquatus*, *Aotus vociferans*, *Alouatta seniculus*, *Lagothrix lagotricha*, and *Cacajao melanocephalus* (Defler and Defler 1996).

During the year of study, we registered five groups of *S. apella* with a home range, or part of it, located within the MICBS's gridded trail system—which is established every 100 m and marked at 25 m intervals. Group composition varied from four to ten individuals. We selected and habituated a group comprised of five individuals: one adult male, two adult females, one subadult male, and one juvenile female. The group composition did not change over the study. This focal group was followed daily from one sleeping site at dawn to the next sleeping site at dusk, for 5 days per month; completing a total of 708 sampling hours between January and December 1999 (with the exception of March, when the group was not found).

The taxonomy of the tufted capuchins is under evaluation. In this study, we followed the classification of Lynch Alfaro et al. (2012), which is currently accepted by the Colombian Primatological Association: *Sapajus apella.*

12.2.2 Comparison Among Food Availability Periods

Our first approach sought to compare ranging patterns between periods when both ripe fruits and arthropods were abundant, with periods of scarcity of both resources. Then, we estimated fruit and arthropod abundance, collected information for six ranging pattern variables (daily distance traveled, home range, backtracking, turning angles, travel speed, and arthropod-foraging speed), and compared these ranging behavioral responses between three different periods of combined fruit and arthropod availability.

12.2.2.1 Abundance of Food Resources

Ripe Fruit Abundance We established five phenological transects (400 × 5 m) within the study group's home range. In each transect, we counted the number of individuals of all trees greater than 3 m in height with mature fruits and the number of mature fruits per tree. Monthly fruit abundance was calculated as the average of the trees with ripe fruit in the five transects. Differences in monthly fruit availability were analyzed by a Kruskal-Wallis test.

Arthropod Abundance Fifteen randomly distributed plots (5 m^2 × 3 m high) were established, per month, within the study group's home range (Boinski 1987; Boinski and Fowler 1989). In each plot, we collected arthropods manually for 1 h (Boinski and Fowler 1989) by searching and examining specific substrates, such as inside dead trunks, under the bark of trees, in suspended leaf litter, on the upper and lower epidermis of leaves, on the surface of stems, and inside flowers. This method was practiced in the field for a month, prior to starting data collection. We captured arthropods by hand, with the help of a wide-mouth lethal bottle. Sampling was equally distributed every month, between 09:00 and 15:00 h, and we did not conduct it under rainfall. All individuals collected in each plot were counted and identified to order level. Monthly arthropod abundance was calculated as the average of the individuals collected in the 15 plots. Variation in monthly arthropod availability was analyzed by a one-way ANOVA. In this study, manual sampling was complemented with aerial net sampling. However, as the main arthropod groups collected with the net were not consumed by the capuchins (see Gómez-Posada 2012), only the results of the manual sampling were taken into account.

Combined Resource Availability For arthropods, we assigned periods of abundance and scarcity to those months in which the number of arthropods collected was significantly higher or lower than the monthly average (calculated as the average between all sampling months), respectively. Under the same criterion, we identified periods of abundance and scarcity of ripe fruits. Although the methods used to estimate the availability of arthropods and fruits are not comparable, we took into account the months in which their abundance or scarcity matched, in order to obtain a pattern of combined availability (Raemaekers 1980). Therefore, we determined two food availability periods that corresponded to either high or low combined availability: abundance, for those months where both resources were abundant, and scarcity, for those where the abundance of both resources was low. The rest of the months were grouped in a category denominated intermediate period.

12.2.2.2 Ranging Pattern Variables

Daily Distance Traveled We followed the focal group between sleeping sites in 1 day (NRC 1981) and recorded its location every 30 min. All-day location points were drawn on maps and connected in a straight line using AutoCAD 2009 version 17.2 for Windows. These data were used to calculate the monthly and total average of daily path lengths.

Home Range We calculated the monthly and total area used by the group in its normal activities (McNab 2002) by superimposing a grid of 1-ha subplots over the study area map and adding up all quadrants visited by the capuchins at least once (NRC 1981). To analyze if the increase in daily path length was related to home range size, these variables were correlated throughout the year (Spearman r_s).

Backtracking Using the daily paths maps, we registered the number of times the group turned back and recrossed its own route in a day (Robinson 1986).

Turning Angle When registering the group's location every 30 min, we measured turning angles between consecutive location points, taken as vectors in a straight line (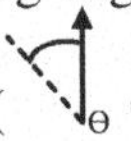 Θ = turning angle), using AutoCAD 2009 version 17.2 for Windows. While a small angle indicated that the group tended to move forward, a greater angle of rotation indicated a tendency to a tortuous movement. When the group did not move, the "0" angle was not included (Robinson 1986). Monthly averages of the angles were calculated with the rectangular coordinates (sine and cosine) (Zar 1996). To test if the angles of rotation followed a random distribution, a Rayleigh test was performed (Robinson 1986).

Travel Speed and Arthropod-Foraging Speed The point of departure and arrival of the group during travel and arthropod-foraging and the time invested in both activities were recorded. Speed was expressed as the number of meters traveled per hour (distances were calculated using AutoCAD 2009 version 17.2 for Windows). Travel speed was registered when the group was moving with a defined direction and without being involved in other activities. Foraging speed was recorded when the group was dispersed to search for invertebrates, moving in a coordinated manner.

12.2.2.3 Ranging Pattern Comparison Among Food Availability Periods

Variation in each ranging behavior variable among periods of high, intermediate, and low food resource abundance was compared by a one-way ANOVA (for average daily path length, travel speed, and arthropod-foraging speed), a Kruskal-Wallis test (for home range and backtracking), and a Watson-Williams test (for turning angles) (Zar 1996).

12.2.3 *Interaction Analysis Between Indicators of Food Resource Supply and the Ranging Patterns of* Sapajus apella

For our second approach, we sought to understand which food supply predictor variables were the ones determining each one of the ranging pattern variables that we assessed (response variables: daily distance traveled, home range, backtracking, turning angles, travel speed, and arthropod-foraging speed), through a generalized linear model analysis. We used two types of predictor variables: a direct estimation of fruit and arthropod abundance (methods explained in the above section) and an indirect approximation of arthropod and fruit availability, using resource use variables that could influence the ranging behavioral responses of *S. apella* (fruit-feeding tree density, fruit-feeding tree dispersion, and arthropod-capture success rate).

12.2.3.1 Use of Food Resources Predictor Variables

Density of Fruit-Feeding Trees On a study area map, we located the trees used by the focal group for consumption of ripe fruit each month. We calculated tree density by dividing the number of trees by the group's home range size each month (Spironello 2001). To analyze if monthly fruit-feeding tree density was related to fruit abundance, these variables were correlated throughout the year (Spearman r_s).

Dispersion of Fruit-Feeding Trees We randomly located three transects of 500 × 20 m, each month, within the study area. In each transect, we established a 10 × 10 m grid and counted the number of trees with fruit of those species that were consumed by the monkeys in each subplot (registered during a diet study using slow scans, see Gómez-Posada 2012). From these data, we determined monthly resource distribution by the Index of Dispersion, ID = s^2/x, where s^2 is the variance and x is the mean (Ludwig and Reynolds 1988). If the sample follows a Poisson distribution, the ID is 1. The difference from 1 can be verified with the test of X^2 and the statistical test $d = \sqrt{2X^2} - \sqrt{(2 - N)} - 1$.

Rate of Capture Success for Arthropod Prey When the group was foraging for arthropods, we estimated the capture success rate, which corresponds to the number of arthropods captured by a capuchin during five continuous minutes of observation, using the focal-animal sampling method (Altmann 1974). Each successful attempt—defined as the jump, movement, or manipulation made by the focal animal toward an arthropod, which ended with its consumption (Melin et al. 2007)—was recorded as a capture event. We alternated observations between the individuals of the group to obtain similar observation times between sexes and ages. To analyze whether the changes in arthropods abundance resulted in a change in capture success, these variables were analyzed with a Spearman correlation throughout the year.

12.2.3.2 Model Analysis of the Interactions Between Food Resource Supply Variables and the Ranging Patterns of *Sapajus apella*

According to our predictions, for each ranging response variable, we constructed different scenarios using generalized linear models (GLM) to test the relationship between each response variable and the predictor ones, evaluating 11 models. Five models used predictor variables independently —one model for each of the predictor variables: fruit abundance, arthropod abundance, fruit tree density, fruit tree dispersion, and capture success rate (these models covered the study's predictions *c* to *f*). Six models combined the effect of two predictor variables: one related to fruit availability and one related to arthropod availability (these models covered predictions *a* and *b*) (Table 12.1). We did not include models with more than two variables, as each model could be over-parameterized regarding the sample size. Model evaluation was based on Akaike's information criterion (AIC) model selection (evaluated using change in the corrected AIC, AICc), with consideration of the weights of models, predictors, and the size of the sample (Burnham et al. 2011). With this information, we calculated the *Akaike weight* (*w*) for each model, which ranges from 0 to 1 and is interpreted as the relative probability that a model is the best across all those evaluated. This means that models with a higher value of *w* will be the most probable model and will have a greater explanatory power of the response factors. For the selected models, we calculated the relation between variables extracting the *P* value and slope from the GLM test. The construction of the different models and their evaluation were performed in R v. 3.3.3 (Dalgaard 2010), using the *GLM* and *model.sel* functions of the *base* and *MuMin* packages (Barton 2018), respectively.

Table 12.1 GLM scenarios to test the influence of predictor variables on each of the ranging behavioral responses of *Sapajus apella*: daily distance traveled, home range, backtracking, turning angle, travel speed, and arthropod-foraging speed

Models		Predictor variables
Single variable effect	M1	Fruit abundance
	M2	Arthropod abundance
	M3	Density of fruit-feeding trees
	M4	Rate of arthropod-capture success
	M5	Dispersion of fruit-feeding trees
Combined variable effect	M6	Fruit abundance + arthropod abundance
	M7	Fruit abundance + rate of arthropod-capture success
	M8	Density of fruit-feeding trees + arthropod abundance
	M9	Density of fruit-feeding trees + rate of arthropod-capture success
	M10	Dispersion of fruit-feeding trees + arthropod abundance
	M11	Dispersion of fruit-feeding trees + rate of arthropod-capture success

12.3 Results

12.3.1 Abundance of Food Resources

Ripe Fruit Abundance The number of trees bearing mature fruits was correlated with the total number of ripe fruits counted per transect each month ($r_s = 0.81$, $n = 12$, $P < 0.01$). Thus, we worked with the number of trees as an indicative of fruit availability. Throughout the year, the availability of fruiting trees varied significantly ($H = 26.9$, DF = 11, $P < 0.01$) (Fig. 12.2a). Some of the most abundant fruit species for capuchins were Amazon grapes (*Pourouma* spp.), different species of guama beans (*Inga* spp.), mamitos (*Virola* spp. and *Iryanthera* spp.), and *Cariniana micrantha* (which was an important resource in months with very low fruit availability). Various tree species fructified abundantly during months of high fruit abundance, such as *Doliocarpus* cf. *dentatus*, *Clusia* sp., *Strychnos* cf. *erichsonii*, *Philodendron* sp., and nuts of cumare palms (*Astrocaryum aculeatum*) (description of the capuchin's diet and fruit availability can be found in Gómez-Posada 2012).

Arthropod Abundance We collected a total of 16,998 individuals (monthly average of 1416.5 ± 274.8; x ± SD, Fig. 12.2a) from 23 groups: 17 orders and 6 categories including other life-cycle stages and families. Formicidae was the most abundant family, followed by Coleoptera, Arachnida, and eggs and larvae of different orders, which are food resources for insectivorous (for arthropod abundance details, see Gómez-Posada 2012). Throughout the year, the arthropods abundance varied significantly ($F = 2.33$, DF = 11, $P < 0.01$) (Fig. 12.2a).

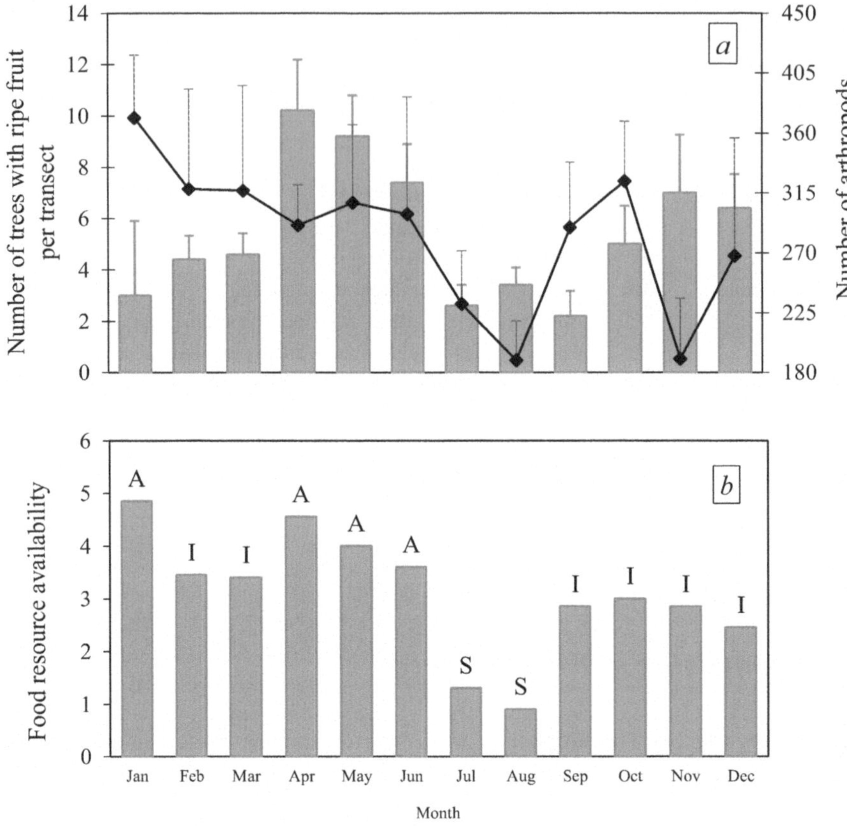

Fig. 12.2 Monthly offer of food resources during 1999 at the Mosiro Itajura-Caparú Biological Station, Colombian Amazon. (**a**) Mean abundance of ripe fruits (bars) and arthropods (line). Standard Error bars are shown. (**b**) Combined availability of food resources to identify periods of abundance (*A*), intermediate offer (*I*), and scarcity (*S*) [By superimposing the abundance graphs of arthropods and fruits (**a**) with the lower and upper points at the same scale on the *Y* axis, a graph of the abundance of the two resources can be obtained in a combined manner (**b**) (Raemaekers 1980)]

Combined Resource Availability When both resources were considered together, January, April, May, and June were the months with combined high abundance (abundance period), while July and August were the months with combined low abundance (scarcity period) (Fig. 12.2b). The rest of the months were grouped as intermediate abundance. Although in January, few trees with ripe fruits were found in the phenological transects, the month was included in the abundance period due to the high offer of *A. aculeatum* across the capuchins' home range, which was an important food resource for the focal group (Gómez-Posada 2012).

Table 12.2 Ranging and movement variables for *Sapajus apella* among three periods of food resource availability, at the MICBS

Ranging behavior	Food resource availability period			*n*	Test value	*P*
	Abundance	Intermediate	Scarcity			
Home range (ha)	116.2 (±22.1)	107 (±28.3)	118.8 (±8.2)	11	$H = 0.62$	0.7^{ns}
Daily distance traveled (m)	2858.0 (±856.1)	3315.6 (±767.9)	3497.0 (±584.5)	49	$F = 2.6$	0.08^{ns}
Backtracking	1.4	1.7	2.0	49	$H = 0.68$	0.7^{ns}
Turning angle	44.2 (±40.5)	54.2 (±41.2)	49.7 (±42.4)	872	$F = 4.32$	0.01^{**}
Travel speed (m/h)	16.5 (±7.5)	13.5 (±5.4)	16.6 (±7.3)	186	$F = 2.7$	0.07^{ns}
Arthropod-foraging speed (m/h)	3.8 (±1.8)	4.5 (±2.4)	4.4 (±2.3)	175	$F = 2.66$	0.07^{ns}

Values are average (±SD) per period. *P* values were calculated from one-way ANOVA (daily path length, travel speed, arthropod-foraging speed), Kruskal-Wallis test (home range, backtracking), and Watson-Williams test (turning angle)
ns nonsignificant differences; **Significant differences

12.3.2 Comparison Among Food Availability Periods

Home Range Size The home range of the focal group was 320 ha during the study period, ranging monthly from 84 to 133 ha. Based on observations made after the year of study, a home range of 367 ha was estimated. Home range size was similar among the tree food availability periods (Table 12.2).

Daily Distance Traveled The average daily path length was 3181 ± 800.4 m ($x \pm$ SD), ranging from 1325.3 to 5243.8 m. When comparing daily path lengths among the three food availability periods, there was a tendency to travel greater distances when resources were scarce, but the differences were marginally significant (Table 12.2). Then, we compared travel distances only between abundance and scarcity periods, as we expected that a lower availability of food resources influenced larger routes. Results indicated that the capuchins moved a greater distance when both fruits and arthropods were in low availability ($t = 2.1$, DF = 25, $P = 0.05$). Larger routes were not related to an increase in the home range ($r_s = 0.37$, $N = 11$, $P = 0.3$).

Backtracking During the study period, the number of times that the capuchin focal group recrossed their own path in a day was 1.7 ± 1.3 ($x \pm$ DS), and all of them were related to the revisit of a fruiting tree, usually the most important in the diet of each month. Backtracking numbers did not vary among the three food availability periods (Table 12.2).

Turning Angles The angles during the capuchins' travels had an average of 48.6° ± 41.4° ($x \pm$ DS) and were not randomly distributed (Rayleigh test <0.01), indicating that the group's movements were not random—they had a direction. Most of the turning angles were less than 90°. When we compared turning angles during the three food availability periods, they were lower (less tortuous paths) during months of high food abundance than during the scarcity period (Table 12.2).

Travel Speed Capuchins traveled at an average speed of 15.8 ± 7.1 m/h (x ± SD). In general, capuchins did not search or consume food during travel events; these movements were generally directed toward a fruiting tree. However, capuchins not only moved to look for food. For example, at the end of the afternoon, they traveled in search for sleeping sites, mostly palms, where they could spend the night. Although they did not have fixed sleeping trees, they frequented sectors where palms were abundant. They also moved when escaping from predators, as a case of a harpy eagle (*Harpia harpyja*) that followed the focal group for about 2 days (escaping speed was not included in the analysis).

Arthropod-Foraging Speed The speed of foraging for arthropods was 4.2 ± 2.1 m/h (x ± SD) during the study year. During these foraging events, the capuchin group dispersed, and the monkeys moved slowly, looking for invertebrates. Foraging was individual, but the movement was coordinated by constant vocalizations of all the members of the group. The group tended to travel and forage for arthropods faster during periods of low food availability, but the differences were only marginally significant (Table 12.2).

12.3.3 Interactions Between Indicators of Food resource Supply and the Ranging Patterns of Sapajus apella

12.3.3.1 Use of Food Resource Variables

Fruit-Feeding Tree Dispersion The spatial distribution of the capuchin's fruit-feeding tree species was variable throughout the study. The relation between tree dispersion and the monthly availability of fruit was not significant ($r_s = 0.42$, $n = 11$, $P = 0.18$); however, months with the highest fruit supply presented a clumped distribution (Table 12.3).

Fruit-Feeding Tree Density The group visited 481 fruit-feeding trees throughout the study period, indicating a density of 1.5 trees/ha in their home range. Months with high fruit availability marginally tended to have a higher density fruit-feeding trees ($r_s = 0.51$, $n = 11$, $P = 0.06$) (Table 12.3).

Rate of Capture Success for Arthropod Prey While foraging for arthropods, the capuchin's arthropod-capture rate was 2.5 ± 3 (x ± SD) individuals per minute. These events of continuous search for arthropods were not long (51.7 ± 29.9 s, with a maximum of 303 s). The monthly abundance of arthropods was correlated with the capture rate ($r_s = 0.76$, $n = 11$, $P = 0.01$) (Table 12.3).

Table 12.3 Monthly fruit-feeding tree dispersion (C: clumped; R: random), density of capuchin's fruit-feeding trees, and arthropod-capture success rate at the MICBS during 1999

Month	Dispersion index ($ID = s^2/x$) $^*P < 0.05$		Number of fruit-feeding trees per hectare	Number of arthropods captured per minute ($x \pm SD$, n)
Jan	3.33*	C	0.65	4.2 ± 0.6, 31
Feb	0.86	R	0.40	2.8 ± 0.4, 20
Apr	1.53*	C	0.36	2.5 ± 0.7, 18
May	2.10*	C	0.63	2.4 ± 0.6, 54
Jun	1.97*	C	0.38	3.2 ± 0.2, 40
Jul	1.03	R	0.10	1.8 ± 0.9, 16
Aug	0.82	R	0.17	1.9 ± 0.1, 15
Sep	0.91	R	0.05	2.4 ± 0.3, 13
Oct	1.03	R	0.20	2.2 ± 1.0, 17
Nov	1.28*	C	0.23	1.8 ± 0.6, 12
Dec	2.96*	C	0.17	2.0 ± 0.4, 10

12.3.3.2 Models Results of the Interaction Between Food Resource Supply Variables and the Ranging Patterns of *Sapajus apella*

Models with only one predictor variable had enough explanatory power to indicate which ones determined each of the response variables (Fig. 12.3). However, none of the combined models that included one variable related to fruit supply and one related to arthropod supply had prediction power (models M6–M11, Table 12.1), probably due to the low sample size (11 months) that might penalize more complex models. The analysis of one-variable models (M1–M5, Table 12.1) showed that food resource availability influenced the group's ranging patterns (Fig. 12.3). For example, the abundance of ripe fruits strongly influenced their home range size, being larger during the high supply period. The density of fruit-feeding trees was negatively related to both the daily path length and the arthropod-foraging speed. Arthropod-capture success rate negatively influenced turning angles and the speed of foraging for arthropods. Travel speed was marginally predicted by fruit-feeding tree dispersion. None of the predictor variables determined backtracking (Fig. 12.3).

12.4 Discussion

Our results showed that the study group modified its ranging behavior according to both fruit and arthropod availability. In general, capuchins increased their daily path length, had more tortuous movements, and tended to move faster during the combined scarcity period. Fruit availability determined the length of the daily movements and, marginally, the travel speed; arthropod availability determined the turning angles—how animals move throughout their home range; and both fruit and

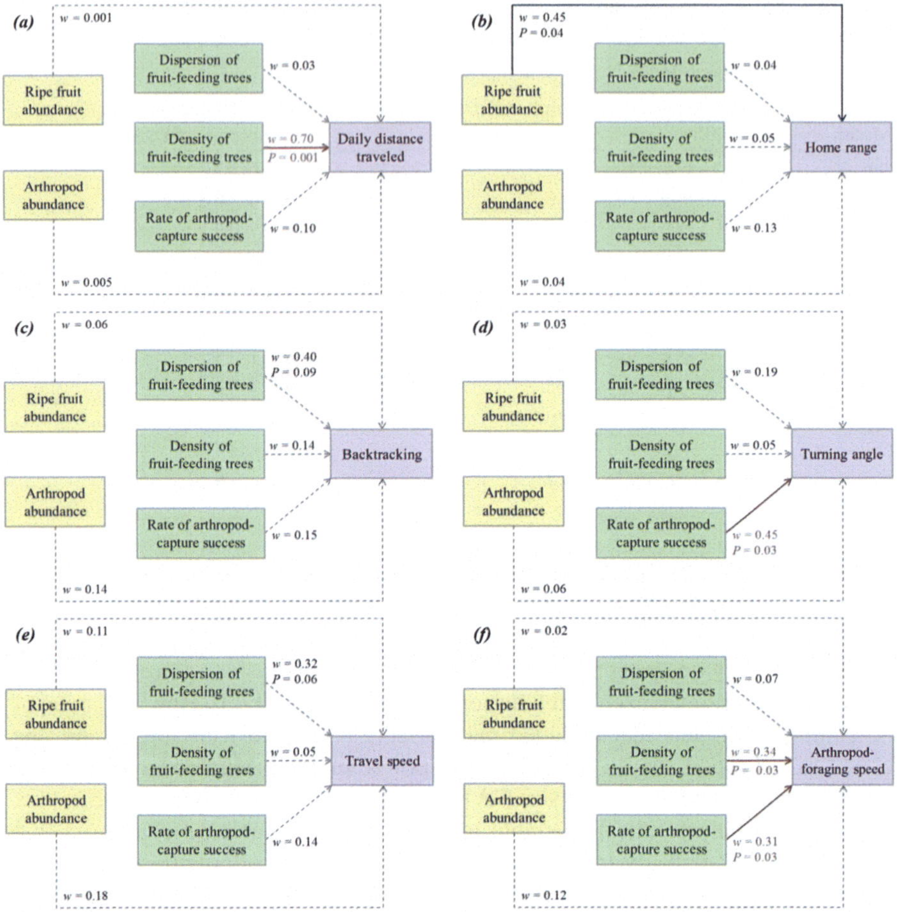

Fig. 12.3 Schematic representation of GLM models analyzing the effects of food resource variables (fruit abundance, arthropod abundance, dispersion and density of fruit-feeding trees, arthropod-capture success rate) on the ranging patterns of *Sapajus apella* at the Mosiro Itajura-Caparú Biological Station, Colombian Amazon: (**a**) daily distance traveled, (**b**) home range, (**c**) backtracking, (**d**) turning angle, (**e**) travel speed, and (**f**) arthropod-foraging speed. The *Akaike weight* (w) is given for all relationships. P value is included for response factors with greater explanatory power ($w > 30\%$). Continuous lines indicate significance level ($P < 0.05$) of positive (blue) or negative (red) relationships

arthropod availability determined the arthropod-foraging speed. There were not conclusive results for backtracking and home range size; the last one was not different between food availability periods but was determined by fruit abundance.

Tropical forests experience seasonal fluctuations in phenological patterns, resulting in resource bottleneck periods for frugivorous and insectivorous animals (Fragaszy et al. 2004; Williams and Middleton 2008). In the study area, periods of high fruit availability were widely abundant, with many species producing fruits synchronically, determining a high density of fruiting trees and a clumped distribution of this food resource. On the contrary, fruit scarcity periods are drastic for large primates in the area, due to the low quality/quantity of fruit production caused by low soil fertility (Defler and Defler 1996; Defler 2013). In consequence, primates must migrate or search for alternative food resources, which are generally of low energetic reward and difficult to obtain (Moegenburg and Levey 2003). Similarly, periods of high arthropod availability were widely abundant and, in general, corresponded with the rainy season (Gómez-Posada 2012; Boinski and Fowler 1989); during low rainy periods, arthropod numbers decreased drastically. The spatiotemporal variation in invertebrate food resources can cause stress as well as dietary and behavioral changes in insect-eating species (Gursky 2000; Rímoli et al. 2008; Terborgh 1983). For example, in this study, the capuchins' arthropod-capture rate was low when arthropod abundance was low. At the MICBS, during periods of fruit and arthropod scarcity, the consumption of both items was not high, and capuchins complemented their diet with plant parts (such as bark, petioles, leaves, roots) and other items such as termite soil, honey, and small vertebrates (Gómez-Posada 2012). In these scarcity periods, capuchins increased the percentage of time spent in foraging and traveling activities (Gómez-Posada 2009).

Capuchins search more often for preys hiding in tough substrate than for mobile preys or the ones from exposed substrates (Janson and Boinski 1992). This species carefully examines specific substrates, due to the cryptic nature of its prey (Fragaszy et al. 2004); thus, the signal of prey presence is not the arthropod but the substrate itself (Janson and Boinski 1992; Melin et al. 2007). In this study, the majority of arthropod orders we collected were those consumed most frequently by the capuchins (Gómez-Posada 2012), showing a correlation between arthropod abundance and the capuchins' arthropod-capture success rate. Therefore, a manual sampling searching substrates can be an effective method to collect the types of prey capuchins consume (Boinski and Fowler 1989; Melin et al. 2007) and a good indicator of prey availability for these primates.

With such a strong fluctuation on the abundance of arthropods and fruits, we expected that capuchins would increase their overall ranging patterns during periods where both items were scarce, in order to find enough resources to fulfill their energy requirements (Janson 2000; Zhang 1995). Our results partially support this hypothesis, as daily path length and turning angles were larger during the scarcity period, travel and arthropod-foraging speeds were marginally larger, and backtracking and home range size did not vary. Explanations for each of these ecological responses are detailed in the following paragraphs.

Our results did not support our prediction of a significant effect of both arthropod and fruit availability on the daily distance traveled by the focal group (prediction *a*). Ripe fruit abundance alone was the main factor determining the daily path length for the study troop. This is not rare as capuchins establish their daily routes according to the location of their main fruit food sources and move between food patches to search for arthropods (Janson 2000). The distances traveled by the focal group on a daily basis decreased in periods of high fruit-feeding tree density, consistent with the capuchins' tendency of traveling to the nearest accessible food source (Garber 2000).

If fruiting trees are at high densities and aggregated in space during abundance periods, it is highly probable to find food patches at a close distance, resulting in shorter distances traveled in a day (Terborgh 1983; Zhang 1995; Di Bitetti 2001). According to the optimal foraging theory, when feeding patches are abundant, grouped, and at high densities, the distance to the nearest food source is short, the energetic cost of traveling is low, and the reward is guaranteed (Chapman and Chapman 2000; Janson 2000).

Correspondingly, the daily paths of the focal group were longer when the density of fruiting trees decreased. During the fruit and arthropod scarcity period, capuchins at the MICBS tended to increase their daily path length, in order to supply their nutritional requirements and maintain a minimum diet quality (Terborgh 1983; Di Bitetti 2001). During this period, the focal group reduced the consumption of ripe fruits and arthropods, increasing the consumption of alternative plant parts and other items (Gómez-Posada 2012). Then, the increase in the daily routes during fruit scarcity periods could be also of a result of the search for plant parts and other food items that are difficult to obtain and dispersed across their home range, than only a direct effect of fruit scarcity (Terborgh 1983). This response to food shortage is similar to those found for capuchin species in Peru (Terborgh 1983) and Argentina (Di Bitetti 2001).

Our second prediction suggested that the group's home range size would be determined by both arthropod and fruit supply (prediction *b*). Our results for home range size were somehow contradictory, as there were no differences among food availability periods, but it was positively related to fruit abundance. Regarding the last one, we could argue that during fruit scarcity periods, capuchins decreased their home range in order to minimize energy requirements (Zhang 1995). However, we found shorter daily path lengths during food abundance periods, related to a high density of fruit-feeding trees. Consequently, we did not find a relation between home range size and daily path length. For example, we found contrasting average path lengths estimated in months where the home range size was relatively similar and with different food availability (January: 1739.8 m, 84 ha; August: 3269.9 m, 87 ha). For some primates, home range size is more variable than daily path length, and these are not necessarily correlated. Thus, the size of a supplying area on a long-term basis is not a direct reflection of its use on a daily basis (Crockett and Eisenberg 1987). Home range size depends on the interplay of various factors, such as group mass, temperature, seasonal climatic and phenological fluctuations, dispersion and density of food resources, dispersion and availability of other resources as sleeping sites, intra- and interspecific competition, and population density, which could explain differences in home range size between seasons, groups, species, and populations (Di Bitetti 2001; Campos et al. 2014). To explain our results, we probably need to consider in the analysis data on a longer time frame, and to include different predictor variables, such as temperature variations, and availability of other resources, among others.

We expected that when fruiting trees were scattered and not abundant, capuchins would tend to return often to the closest feeding tree, instead of visiting farther ones (Garber 2000), which would result in a higher chance of the group recrossing their own path in a day (prediction *c*). However, we did not find differences in the number of backtrackings among food availability periods, and the model only suggested that backtracking could be negatively related to the dispersion of fruit-feeding trees ($w = 0.40$), but this relation was not significant ($P = 0.09$). By foraging frequently in the same area, food items could decrease to the point of being so scarce that it is difficult to obtain them (functional depletion) (Charnov 1976; Chapman and Chapman 2000). Thus, once food patches became functionally or totally depleted, the group had to look for other feeding sources, avoiding revisiting the depleted ones (Chapman and Chapman 2000). This might explain why we did not find a significant effect of the dispersion of fruit-feeding trees on the group's tendency to recross their route paths.

We expected that travel speed would be low when fruits are abundant and aggregated (prediction *d*). For capuchin monkeys, travel speed is considered to be faster when moving toward known and more rewarding food patches than toward unfamiliar ones (Tujague and Janson 2017). This notion of known/unknown feeding sites has been linked to the monkeys' ability to recall the location of food resources, as a result of previous experiences (Janson 2000; Janson and Byrne 2007). In addition, it has been related to their capacity of perceiving *a priori* the occurrence and productive condition of important fruit-feeding trees, which might be enhanced by sensory (visual, olfactory, auditory) cues (Tujague and Janson 2017). Despite the approach, it is evident that from a specific distance, capuchins can detect the presence, location, and even value (amount and quality) of resources (Tujague and Janson 2017). Therefore, a traveling event between two feeding sources will not be a constant movement but a dynamic one, in relation to spatial attributes. Correspondingly, the study group at the MICBS displayed the common movement pattern observed in capuchins of traveling at a relatively slow speed when leaving a food source and accelerating when approaching the subsequent resource patch (per. obs.) (Janson and Di Bitetti 1997; Tujague and Janson 2017). This behavior is proposed as an individual's response to spatial food competition (scramble competition), for which it will travel faster and race other group members in order to arrive first to the food goal (Janson and Byrne 2007; Tujague et al. 2016). Our results suggest that travel speed tended to decrease when food resources were clumped ($w = 0.32$, $P = 0.06$). As the clumped distribution of fruit-feeding trees corresponded to the abundance period of ripe fruits, the spatial offer of this resource was not rare on a patch scale. Therefore, capuchins did not need to move fast and invest energy in accessing other feeding sites, which resulted in an energetic saving that exceeded their ability to obtain it (Janson 2000). Moreover, as the resources were clumped, scramble competition could be relaxed (Tujague and Janson 2017). However, as we did not consider differences among group members nor the spatiotemporal fluctuations of the movement—being higher in the morning, declining around midday, and increasing later in the evening (Tujague and Janson 2017)—we cannot support our hypothesis and state which factor is conclusive.

During the scarcity period, capuchins tended to move slower during invertebrate foraging; this speed was determined by the arthropod-capture success rate (as expected in prediction *e*) and, surprisingly, also by the fruit-feeding tree density.

Regarding capture rate, capuchins search more often for preys hiding in substrates that require forcefully pulling with the teeth, arms, or even the whole body (Janson and Boinski 1992) and perform different foraging manual actions on the substrate, such as tapping, pounding, ripping, and probing, among others (Pouydebat et al. 2014), which take time. Then, when the arthropod-capture rates are high, the capuchins would move slowly in order to capture, manipulate, and consume the preys. However, if the capture rates were low, the capuchins would search the substrate and, when finding no prey, would continue the search, resulting in a faster moving rate than during periods of abundance. We could not explain why the fruit-feeding tree density also determined the arthropod-foraging speed. This result could be due to a coincidence between high fruit and high arthropod availability periods—capuchins move between fruiting trees, while slowly foraging arthropods among them. More analysis is needed to explain this pattern.

Capuchin at the MICBS traveled more tortuous paths (greater turning angles) during the food scarcity period, and this response was also determined by the arthropod-capture success rate (prediction *f*). As path tortuosity reflects the foraging intensity of an animal (Calengea et al. 2009), the capuchins' tortuous movements indicate a more meticulous and fine-scaled search, allowing a wide scanning of an area (Phiphatsuwannachai et al. 2018). Covering a greater area with this foraging strategy would result in more opportunities to find arthropods during the season of low availability.

The strategies developed by capuchins are widely versatile due to its frugivorous-insectivorous diet. In some studies, fruit availability during periods of scarcity has been proposed as the driver of variations in the genus' ranging behavior; for example, reduction or expansion of the home range area (Peres 1994; Spironello 2001; Terborgh 1983; Zhang 1995; Stevenson et al. 2000). However, based on our results, we could suggest that the contrasting ranging behavioral responses found for *S. apella* in different types of forest are due not only to the distribution and abundance of fruits but also to the availability of arthropods, which is not usually measured. If fruit scarcity in a forest does not happen at the same time as arthropod scarcity, the strategy will be different and probably opposite to the one that is found when the two main resources decrease at the same time. Evidently, while the capuchins' large-scale travels throughout its home range reflect a frugivorous diet, their small-scale movements reveal an insectivorous one. Current hypotheses concerning primate ranging behavior suggest that frugivorous species are expected to have greater day ranges and larger home ranges than folivorous species, and invertebrate consumption is expected to produce a more wide-ranging pattern (Nunn and van Schaik 2002).

Our study illustrates some of the complex ecological processes that determine movement behavior in a capuchin monkey with an insectivorous-frugivorous diet. Fluctuations in the availability and distribution of major food resources shape primate ranging patterns (Gursky 2000); hence, in the case of *S. apella*, both fruits and invertebrates supply should be taken into account when predicting ranging behaviors. The insights gained from these kinds of predictions can be used to understand how animals will respond to a diversity of disturbance and changing scenarios. Therefore, the field of movement ecology is determinant for conservation management as it approaches to not only an animal's motivation to move (e.g., finding food sources, breeding, avoiding adverse conditions or predators) but to the external abiotic and biotic factors influencing its movement (Nathan et al. 2008).

Acknowledgments Research funding was provided by Fundación Natura/COLCIENCIAS. Special thanks to Thomas Defler, director of this project, for his constant support and advice and for the tireless dedication to the study of Colombian primates. To Michael Alberico (†), co-director of this work, for his advice and help. To Conservation International Colombia, for the support provided, especially to Erwin Palacios and Adriana Rodríguez. To Marcelino Días and family, Jaime Castillo, and Ángela Maldonado, for their support in the field. To Isidoro Cabrera and Rodrigo Botina, for the identification of the plant material. To Marcia Muñoz Neyra, for her collaboration in the revision of this document.

References

Altmann J (1974) Observational study of behavior: sampling methods. Behavior 49:227–265

Barton K (2018) MuMIn: multi-model inference. R package version 1.40.4. Available at: https://CRAN.R-project.org/package=MuMIn

Boinski S (1987) Habitat use by squirrel monkeys (*Saimiri oerstedii*) in Costa Rica. Folia Primatol 49:151–167

Boinski S, Fowler N (1989) Seasonal patterns in a tropical lowland forest. Biotropica 21:223–233

Brown A, Zunino G (1990) Dietary variability in *Cebus apella* in extreme habitats: evidence for adaptability. Folia Primatol 54:187–195

Burnham KP, Anderson DR, Huyvaert KP (2011) AIC model selection and multimodel inference in behavioral ecology: some background, observations, and comparisons. Behav Ecol Sociobiol 65:23–35

Calengea C, Draya S, Royer-Carenzia M (2009) The concept of animals' trajectories from a data analysis perspective. Eco Inform 4:34–41

Campos FA, Bergstrom ML, Childers A, Hogan JD, Jack KM, Melin AD, Mosdossy KN, Myers MS, Parr NA, Sargeant E, Schoof VAM, Fedigan LM (2014) Drivers of home range characteristics across spatiotemporal scales in a Neotropical primate, *Cebus capucinus*. Anim Behav 91:93–109

Chapman C (1988) Patterns of foraging and range use by three species of neotropical primates. Primates 29:177–194

Chapman CA, Chapman LJ (2000) Determinants of group size in primates: the importance of travel costs. In: Boinski S, Garber PA (eds) On the move: how and why animals travel in groups. University of Chicago Press, Chicago, pp 24–41

Chapman CA, Bortolamiol S, Matsuda I, Omeja PA, Paim FP, Reyna-Hurtado R, Sengupta R, Valenta K (2018) Primate population dynamics: variation in abundance over space and time. Biodivers Conserv 27:1221–1238

Charnov EL (1976) Optimal foraging: the marginal value theorem. Theor Popul Biol 9:129–136

Crockett CM, Eisenberg JF (1987) Howlers: variations in group size and demography. In: Crockett CM, Eisenberg JF, Smuts BB (eds) Primate societies. The University of Chicago Press, Chicago, pp 54–68

Defler TR (2010) Historia Natural de los Primates Colombianos. Universidad Nacional de Colombia, Bogota

Defler TR (2013) Species richness, densities and biomass of nine primate communities in eastern Colombia. Rev Acad Colomb Cienc Exactas Físicas Natur 37(143):253–262

Defler TR, Defler SB (1996) Diet of a group of *Lagothrix lagotricha lagotricha* in southeastern Colombia. Int J Primatol 17(2):161–190

Di Bitetti M (2001) Home range use by the tufted capuchin monkey (*Cebus apella nigritus*) in a subtropical rainforest of Argentina. J Zool Lond 253:33–45
Dalgaard P (2010) R Development Core Team: R: A language and environment for statistical computing. Computer programme, Retrieved from http://www.R-project.org/
Fragaszy DM, Visalberghi E, Fedigan LM (2004) The complete capuchin: the biology of the genus cebus. Cambridge University Press, Cambridge, UK
Garber PA (2000) Evidence for the use of spatial, temporal, and social information by primate foragers. In: Boinski S, Garber PA (eds) On the move: how and why animals travel in groups. University of Chicago Press, Chicago, pp 261–298
Gómez-Posada C (2003) Estrategias de forrajeo de un grupo provisionado de mico maicero, *Cebus apella* (Primate: Cebidae) en el Parque Nacional Natural Tinigua, Colombia. In: Pereira V, Nassar F, Savage A (eds) Primatología del Nuevo Mundo: Biología, medicina, manejo y conservación. Centro de Primatología Araguatos, Bogotá, pp 136–146
Gómez-Posada C (2009) *Cebus apella*: variación del patrón de actividad de acuerdo a la oferta de frutos y artrópodos. In: Alarcón-Nieto G, Palacios E (eds) Estación Biológica Mosiro Itajura-Caparú. Conservación Internacional Colombia, Bogotá, pp 179–185
Gómez-Posada C (2012) Dieta y comportamiento alimentario de un grupo de mico maicero *Cebus apella* de acuerdo a la variación en la oferta de frutos y artrópodos en la Amazonía colombiana. Acta Amazon 42(3):363–372
Gursky S (2000) Effect of seasonality on the behavior of an insectivorous primate, Tarsius spectrum. Int J Primatol 21(3):477–495
Janson CH (2000) Spatial movement strategies: theory, evidence, and challenges. In: Boinski S, Garber PA (eds) On the move: how and why animals travel in groups. University of Chicago Press, Chicago, pp 165–203
Janson CH, Boinski S (1992) Morphological and behavioral adaptations for foraging in generalist primates: the case of the Cebinae. Am J Anthropol 88:483–498
Janson CH, Byrne R (2007) What wild primates know about resources: opening up the black box. Anim Cogn 10(3):357–367
Janson CH, Di Bitetti MS (1997) Experimental analysis of food detection in capuchin monkeys: effects of distance, travel speed and resource size. Behav Ecol Sociobiol 41:7–24
Ludwig J, Reynolds J (1988) Statistical ecology: a primer on methods and computing. Wiley, New York
Lynch Alfaro JW, Silva JdSE Jr, Rylands AB (2012) How different are robust and gracile capuchin monkeys? An argument for the use of *Sapajus* and *Cebus*. Am J Primatol 74:273–286
McNab BK (2002) The physiological ecology of vertebrates: a view from energetics. Cornell University Press, Ithaca
Melin AD, Fedigan LM, Hiramatsu C, Sendall CL, Kawamura S (2007) Effects of colour vision phenotype on insect capture by a free- ranging population of white-faced capuchins, *Cebus capucinus*. Anim Behav 73:205–214
Moegenburg S, Levey D (2003) Do frugivores respond to fruit harvest? An experimental study of short-term responses. Ecology 84(10):2600–2612
Nathan R, Getz WM, Revilla E, Holyoak M, Kadmon R, Saltz D, Smouse PE (2008) A movement ecology paradigm for unifying organismal movement research. Proc Natl Acad Sci U S A 105(49):19052–19059
NRC: National Research Council (1981) Techniques for the study of primate population ecology. National Academy Press, Washington, DC
Nunn CL van Schaik CP (2002) A comparative approach to reconstructing the socioecology of extinct primates. Plavcan JM, Kay RF, Junger WL van Schaik CP Reconstructing behavior in the primate fossil record. Kluwer Academic/Plenum, New York, 159–215
Peres C (1994) Primate responses to phenological changes in an Amazonian terra firme forest. Biotropica 25(1):98–112
Phiphatsuwannachai S, Westcott DA, McKeown A, Savini T (2018) Inter-group variability in seed dispersal by white-handed gibbons in mosaic forest. Biotropica 50:106–115

Pouydebat E, Borel A, Chotard H, Fragaszy D (2014) Hand preference in fast-moving versus slow-moving actions in capuchin, *Sapajus* spp., and squirrel monkeys, *Saimiri sciureus*. Anim Behav 97:113–123

Raemaekers J (1980) Causes of variation between months in the distances traveled daily by gibbons. Folia Primatol 34:46–60

Reyna-Hurtado R, Teichroeb JA, Bonnell TR, Hernández-Sarabia RU, Vickers SM, Serio-Silva JC, Sicotte P, Chapman CA (2018) Primates adjust movement strategies due to changing food availability. Behav Ecol 29(2):368–376

Rímoli J, Strier K, Ferrari S (2008) Seasonal and longitudinal variation in the behavior of free-ranging black tufted capuchins *Cebus nigritus* (Goldfuss, 1809) in a fragment of Atlantic Forest in Southeastern Brazil. In: Ferrari S, Rímoli J (eds) A Primatologia no Brasil (9). Sociedade Brasileira de Primatologia, Aracaju, pp 130–146

Robinson JG (1986) Seasonal variation in use of time and space by the wedge-capuchin monkey, *Cebus olivaceus*: implications for foraging theory. Smithson Contrib Zool 431:1–60

Spironello W (2001) The brown capuchin monkey (*Cebus apella*): ecology and home range requirements in central Amazonia. In: Bierregard RO, Gascon C, Lovejoy TE, Mesquita R (eds) Lessons from Amazonia. Yale University Press, New Haven, pp 271–283

Stevenson P, Quiñones M, Ahumada J (2000) Influence of Fruit Availability on Ecological Overlap among Four Neotropical Primates at Tinigua National Park, Colombia. BIOTROPICA 32(3):533–544

Terborgh J (1983) Five New World Primates: a study in comparative ecology. Princeton University Press, New Jersey

Tujague MP, Janson CH (2017) Wild capuchin monkeys anticipate the amount of ripe fruit in natural trees. Anim Cogn 20:841–853

Tujague MP, Bacigalupe MA, Lahitte HB, Janson CH (2016) Memoria espacial en monos capuchinos de Argentina: un estudio observacional en vida silvestre. Rev Argent Antropol Biol 18(1):1–13

Vasconcellos A, Andreazze R, Almeida AM, Araujo HFP, Oliveira ES, Oliveira U (2010) Seasonality of insects in the semi-arid Caatinga of northeastern Brazil. Rev Bras Entomol 54:471–476

Wallace RB (2008) Factors influencing spider monkey habitat use and ranging patterns. In: Campbell C (ed) Spider monkeys: behavior, ecology, and evolution of the genus Ateles. Cambridge University Press, Cambridge, UK, pp 138–154

Williams SE, Middleton J (2008) Climatic seasonality, resource bottlenecks, and abundance of rainforest birds: implications for global climate change. Divers Distrib 14:69–77

Winkler DW, Jørgensen C, Both C, Houston AI, McNamara JM, Levey DJ, Partecke J, Fudickar A, Kacelnik A, Roshier D, Piersma T (2014) Cues, strategies, and outcomes: how migrating vertebrates track environmental change. Mov Ecol 2:10

Zar J (1996) Biostatisical analysis. Prentice Hall, New Jersey

Zhang S (1995) Activity and ranging patterns in relation to utilization by brown capuchins (*Cebus apella*) in French Guiana. Int J Primatol 16:489–507

Chapter 13
Insights of the Movements of the Jaguar in the Tropical Forests of Southern Mexico

J. Antonio de la Torre and Marina Rivero

13.1 Introduction

Movement is a critical animal behavior which reflects the animal response to its current biological needs and to its environment. Movement can maximize probability of survival in animals, because it can ensure access to resources such as food and water, the evasion of predators and competitors, and to find mates in reproduction (Morales et al. 2010). Understanding how and why animals use the available space and the underlying drivers of animal movements is essential to the management and conservation for both the species and the ecosystems where they inhabit (Allen and Singh 2016). The disruption of the natural patterns of animal movement can have important implications for ecosystem dynamics and function, including disturbances to trophic chains and nutrients cycling process (Ning et al. 2016; Tucker et al. 2018). Studies about movement ecology have grown in recent years with the improvement of tracking devices and the analytical tools, and these had provided unprecedented insights to understand where, when, and why the species move across the landscapes (Cagnacci et al. 2010a, b; Tomkiewicz et al. 2010).

Most apex predators in terrestrial ecosystems are facing dramatic population declines and range reduction (Ripple et al. 2014, 2016; Treves and Bruskotter 2014). Understanding the details of how these species move and use the space is essential to develop and implement effective conservation plans to ensure their long-term

J. A. de la Torre (✉)
Instituto de Ecología, UNAM, Laboratorio de Ecología y Conservación de Vertebrados Terrestres, Ciudad de México, Mexico

Bioconciencia A.C., Ciudad de México, Mexico
e-mail: adelatorre@iecologia.unam.mx

M. Rivero
Bioconciencia A.C., Ciudad de México, Mexico

El Colegio de la Frontera Sur, Campeche, Campeche, Mexico

R. Reyna-Hurtado, C. A. Chapman (eds.), *Movement Ecology of Neotropical Forest Mammals*, https://doi.org/10.1007/978-3-030-03463-4_13

conservation (Chetkiewicz et al. 2006; Lehmann et al. 2008; Marker et al. 2008; de la Torre et al. 2017a; Tucker et al. 2018). The jaguar (*Panthera onca*) is the largest felid in the Neotropics and probably the least studied species of the *Panthera* genus (de la Torre and Medellín 2011). Historically, jaguars ranged from southwestern United States to central Argentina (Seymour 1989; Sanderson et al. 2002). However, this species has been extirpated from more than half of its original range over the last 100 years, and recent conservation assessments have concluded that jaguars are declining in much of its remaining range (Swank and Teer 1989; Rabinowitz and Zeller 2010; Medellín et al. 2016; de la Torre et al. 2018). Though the jaguar is not endangered as the same level of the other species of large felids such as lions, tigers, and snow leopards, the most current evaluations of the conservation status of this species have shown that most of its populations are threatened (de la Torre et al. 2018). The jaguar currently is listed as Near Threatened by the IUCN Red List (Quigley et al. 2017). However, fragmentation, habitat conversion, and poaching in retaliation to cattle predation are severe threats for the species in most of its distribution range (Nowell and Jackson 1996; Sanderson et al. 2002; Haag et al. 2010; Medellín et al. 2016; Quigley et al. 2017).

Diverse factors determine the patterns of movement and space utilization of solitary large felids such as the jaguar. For instance, home range sizes and patterns of space use in these animals are usually determined by their body size (Gittleman and Harvey 1982; Jetz et al. 2004; Carbone et al. 2005, 2007; Macdonald et al. 2010), prey availability across the space and time (Herfindal et al. 2005; Odden and Wegge 2005; Marker et al. 2008; Schmidt 2008), habitat suitability (Spong 2002; Broomhall et al. 2003; Conde et al. 2010), and sociality (Logan and Sweanor 2001; Azevedo and Murray 2007; Cavalcanti and Gese 2009; Goodrich et al. 2010). Furthermore, other factors such as interactions with other large carnivores can influence patterns of space use in large felids (Scognamillo et al. 2003; Harmsen et al. 2009; Odden et al. 2010; Bhattarai and Kindlmann 2012; Noss et al. 2012; Foster et al. 2013). Landscape configuration and connectivity are critical aspects that determine the space use in large felids, as well (Colchero et al. 2009; Conde et al. 2010; de la Torre et al. 2017a).

Due to its natural history and secretive behavior, studying the movements of large felids is expensive and logistically challenging, especially in the tropical forests. Only a few studies aimed at large felids in tropical forests have been able to capture and monitor several individuals to give relevant conclusions (Rabinowitz and Nottingham 1986; Jenny 1996; Karanth and Sunquist 2000; Ceballos et al. 2002; Odden and Wegge 2005; Nuñez 2006; Figueroa 2013; de la Torre et al. 2017b). Most of the former studies focused on evaluating jaguar spatial ecology have tried to describe the home range areas and spatial organization (Schaller and Crawshaw 1980; Rabinowitz and Nottingham Jr. 1986; Crawshaw and Quigley 1991; Ceballos et al. 2002; Cavalcanti and Gese 2009; Chávez 2009; Morato et al. 2016; de la Torre et al. 2017b) and jaguar habitat use (Conde et al. 2010; Cullen et al. 2013; de la Torre et al. 2017a). Other aspects of jaguar movement ecology that have been previously addressed are the assessment of their movement rates with respect to different variables such as time of day, moon phases, sex, etc. (Cavalcanti and Gese 2009; de la Torre et al. 2017b), as well as understanding the movement decisions of jaguars in order to predict the most appropriate areas to maintain the connectivity between populations or patches of suitable habitat (Colchero et al. 2011; de la Torre et al. 2017).

With the aim to contribute in the knowledge of the spatial ecology of the jaguar in the tropical forests, in this chapter we summarize all information that has been obtained about the movements of this species for the Greater Lacandona Ecosystem, Southern Mexico. This jaguar population is one of the most studied in Mexico (de la Torre and Medellín 2011; de la Torre and Rivero 2017; de la Torre et al. 2017a, b; Towns et al. 2017). Formerly, we published in different sources some information related to movements of these jaguars tracked in this region. We analyzed these data comparing the spatial ecology of jaguars and pumas in our study area with the aim to compare the spatial requirements of these species in the tropical forest of Southern Mexico (de la Torre et al. 2017b). Additionally, we identified the suitable habitat and the functional corridors for jaguars to design a strategy to maintain connectivity in the Southern Mayan Forest, which spans the border of Mexico and Guatemala (de la Torre et al. 2017a). With this analysis we gave recommendations to decision-makers in Mexico and Guatemala to invest in specific sites for conservation, management programs, and habitat restoration to ensure the long-term viability of the jaguar population of the Mayan Forest (de la Torre et al. 2017a).

In order to explore other aspects of the movement ecology of jaguars in tropical forests, this chapter had the following objectives: (1) to evaluate jaguar home range sizes and movement rates in the tropical rainforests of Southern Mexico using new methods; (2) to evaluate the periodicity of jaguar movement, analyzing the frequency and the intensity with which they visit certain sites within their home range; and (3) to evaluate movement decisions of jaguars across different landscape features. The questions that we intended to answer are: (1) How do the jaguars use the space within their home range area in the Southern Mexico? and (2) Does the landscape configuration influence the behavior and movement patterns of jaguars? We expect that this analysis help to improve the understanding of the movement ecology of jaguars in tropical forests and increase the knowledge of this species to advance in its conservation throughout the tropical forest of Central America.

13.2 Methods

13.2.1 Study Area

Our study area was located in the Greater Lacandona Ecosystem in southeastern Mexico. This region contains the largest remaining portion of tropical humid rainforest in Mexico and is considered one of the most biodiverse regions of the country. The Greater Lacandona Ecosystem is part of what is called the "Mayan Forest," a trinational forest shared by Mexico, Guatemala, and Belize that holds important cultural and natural associations (Medellín 1994; de la Torre and Medellín 2011). The Mayan Forest is crucial for the conservation of different species of large wildlife, because it is one of the few landscapes in Mesoamerica that is large enough to maintain viable populations of large mammals such as jaguars, white-lipped peccaries (*Tayassu pecari*), and Baird's tapirs (*Tapirus bairdii*) (March 1993; Medellín 1994; Matola et al. 1997; Sanderson et al. 2002; Naranjo et al. 2015; Reyna-Hurtado et al. 2017). Despite its importance for the conservation of numerous species, the Greater Lacandona Ecosystem continues to be subjected to anthropogenic destruction. Of its original 1,500,000 ha of rainforest, 2/3 has been lost due to human impact in the past 40 years (Mendoza and Dirzo 1999; Jong et al. 2000). The main threats for this ecosystem are the rapid growth of human populations, deforestation, unregulated extraction of flora and fauna, and the illegal use and extraction of natural resources from nature reserves (Medellín 1994; Davis et al. 1997; Mendoza and Dirzo 1999).

There are seven natural protected areas within the Greater Lacandona Ecosystem. These cover the largest forested areas of this region: Montes Azules (3312 km^2), Lacantún (619 km^2), Bonampak (48 km^2), Yaxchilán (26 km^2), Chan-kin (122 km^2), Naha (38 km^2), and Metzabok (33 km^2). Our study area was located in the north of Greater Lacandona Ecosystem and covered approximately 2500 km^2, including Yaxchilán (Usumacinta River as the border with Guatemala on the north), Bonampak, Lacantún, and Montes Azules Natural Protected Areas (Fig. 13.1). The study area also encompassed the Sierra la Cojolita, a tract of natural habitat in the

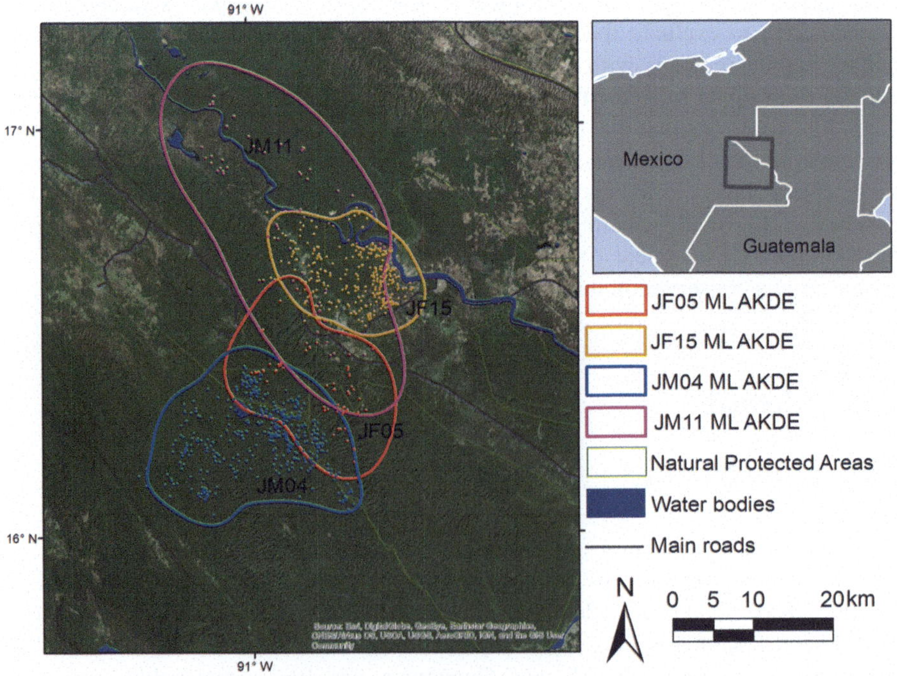

Fig. 13.1 Location of our study area and home ranges of jaguars in Southern Mexico using the autocorrelated kernel density estimation

north of Greater Lacandona Ecosystem that is protected by local communities. Three indigenous communities inhabit the Mexican part of our study area: Lacanja Chansayab (~1000 people), Nueva Palestina (~20,000), and Frontera Corozal (~15,000). The main economic activities in this region include ranching, farming, and ecological and archeological tourism. Thus, the landscape outside the protected areas is composed of a matrix of primary forest surrounded by secondary forest, agricultural fields, and managed grasslands for livestock. The climate is hot (mean of 25 °C), and the mean rainfall per year is 2800 mm, with the greatest concentration in June–September and the lowest in March–April (reviewed by O'Brien 1998).

Among the species that potentially are important prey for jaguars in the area are collared peccary (*Pecari tajacu*), white-lipped peccary, the brocket deer (*Mazama temama*), white-tailed deer (*Odocoileus virginianus*), agouti (*Cuniculus paca*), paca (*Dasyprocta punctata*), and the nine-banded armadillo (*Dasypus novemcinctus*). Our field base camps to implement this study were located in the community of Lacanja Chansayab and in the CONANP (National Commission of Natural Protected Areas) field station at Yaxchilán Natural Protected Area.

13.2.2 Data Collection

We captured jaguars using foot snares (Frank et al. 2003). To respond immediately to any animal capture, we also placed a VHF radio transmitter at the site of each snare trap to monitor if traps were triggered (Halstead et al. 1995). We checked the traps 4 h throughout the night and, depending on weather conditions, several times during the day. We conducted three trapping seasons, one in Bonampak (November 2011 to January 2012) and two in Yaxchilán (July to September 2012 and February to April 2013). In Bonampak, we covered an area of approximately 60 km^2 with foot snares, and in Yaxchilán we covered an area of 25 km^2. We used 8–15 foot snares during each trapping day. All capture and handling protocols followed the Institutional Animal Care and Use Committees guidelines approved by the American Society of Mammalogists (Sikes et al. 2011).

Upon capture, we immobilized jaguars using medetomidine (0.08 mg/kg) combined with ketamine (5 mg/kg) using a dart fired from a CO_2 pistol or rifle. While immobilized, we examined individual body condition and determined sex. We estimated age based on coat color, tooth wear (Stander 1997), and gum-line recession (Laundré et al. 2000). Body mass and linear measurements were recorded (de la Torre and Rivero 2017). Weight was recorded using a portable scale.

Captured jaguars were fitted with a satellite GPS collar (Telonics®, GEN IV, model TGW-4580). This model included a component for sending information through the ARGOS satellite system. We programmed the global positioning system collars to acquire a location every 4.8 h (four locations/day) and to send data packets through the ARGOS system every 4 days. All collars included a programmable release mechanism (model CR-2a, Telonics®), and we scheduled the release of the collars 12–14 months after jaguars were captured. The collars were recovered, when possible, using the locations obtained after their release through the ARGOS system and searching the VHF pulses using a receiver (model TR4, Telonics®) and antennas of two and three elements (models RA23K and RA3, Telonics®). For all analyses, we used only the 3D GPS fixes that we obtained from collared animals that were calculated from four or more GPS satellites that provided a location estimate with a typical accuracy of 2–10 m.

Our trapping capture rate of jaguars was 2.5 captures per 1000 trap night (total trapping effort = 1,934 trap nights). We captured three females and two males. The female jaguars weighted in average 34.3 ± 1.5 kg and the males 52.5 ± 3.5 kg (de la Torre and Rivero 2017). We successfully deployed satellite global positioning system collars on these five jaguars. A total of 1134 3D GPS fixes were obtained from the collared animals and mean horizontal error was 5.01 ± 4.61 m. However, the fix success of the GPS collars was very low for most of the jaguars probably due to the closed forest canopy, since other studies (Gitzen et al. 2013; Camp et al. 2016) have shown that the performance of the GPS radio collars devices diminishes in tropical forests with multiple canopy layers or in areas with steep terrain. Furthermore, we only recovered three of the collars deployed (JF12, JF15, and JM04), and we could not recover one collar (JF05), because after the release mechanisms were activated, we did not obtain fixes through the ARGOS satellite system to search for this collar. We also never found it using the VHF signal in the field to search for this collar after the release mechanism was activated. Additionally, we lost the signal of jaguar male JM11 after 5 months of tracking through the ARGOS satellite system. Thus, data shown for animals JF05 and JM11 corresponded only to data recovered through the ARGOS satellite system (Table 13.1). We obtained a very few data of the female jaguar JF12, because she was shot 20 days after the capture (Table 13.1).

Table 13.1 Movement parameters and home range sizes for the jaguars tracked in the Greater Lacandona Ecosystem, southern Mexico

Animal ID	Sex	Age class	Weight	Number days/ fixes	Home range crossing time (day)	Velocity autocorrelation time scale (h)	Average distance traveled (km/day)	95% KDE (km²)	AKDE (km²) (95% CI)
JF05	♀	Adult	34	331/88	9.9 (3.6–27.2)	1.65 (0.31–8.5)	7.2 (5.1–9.3)	211.6	297.2 (163.4–470.4)
JF15	♀	Adult	33	350/395	4.6 (3.1–6.7)	2.7(2.0–3.6)	6.9 (6.3–7.5)	198.8	203.4 (151.0–263.4)
JM04	♂	Adult	50	367/578	4.4 (3.2–5.9)	4.2 (3.4–5.1)	8.1 (7.6–8.5)	322.8	375.7 (294.8–466.4)
JM11	♂	Adult	55	133/72	4.7 (1.8–12.2)	5.2 (2.9–9.2)	9.6 (8.2–11.0)	581.9	714.1 (412.3–1097.4)

Home ranges were estimated using 95% kernel density estimator (KDE) and autocorrelated kernel density estimator (AKDE)

13.3 Analysis

13.3.1 Home Range Sizes and Movement Rates of Jaguars in Southern Mexico

One of the basic issues in the study of movement patterns of the wild animals is to define the area used by individuals in its normal activities for food acquisition, mating, and rearing progeny. This area is known as "home range" and generally is estimated aggregating the telemetry locations to evaluate the total use of the space by an animal (Powell and Mitchell 2012). To estimate home range sizes of jaguars in the Greater Lacandona Ecosystem, we used continuous-time stochastic movement models and the autocorrelated kernel density estimation (AKDE), which allows more accurate home range estimation, even if the data are strongly correlated. We calculated variograms, movement models, and estimated home range sizes using the *ctmm* package of the statistical software R (R Core Team 2016). Using the command *variogram*, we plotted the estimated semi-variance function for each animal to visually inspect the autocorrelation structure. We also evaluate the residence of the individuals tracked using the variograms. We expected that resident individuals reached an asymptote on timescale that roughly corresponds to the home range crossing time, and individuals whose plotted semi-variance did not reach an asymptote were not assumed to be home range residents (Morato et al. 2016).

The *ctmm* package allows to fit several movement models and select the best that describe the movement patterns of the individual studied. The *ctmm* package includes four basic movement models: (1) Brown motion (BM), which is based on the regular diffusion, where the velocities are uncorrelated and the space use is not limited. This model is used for data where the time between locations is very broad so it cannot reveal the autocorrelation velocity and too brief to show the residency areas. (2) Ornstein-Uhlenbeck (OU) movement model is a combination between the BM, but in this case the movement tends to remain in a certain area. The OU model fit to data that has no evidence of directional persistence; nevertheless the use of space is restricted. (3) Integrated OU (IOU) model is similar to BM in that it is a process that diffuses continuously, but it has autocorrelated velocities in short time scales. This model is better for data that show directional persistence, but the time is not enough to show a range residency. Finally, the Ornstein-Uhlenbeck Foraging (OUF) model is a combination between the OU and the IOU; this means that the movement tends to have correlated velocities and restricted use of space. This model is appropriated for data that has a very detail information to show velocity autocorrelation and long enough to show range residency (Calabrese et al. 2016).

OU models are described by two parameters, which are home range crossing time (days) and variance (km^2), and it allows to estimate the home range and home range crossing time, while the OUF models are described by three parameters which are home range crossing time, velocity autocorrelation (hours), and variance (km^2) and allow to estimate home range, home range crossing time, velocity autocorrelation time scale, and average distance traveled for each individual. We fitted the models using the function *ctmm.fit*, which use maximum likelihood, and we ranked the models based on AICc using the function *ctmm.select*. We estimated the home range based on the best fitted model for each individual using the function *akde*.

For comparative proposes, we estimated home ranges using conventional kernel density estimators at 95% for each animal (Worton 1989). For this we also fit in the *ctmm* the independent identically distributed (IID) model, which by definition ignores autocorrelation in the data, and then we adjusted the fitted model to the *akde* function. We compared our jaguar home range estimations with those obtained in other studies. To compare the movement rates between female and male jaguars, we used the distance average traveled by each individual estimated using the OUF models (Morato et al. 2016).

13.4 Frequency and Intensity of Visits of Jaguars to Certain Sites Within Home Ranges

Traditionally, the space use has been studied through classical kernel methods for estimating the utilization distribution within an animal home range (Worton 1989). These methods consider locations as static and independent points rather than a movement process. However, the use of GPS radio tracking devices on wild animals provides information of sequential and correlated localizations that can be used to have a more dynamic approach to describe the space use and habitat selection. The movement-based kernel approach incorporates movement information and considers the time spent between successive relocations to analyze the trajectories of the individuals. This approach seeks to describe the intensity in which the individual actively uses the habitat patches within its home range, rather than the simple area occupied (Benhamou 2011; Benhamou and Riotte-Lambert 2012). Therefore, the movement-based kernel approach is focused in estimating the utilization distribution (UD) and additionally considers that the intensity of the space use is proportional to the activity time spent by unit area (Benhamou and Cornélis 2010).

The movement-based kernel approach can be analyzed under the biased random bridge (BRB) model. This BRB method is based on the biased random walk model, and its main characteristic is that it not only considers the diffusive movement but also includes the advection component; this means that the model considers the drift between successive relocations in the trajectory analysis (Benhamou 2011). Additionally, the BRB model decomposes the use of space in two components, since it identifies the areas that the animal might visit only once but stays for a long time, the intensity distribution (ID), and the areas that the animal visits repeatedly, the recursion distribution (RD), inside their home range. These components can be very helpful to understand how the animals select and make use of space within their home range areas (Benhamou and Cornélis 2010; Benhamou 2011; Benhamou and Riotte-Lambert 2012).

Using the BRB approach, we estimated the space use of two individuals of jaguar (JF15 and JM04) for our study area. We only used these individuals because they were the ones which we obtained a larger number of fixes. The analysis was conducted using adehabitatHR (Calenge 2013a) of the statistical software R (R Core Team 2016). To calculate the UD, ID, and RD, we first estimated the diffusion coefficient (*D*) (Benhamou 2011). The *D* determines the smallest distance below which we consider that the animal is not moving; for our data we used a maximum time (T_{max}) = 24 h and a minimum distance (L_{min}) = 200 m to calculate *D*. The minimum smoothing parameter h_{min} was set to 100 m, and the radius of circle (radius) of the resident time and the number of visit used for the model was set to 1000 m (for more description regarding the model, see Calenge 2013a). We defined up to 95% cumulative frequency isopleths for UD, because it is better for analyzing the global home range area use, and 30% of isopleth for the ID and RD, because we were interested in identifying the highest density values for these distributions (intensively exploited/repeatedly visited; Benhamou and Riotte-Lambert 2012).

13.5 Movement Decisions of Jaguars Across Different Landscape Features

To understand how the landscape structure of our study area affects jaguar movements, we analyzed the sequential locations obtained from the jaguars tracked in our study using step selection functions (Fortin et al. 2005; Thurfjell et al. 2014; Ziółkowska et al. 2016). Step selection functions are relative new useful models for studying resource selection by animals moving through the landscape, and these compare environmental attributes of observed steps (the linear segment of two consecutives fixes) with alternative random steps taken from the same starting point. This approach has been used to study habitat selection, human wildlife interactions, movement corridors, and dispersal behaviors in carnivore and ungulate species (Thurfjell et al. 2014). With the aim to evaluate how the landscape configuration of the Greater Lacandona Ecosystem affects the movement decision of jaguars, we compiled a geospatial dataset using a suite of environmental and anthropogenic variables for our study area (Table 13.2). Then, for each observed step, we calculated its length (*d*) and turning angle (α) using the package "adehabitatLT" (Calenge 2013b) from R 3.1.1 (R Core Team 2016). Steps were divided into "active" and "passive," based on step length (Ziółkowska et al. 2016), with all steps ≥500 m constituting active steps. We only used the active steps for implementing the step selection function analysis.

Each active step was paired with 100 control steps that shared the same starting point but differed either in length, direction, or both. The length and turning angles of control steps of a given individual jaguar were sampled from those observed of the other individuals to avoid problems of circularity (Fortin et al. 2005). We used the command "movement.ssfsamples" of the "Geospatial Modelling Environment" package to generate the control steps (Beyer 2012). For each observed and control

Table 13.2 Landscape variables used to evaluate the potential habitat for jaguars in the Greater Lacandona Ecosystem through resource selection probability functions

Type	Variable name	Abbreviation	Description	Units	Justification
Forest cover	Forest cover 30 m	FCov-30	Percentage of forest coverage around the pixel	%	Jaguars in tropical forest are restricted mostly to areas of primary forest where they find their natural prey and refuge (Rabinowitz and Nottingham 1986; Conde et al. 2010; Cullen et al. 2013). To appraise habitat quality, we used the high-resolution global forest map (Hansen et al. 2013). This dataset includes the extent of global tree cover density, which is defined as vegetation higher than 5 m and is expressed as a percentage (Hansen et al. 2013). Given that habitat use and movement by jaguars might be limited by the amount of surrounding forest, as it is by other large predator species (Beier 1995; Dickson et al. 2005), we estimated forest cover in a neighborhood of 240 m (FCov-240) and 510 m (FCov-510) around each raster pixel
	Forest cover 240 m	FCov-240	Percentage of forest coverage 240 m around the pixel	%	
	Forest cover 510 m	FCov-240	Percentage of forest coverage 510 m around the pixel	%	
Terrain	Topographic position index	TPI	Classification of landscape according to the slope position within a different radius. This index incorporated richness and evenness into a single measure	–	Jaguar habitat use and movement are affected by different terrain conditions. Jaguar mobility would be facilitated in areas of high topographic complexity, especially if the flat areas of suitable habitat have been cleared (Dickson et al. 2005; Morato et al. 2014). However, jaguars are frequently associated with lowland areas, and jaguar occupancy and movements would be hampered by the mountain ranges at higher altitudes (Rabinowitz and Nottingham 1986; Rabinowitz and Zeller 2010; Zeller et al. 2011). Topographic position index was used to characterize the slope's position in the terrain and landform. This index incorporated richness and evenness into a single measure. Using a 30 m digital elevation model, we obtained elevation ranges. Shannon index was used to evaluate differences in heterogeneity for ruggedness. In general, higher values reflected more diversity and a better balance among unique landform values (Riley et al. 1999)
	Elevation	ELE	Elevation	m.a.s.l.	
	Shannon topographic index	SHAN	Differences of ranges of elevation values within a different radius	–	

Water runoffs	Distance to water courses	DistW	The minimum distance to the nearest water runoff	km	Jaguars are known to use riparian habitats to move through landscapes and use sites with permanent water more frequently (Schaller and Vasconcelos 1978; Rabinowitz and Nottingham 1986; Emmons 1987; Nuñez 2006). We constructed this raster using a layer of the hydrological drainage system of the entire study area (Tapia and Nuñez 2008)
Human	Distance to towns	DistT	The minimum distance to the nearest town	km	Human activity affects habitat use by jaguars negatively due to disturbance and persecution (Conde et al. 2010; Colchero et al. 2011; Espinosa et al. 2018). Information of towns and roads was obtained from INEGI (Instituto Nacional de Estadistica e Informatica) for Mexico and IGN-SEGLAPAN (Instituto Geográfico-Secretaria de Planificación y Programación de la Presidencia) for Guatemala. To evaluate the distances that jaguars use from the boundaries that separated the forest from open modified areas, we used a raster layer with the minimum distance to the nearest deforested patches $<1\ km^2$ since the average size of the cleared areas transformed to create grassland for livestock is $1\ km^2$ in our study area (equivalent to 100 ha)
	Distance to paved roads	DistR	The minimum distance to the nearest paved roads	km	
	Distance to deforestation edge	DistD	The nearest distance to the deforested patch $>1\ km^2$	km	

step, we calculated the exact values of the predictors covariables at the endpoint of steps (Thurfjell et al. 2014).

We constructed step selection function models with the package "ResourceSelection" (Lele et al. 2014) from R 3.1.1 (R Core Team 2016) using a conditional regression with log link, which is adequate to construct this kind of models (Thurfjell et al. 2014). The step selection functions were fit using the observed steps matched to their respective control steps (Lele and Keim 2006; Northrup et al. 2013). We tested 54 step selection function models with different combinations of explanatory variables (Table 13.5). Then, we used the Akaike information criterion (AIC) to identify the best step selection function models (Burnham and Anderson 2002). We considered models comparable if ΔAIC was <2.0, and we compared the AIC weights (wi) to determine the most appropriate models that described jaguar movements.

13.6 Results

13.6.1 Home Range Sizes and Movement Rates of Jaguars in Southern Mexico

After the preliminary analysis with the variograms, we only calculated the home ranges of four jaguars tracked (2 ♀, 2 ♂), because the variogram of the female jaguar JF12 did not reach the asymptote. The best movement model, for the four jaguars that we estimated the home ranges, was the OUF. The examination of our home range estimations showed that jaguars use vast annual areas in the tropical forests in Southern Mexico. Mean annual home range for female ranges from 203 to 297 km^2 and for male jaguars from 375 to 714 km^2 (Fig. 13.1 and Table 13.1). Home range sizes of jaguars in Southern Mexico were similar to other studies carried on in tropical rainforests, which ranged from 10 to 268 km^2 for females and from 34 to 624 km^2 for males (Table 13.3). However, some of these estimations were obtained using VHF tracking technique and conventional tools to estimate home ranges such as minimum convex polygons and kernel density estimator which suggest that some of these were underestimated (Morato et al. 2016).

13.6.2 Frequency and Intensity of Visits of Jaguars to Certain Sites Within Home Ranges

The results of the BRB model allowed identifying the areas more intensively used and the areas more frequently visited within the home range areas of the female jaguar JF15 and the male jaguar JM04. It is plausible that areas intensively used by both jaguars were areas associated with a killing of large prey event, since jaguars

Table 13.3 Jaguar home range estimates (in km^2) obtained from different studies using VHF and GPS tracking techniques and the autocorrelated kernel density estimation (AKDE), kernel density estimator (KDE), and minimum convex polygon (MCP). Also are included the countries, regions, and habitat types

Study area	Country	Habitat type	Tracking technique	Estimation method	Males	Females	♂/♀	References
Mexican Pacific	Mexico	Tropical dry forest	VHF	MCP 100%	92.7 (2)	38.0 (4)	2.4	Nuñez (2006)
Calakmul	Mexico	Tropical rainforest	VHF	MCP 100%	36.9 (2)	45.5 (2)	0.8	Ceballos et al. (2002)
Calakmul	Mexico	Tropical rainforest	GPS	Kernel 95%	624.95 (3)	155.42 (7)	4.0	Chávez (2009)
Lacandona	Mexico	Tropical rainforest	GPS	AKDE	544 (2)	250 (2)	2.3	This study
Center of Belize	Belize	Tropical rainforest	GPS	Kernel 95%	264.3 (6)	169.3 (1)	1.5	Figueroa (2016)
Cockscomb	Belize	Tropical rainforest	VHF	MCP	34.4 (4)	10 (2)	3.4	Rabinowitz and Nottingham (1986)
Los Llanos	Venezuela	Tropical savannas	VHF	MCP	130 (1)	49 (2)	2.6	Scognamillo et al. (2003)
Dry Chaco	Paraguay	Tropical dry forest	GPS	AKDE	727(5)	255 (2)	2.8	McBride and Thompson (2018)
Humid Chaco	Paraguay	Tropical savannas	GPS	AKDE	818(4)	237 (8)	3.4	McBride and Thompson (2018)
Amazon	Brazil	Tropical rainforest	GPS	AKDE	211 (4)	68 (6)	3.1	Morato et al. (2016)
Cerrado	Brazil	Tropical savannas	GPS	AKDE	1268	–	–	Morato et al. (2016)
Cerrado	Brazil	Tropical savannas	VHF	MCP 80%	265 (1)	228 (1)	1.1	Silveira (2004)
Atlantic Forest	Brazil	Tropical rainforest	GPS	AKDE	462 (5)	268 (5)	1.7	Morato et al. (2016)
Pantanal	Brazil	Tropical savannas	GPS	AKDE	144 (6)	52 (10)	2.7	Morato et al. (2016)
Pantanal	Brazil	Tropical savannas	GPS	Kernel 95%	154 (6)	62.5 (4)	2.4	Cavalcanti and Gese (2009)
Pantanal	Brazil	Tropical savannas	VHF	Kernel 95%	67.3 (3)	38.2 (5)	1.7	Azevedo and Murray (2007)
Pantanal	Brazil	Tropical savannas	VHF	MCP 100%	152.4 (1)	139.5 (4)	1.1	Crawshaw and Quigley (1991)
Pantanal	Brazil	Tropical savannas	VHF	MCP 100%	80 (1)	31.5 (2)	2.5	Schaller and Crawshaw (1980)

can spend several days feeding of a carcass of a large prey and they can stay around killing sites from 1 to 21 days (Cavalcanti and Gese 2010). In Figs. 13.2 and 13.3 is shown the intensity distribution of both animals using an isopleth of 30%. This intensity distribution at 30% of isopleth covered 23 km² for the female jaguar JF15 and 40 km² in the male jaguar JM04.

On the other hand, areas frequently visited might be associated with refuge areas or areas used frequently as foraging sites. Contrasting the recursion utilization at 30% of isopleth with the core areas estimated through 50% fixed kernel (utilization distribution), we observed that these areas were very similar in both individuals.

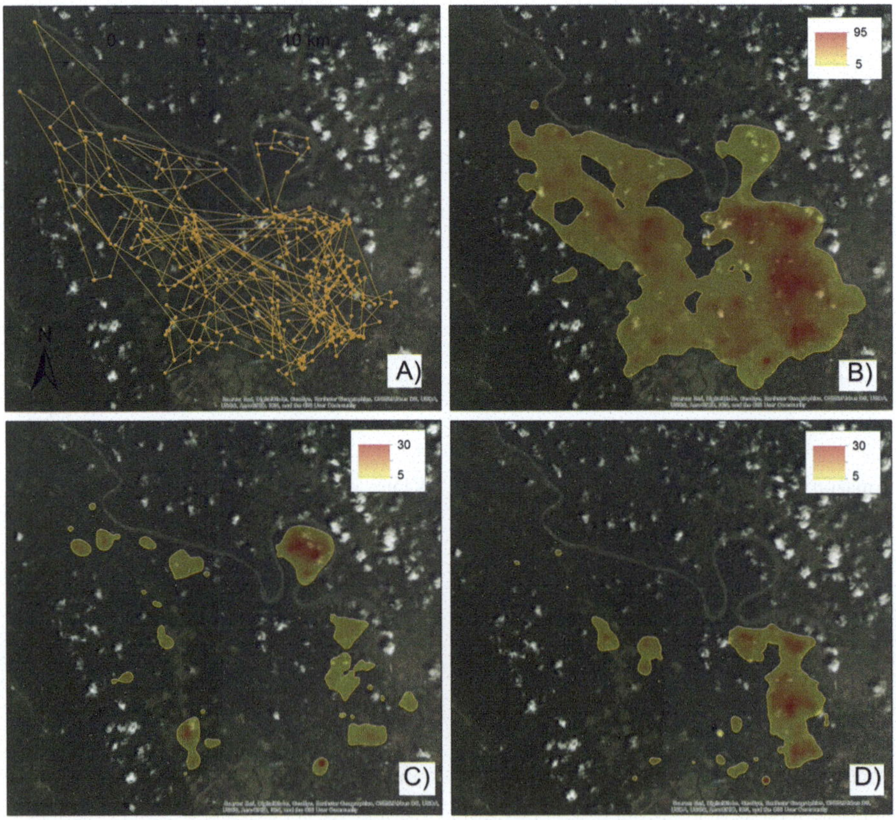

Fig. 13.2 Home range polygons obtained using the biased random bridge model for the jaguar female JH15. (**a**) Locations and trajectories obtained from the individual followed by the GPS collar. (**b**) Utilization distribution using an isopleth of 95%. (**c**) Intensity distribution using an isopleth of 30%. (**d**) Recursion distribution using an isopleth of 30%

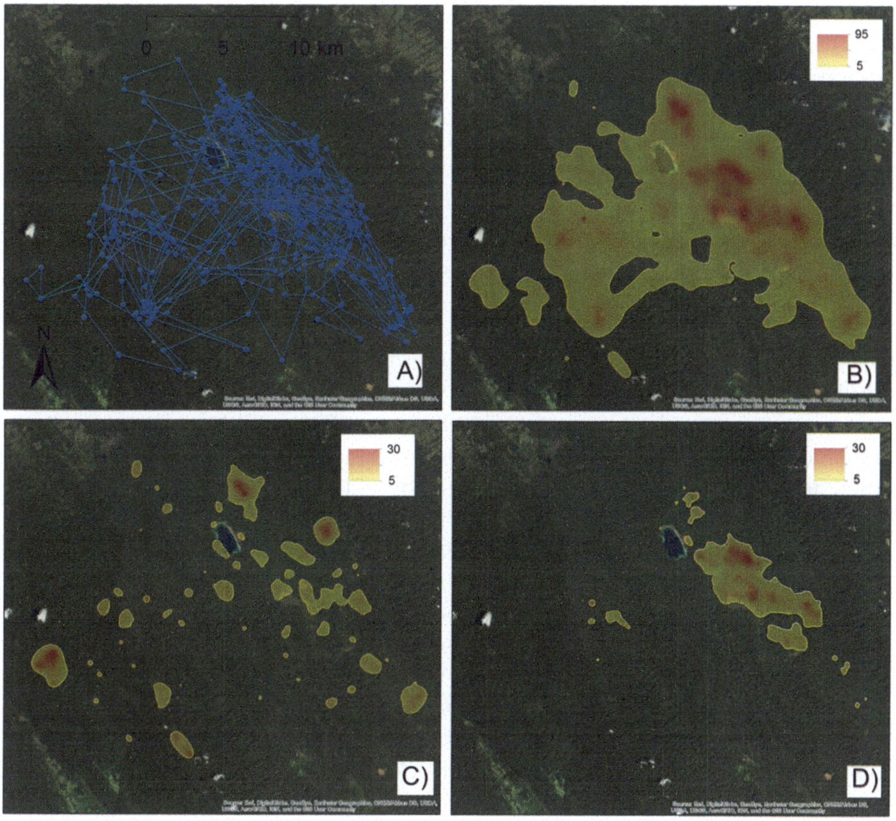

Fig. 13.3 Home range polygons obtained using the biased random bridge model for the jaguar male JM04. (**a**) Locations and trajectories obtained from the individual followed by the GPS collar. (**b**) Utilization distribution using an isopleth of 95%. (**c**) Intensity distribution using an isopleth of 30%. (**d**) Recursion distribution using an isopleth of 30%

13.6.3 Movement Decisions of Jaguars Across Different Landscape Features

We analyzed a total of 452 movement steps, of which 43.5% were classified as active. Our best step selection function model included the forest cover in a neighborhood of 240 m (FCov-240), the TPI, and elevation (ELEV) as the variables associated most strongly with movement probability by jaguars in the Greater Lacandona Ecosystem landscape (Table 13.4). For this model all the variables were informative, and the 95% CI did not overlap 0 (Table 13.4). However, this model only had a ΔAIC value of 0.54 for the second best ranked model (Table 13.5). This second ranked model included the same variables than the first ranked one but as well including the nearest distance to deforested patches that were >1 km^2 (DistD)

Table 13.4 Estimated coefficients (β), standard errors (SE), 95% confidence intervals (95% CI), z values (z), and P values (P) for the best step selection functions for movements of jaguars in the Southern Mayan Forest

Covariate	β	SE	95% CI	z	P
FCov-240	0.0335	0.0079	0.0199 to 0.0527	4.241	<0.0001
TPI	0.0379	0.0109	0.0166 to 0.0595	3.475	<0.001
ELEV	−0.0070	0.0015	−0.0103 to −0.0049	−4.680	<0.0001

We estimated the coefficients and the standard errors for percentage of forest cover in a neighborhood of 240 m (FCov-30), the topographic position index (TPI), and the elevation (ELEV) by bootstrapping ($B = 5000$)

Table 13.5 The ten best step selection function models tested to understand how the landscape structure of the Greater Lacandona Ecosystem affects jaguar movements

Model	df	LogLik	AICc	ΔAIC	Weight	Rank
FCov-240 + TPI + ELEV	3	−893.414	1793.00	0	0.355	1
FCov-240 + DistD + TPI + ELEV	4	−892.642	1793.50	0.54	0.271	2
FCov-510 + TPI + ELEV	3	−893.771	1793.70	0.71	0.248	3
FCov-510 + DistT+ TPI+SHANN	4	−894.667	1797.50	4.59	0.036	4
FCov-240 + DistD + DistT + ELEV	4	−895.108	1798.40	5.47	0.023	5
FCov-510 + TPI + SHANN	3	−896.844	1799.80	6.86	0.011	6
FCov-240 + DistT + ELEV	3	−896.919	1800.00	7.01	0.011	7
FCov-510 + DistT + ELEV	3	−897.367	1800.90	7.91	0.007	8
FCov-240 + DistD + ELEV	3	−897.409	1800.90	7.99	0.007	9
FCov-30 + DistD + TPI + ELEV	4	−896.446	1801.10	8.15	0.006	10
Null model	1	−910.457	1822.90	29.98	0	55

variable. However this last variable was not informative because the 95% CI overlapped 0. The third best ranked model was very similar to our best model (wi = 0.24) and included the forest cover in a neighborhood of 510 m, TPI, and elevation (ELEV). Since the three ranked models were very similar, we used the first ranked model to explain the movement decisions of jaguars across the different landscape features of our study area.

13.7 Discussion

13.7.1 Home Range Sizes and Movement Rates of Jaguars in Southern Mexico

Comparing our home range sizes, estimations with other studies where home ranges were estimated using GPS tracking technique suggest that the areas used by the jaguars in the tropical rainforest of Central America are similar (Table 13.3). This demonstrates that jaguars require extensive areas to meet their spatial requirements in this ecosystem type. On the other hand, it appears that in other kinds of ecosystems such as tropical savannas, such as in the Pantanal, Brazil, jaguars use smaller home ranges than in the tropical rainforests (Table 13.3).

Contrasting the home range sizes between sexes, male jaguars had 2.1 times larger home range than those of female jaguars. Jaguars, as most of the felid species, are solitary except for periods of mating and breeding. Usually, females have smaller home ranges than males, and the home ranges of males include and overlap with the home ranges of several females (Schaller and Crawshaw 1980; Rabinowitz and Nottingham Jr. 1986; Cavalcanti and Gese 2009). Home ranges of females are shaped by the abundance and distribution of prey, which is critical to meet the energetic requirements that raising cubs demands (Eisenberg 1986; Sandell 1989; Logan and Sweanor 2001). Another factor that determines space use in solitary female felids is the availability of safe refuges for offspring (Nuñez 2006; Steyn and Funston 2009). Larger home ranges of males are determined by the spatial distribution of females and have to be large enough to cover several females to increase mating opportunities. Other factor that determines the space in males is defending their territory from other males, which would compete for the mates or would kill their cubs (Eisenberg 1986; Cavalcanti and Gese 2009; Ferguson et al. 2009; Goodrich et al. 2010).

Movement rates also varied between male and female jaguars. Male movement paths, represented by velocity autocorrelation time scale and distance traveled by day, were larger compared with female jaguars (Table 13.1). However, home range crossing time was variable between individuals. These results are similar to previous studies of jaguar movements which have reported that female jaguar movement paths are proportionally more erratic or tortuous with smaller distances traveled per day (Cavalcanti and Gese 2009; Morato et al. 2016).

13.7.2 Frequency and Intensity of Visits of Jaguars to Certain Sites Within Home Ranges

The results obtained from the analysis suggest that jaguars exhibited a selective behavior in certain areas within their home ranges, and probably these provide significant habitat component for their survival or reproduction (Samuel et al. 1985; Bingham and Noon 1997; Plowman et al. 1998). For instance, areas frequently used by female jaguar JF15 were associated with rugged terrain across a small mountain range covered with forest, but that has been deforested around it in the flat areas. In the case of female jaguars, rugged terrain would provide den and refuge when they are rearing cubs (Steyn and Funston 2009; White et al. 2015) and could provide protection because these areas maintain more forest cover in contrast with flat areas which are more susceptible to deforestation and other human activities.

Frequently visited areas also might be associated with areas of exclusive use. Other studies have shown that jaguars have little spatial overlap in core areas (Azevedo and Murray 2007; de la Torre et al. 2017b). This spatial segregation observed in jaguars and other large felids implies territorial behavior because core areas have been frequently associated with regions of resource concentration (Samuel et al. 1985; Plowman et al. 1998; Nuñez 2006; Azevedo and Murray 2007). Spatial segregation of frequently visited areas or core areas could facilitate the coexistence between individuals through resource partitioning (Scognamillo et al. 2003; Azevedo and Murray 2007). Communication between individuals through scent marking using scats, scrapes, or spray urine may provide spatial and temporal information to reduce confrontation and may help to maintain the exclusive territories (Harmsen et al. 2010; Towns et al. 2017).

13.7.3 Movement Decisions of Jaguars Across Different Landscape Features

The step selection function analysis indicates that the probability of movement of jaguars in the Greater Lacandona Ecosystem is positively associated with a high percentage of forest cover in a neighborhood of 240 m. This result suggest that despite the landscape configuration, jaguars prefer to move across forested areas in the tropical forest of south of Mexico. Additionally our model indicates that probability of movement for jaguars was associated with positive values of TPI, which indicated that jaguar movements were facilitated by sites with medium to moderate slopes and flat valleys (Dickson et al. 2005). This is especially plausible if the flat areas had been cleared for cattle pastures or crops, which is a common situation in our study area (Morato et al. 2014). Our best model also indicates that jaguars avoided moving through the ridge tops of mountain ranges, since the probability of movement by jaguars decreased with elevation. Other studies also have suggested that movements of jaguars would be hampered by mountain ranges at higher altitudes (Rabinowitz and Zeller 2010; Zeller et al. 2011); however, it is necessary to track some individuals in forested areas located at higher altitudes to have conclusive results and to improve the knowledge of jaguar movement behavior in areas of higher altitudes and rugged terrain.

13.8 Conclusions

In this study we illustrate that the extensive area requirements of jaguars have crucial implications for the conservation of the species at the landscape scale in the tropical rainforests of Central America; therefore, forest transformation and fragmentation would have severe effects for the species. Although jaguars do not necessarily use areas which are covered completely with forest, several studies have shown that they preferred primary forest in contrast to modified environments associated with human activities (Conde et al. 2010; Foster et al. 2010; Colchero et al. 2011; Cullen et al. 2013; de la Torre et al. 2017a).

Our study demonstrates that to ensure the long-term conservation of jaguars in the tropical rainforest of Central America, large extensions of primary forest should be preserved, for instance, our estimations of jaguar home ranges indicate that it is necessary to conserve connected parches of tropical forest of at least 200 km^2 to guarantee the annual spatial requirements of a single female jaguar in the south of Mexico. This estimation gives an idea of the scale and the magnitude of conservation actions required to preserve jaguar populations in tropical forest areas. Large tropical forest areas provide to jaguars prey availability, dense foliage for stalking prey, and refuge for rearing cubs (Rabinowitz and Nottingham 1986; Crawshaw and Quigley 1991; Conde et al. 2010; Foster et al. 2010; Cullen et al. 2013).

The movement probability model also demonstrates the importance to conserve areas with primary forest to allow the movement of the jaguars across fragmented landscapes, because the probability of movement of jaguars was greater in areas with a forest cover in a neighborhood of 240 m. This outcome has important implications for the design of corridors focused on maintaining the connectivity between suitable habitat patches for the species, because our results suggest that the minimal width at which corridors could be functional for jaguars is 240 m. This result is similar to studies using other species of large felids such as pumas, which indicates that corridors >400 m wide covered with forest or other coverage were functional for dispersal (Beier 1993, 1995).

Our results also showed that BRB model is a new alternative to the traditional home range estimators to study the space use of large carnivores in tropical forests. This model allowed us to decompose the home range of the jaguars into two components, intensity of use and frequency of visit. The results obtained from the estimation of the intensity distribution (ID), which are the areas where the jaguars spend prolonged periods of time, but only visited once, are probably associated with hunting sites. On the other hand, the estimation of the recursion distribution (RD) indicated areas that are frequently visited by jaguars so it can be probably associated with resting or refuge sites. Together, both distributions could represent an approximation of the areas identified by the estimation of the fixed kernel at 50%, considered as the core areas; however, this new approach generated from the BRB model allows us to make a more detailed description of the use of space within the home range by jaguars. Future studies could incorporate this approach combined with an intensive ground tracking of animals to give clearer insights regarding the utilization of certain areas within their home range by jaguars in tropical forests.

We recognize that a caveat in our analyses is that the GPS records used to fit all the models come only from two to five animals tracked, and this sample size probably is not representative of the jaguar population of our study area. Other limitation is that these five animals were tracked in a relatively small area which is mainly represented by lowland areas. However, this effort is the first step to understating jaguar movement in the humid tropical rainforest of Southern Mexico and has opened a window to develop sound conservation actions that ensure the connectivity of this jaguar population across the Greater Lacandona Ecosystem landscape (de la Torre et al. 2017a).

The movement ecology is a discipline that in recent times, with the development of new technologies and analytical tools, has allowed making more precise inferences regarding the space use and movement of wildlife animals. We hope that the growing of this discipline allows in the future a better understanding of the ecology and behavior of the species that inhabit the tropical rainforest to generate the information necessary to implement policy conservation actions that ensure the long-term persistence of the wildlife and these ecosystems.

Acknowledgments This work was supported by The Conservation, Food & Health Foundation, Conservation Leadership Programme, Cleveland Metroparks Zoo and Cleveland Zoological Society, Commission of Natural Protected Areas of the Mexican Federal Government (CONANP), the Mesoamerican Biological Corridor-Mexico, and the National Council of Science and Technology of Mexico (CONACyT). J. A. de la Torre acknowledges the scholarship and financial support provided by the National Council of Science and Technology (CONACyT) and UNAM during the implementation of this research. We thank Dr. Rodrigo Medellín for his invaluable support to develop the research. We appreciate the assistance of our field assistants from Lacanja Chansayab (Luisito, Gregorio, Paco, and Rodrigo) and Frontera Corozal (Silvano, Chano, Celestino, Bernabe, Wicho, Arnoldo, and Florencio) communities for their helpful and hard work. We are very grateful to the Commission of Natural Protected Areas of the Mexican Federal Government (CONANP), especially with S. Nañez and V. Silvano for supporting this research. We appreciate the invaluable support of Wildlife Service and their team of veterinarians, especially Ivonne Cassaigne, Susana Ilescas, Octavio Lopez, Dane Hawk, and Dave Kenny. We also thank Cuauhtémoc Chávez, Heliot Zarza, José Gonzalez Maya, Rurik List, Rodrigo Sierra, Eduardo Ponce, Gerardo Ceballos, Enrique Martinez-Meyer, David Valenzuela, Jorge Servin, Jaime Zuñiga, Omar Figueroa, Brian Jansen, Ron Thomson, and Valeria Towns for their helpful suggestions in the development of this study. We thank Ana Ibarra, Alma Moreno, and Luis Viquez for technical support. We would also like to thank to Cecilia Vargas, Aseneth Ureña, and Nora Torres for their helpful assistance during field work. Throughout the analysis, we had the assistance of the Centro de Investigación en Geografía y Geomática "Ing. Jorge L. Tamayo" A.C. We thank the logistical support of Bioconciencia A.C. and Natura Mexicana A.C. Permission to conduct the field captures was granted by the General Office of Wildlife-SEMARNAT Mexico and facilitated by Fernando Sánchez.

References

Allen AM, Singh NJ (2016) Linking movement ecology with wildlife management and conservation. Front Ecol Evol 3:1–13

Azevedo FCC, Murray LD (2007) Spatial organization and food habits of jaguars (*Panthera onca*) in a floodplain forest. Biol Conserv 137:391–402

Beier P (1993) Determining minimum habitat areas and habitat corridors for cougars. Conserv Biol 7:94–108

Beier P (1995) Dispersal of juvenile cougars in fragmented habitat. J Wildl Manag 59:228–237

Benhamou S (2011) Dynamic approach to space and habitat use based on biased random bridges. PLoS One 6:e14592

Benhamou S, Cornélis D (2010) Incorporating movement behavior and barriers to improve kernel home range space use estimates. J Wildl Manag 74:1353–1360

Benhamou S, Riotte-Lambert L (2012) Beyond the utilization distribution: identifying home range areas that are intensively exploited or repeatedly visited. Ecol Model 227:112–116 Elsevier B.V.

Beyer HL (2012) Geospatial modelling environment. Geospatial Model Environ 2010:1–140

Bhattarai BP, Kindlmann P (2012) Interactions between Bengal tiger (*Panthera tigris*) and leopard (*Panthera pardus*): implications for their conservation. Biodivers Conserv 21:2075–2094

Bingham BB, Noon BR (1997) Mitigation of habitat "Take": application to habitat conservation planning. Conserv Biol 11:127–139

Broomhall LS, Mills MGL, du Toit JT (2003) Home range and habitat use by cheetahs (*Acinonyx jubatus*) in the Kruger National Park. J Zool 261:119–128

Burnham KP, Anderson DR (2002) Model selection and multimodel inference: a practical information-theoretic approach, 2nd edn. Springer-Verlag, New York

Cagnacci F, Boitani L, Powell RA, Boyce MS (2010a) Challenges and opportunities of using GPS-based location data in animal ecology. Philos Trans R Soc Lond Ser B Biol Sci 365:2155

Cagnacci F, Boitani L, Powell RA, Boyce MS (2010b) Animal ecology meets GPS-based radiotelemetry: a perfect storm of opportunities and challenges. Philos Trans R Soc Lond Ser B Biol Sci 365:2157–2162

Calabrese JM, Fleming CH, Gurarie E (2016) Ctmm: an R package for analyzing animal relocation data as a continuous-time stochastic process. Methods Ecol Evol 7:1124–1132

Calenge AC [online] (2013a) Package 'adehabitatHR'. <https://trac.faunalia.it/animove/>

Calenge C [online] (2013b) "Package adehabitatLT" analysis of animal movements. CRAN. <https://trac.faunalia.it/animove/>

Camp MJ, Rachlow JL, Cisneros R, Roon D, Camp RJ (2016) Evaluation of global positioning system telemetry collar performance in the tropical Andes of southern Ecuador. Nat Conservacao 14:128–131 Associação Brasileira de Ciência Ecológica e Conservação

Carbone C, Cowlishaw G, Isaac NJB, Rowcliffe JM (2005) How far do animals go? Determinants of day range in mammals. Am Nat 165:290–297

Carbone C, Teacher A, Rowcliffe JM (2007) The costs of carnivory. PLoS Biol 5:e22

Cavalcanti SMC, Gese EM (2009) Spatial ecology and social interactions of jaguars (*Panthera onca*) in the southern Pantanal, Brazil. J Mammal 90:935–945

Cavalcanti SMC, Gese EM (2010) Kill rates and predation patterns of jaguars (*Panthera onca*) in the southern Pantanal, Brazil. J Mammal 91:722–736

Ceballos G, Chávez C, Rivera A, Manterola C (2002) Tamaño poblacional y conservación del jaguar en la Reserva de la Biosfera de Calakmul, Campeche, México. In: Medellín RA, Equihua C, Chetkiewicz CLB, Crawshaw PG Jr, Rabinowitz A, Redford KH et al (eds) El Jaguar en el Nuevo Milenio. Fondo de Cultura Económica, Universidad Nacional Autónoma de México, Wildlife Conservation Society, México, pp 403–418

Chávez C (2009) Ecología y conservación del jaguar (*Panthera onca*) y puma (*Puma concolor*) en la región de Calakmul y sus implicaciones para la conservción de la Peinínsula de Yucatán. Universidad de Granada. Granada, Spain

Chetkiewicz C-LB, Clair CCS, Boyce MS (2006) Corridors for conservation: integrating pattern and process. Annu Rev Ecol Evol Syst 37:317–342

Colchero F, Medellin RA, Clark JS, Lee R, Katul GG (2009) Predicting population survival under future climate change: density dependence, drought and extraction in an insular bighorn sheep. J Anim Ecol 78:666–673

Colchero F, Conde DA, Manterola C, Chávez C, Rivera A, Ceballos G (2011) Jaguars on the move: modeling movement to mitigate fragmentation from road expansion in the Mayan Forest. Anim Conserv 14:158–166

Conde DA, Colchero F, Zarza H, Christensen NL, Sexton JO, Manterola C et al (2010) Sex matters: modeling male and female habitat differences for jaguar conservation. Biol Conserv 143:1980–1988 Elsevier Ltd

Crawshaw PG Jr, Quigley HB (1991) Jaguar spacing, activity and habitat use in a seasonally flooded environment in Brazil. J Zool 223:357–370

Cullen LJ, Sana DA, Lima F, De Abreu KC, Uezu A (2013) Selection of habitat by the jaguar, *Panthera onca* (Carnivora: Felidae), in the upper Paraná River, Brazil. Fortschr Zool 30:379–387

Davis S, Haywood V, Herrera-MacBride O, Villa-Lobos J, Hamilton A (1997) Centre of plants diversity. A guide and strategy for their conservation. The World Wide Fund for Nature, UICN and The World Conservation Union, Cambridge, UK

De Jong BHJ, Ochoa-Gaona S, Castillo-Santiago MA (2000) Carbon flux and patterns of land-use/land-cover change in the Selva Lacandona. Ambio 29:504–511

de la Torre JA, Medellín RA (2011) Jaguars *Panthera onca* in the Greater Lacandona Ecosystem, Chiapas, Mexico: population estimates and future prospects. Oryx 45:546–553

de la Torre JA, Rivero M (2017) A morphological comparison of jaguars and pumas in southern Mexico. Therya 8:117–122

de la Torre JA, Núñez JM, Medellín RA (2017a) Spatial requirements of jaguars and pumas in Southern Mexico. Mamm Biol 84:52–60

de la Torre JA, Núñez JM, Medellín RA (2017b) Habitat availability and connectivity for jaguars (*Panthera onca*) in the Southern Mayan Forest: conservation priorities for a fragmented landscape. Biol Conserv 206:270–282

de la Torre, JA, González-Maya JF, Zarza H, Ceballos G and Medellín RA (2018) The jaguar's spots are darker than they appear: Assesing the global conservation status of the jaguar Panthera onca. Oryx 52:300–315

Dickson BG, Jenness JS, Beier P (2005) Influence of vegetation, topography, and roads on cougar movement in southern California. J Wildl Manage 69, 264–276

Eisenberg JF (1986) Life history strategies of the Felidae: variations on a common theme. In: Miller SD, Everett DD (eds) Cats of the world: biology conservation and management. National Wildlife Federation, Washington, DC, pp 293–303

Emmons LH (1987) Comparative feeding ecology of felids in a neotropical rainforest. Behav Ecol Sociobiol 20:271–283

Espinosa S, Celis G, Branch LC (2018) When roads appear jaguars decline: increased access to an Amazonian wilderness area reduces potential for jaguar conservation. PLoS One 13:1–18

Ferguson AW, Currit NA, Weckerly FW (2009) Isometric scaling in home-range size of male and female bobcats (*Lynx rufus*). Can J Zool 87:1052–1060

Figueroa O (2013) The ecology and conservation of jaguars (*Panthera onca*) in Central Belize: conservation status, diet, movement patterns and habitat use. Gainsville, Florida, USA. 163 pp.

Fortin D, Beyer HL, Boyce WM, Smith DW, Duchesne T, Mao JS (2005) Wolves influence elk movements: behavior shapes a trophic cascade in Yellowstone National Park. Ecology 86:1320–1330

Foster RJ, Harmsen BJ, Doncaster CP (2010) Habitat use by sympatric jaguars and pumas across a gradient of human disturbance in Belize. Biotropica 42:724–731

Foster VC, Sarmento P, Sollmann R, Tôrres N, Jácomo ATA, Negrões N et al (2013) Jaguar and Puma activity patterns and predator-prey interactions in four Brazilian biomes. Biotropica 45:373–379

Frank L, Simpson D, Woodroffe R (2003) Foot snares: an effective method for capturing African lions. Wildl Soc Bull 31:309–314

Gittleman JL, Harvey PH (1982) Carnivore home-range size, metabolic needs and ecology. Behav Ecol Sociobiol 10:57–63

Gitzen RA, Belant JL, Millspaugh JJ, Te Wong S, Hearn AJ, Ross J (2013) Effective use of radiotelemetry for tropical forest carnivores. Raffles Bull Zool 28:67–83

Goodrich JM, Miquelle DG, Smirnov EN, Kerley LL, Quigley HB, Hornocker MG (2010) Spatial structure of Amur (Siberian) tigers (*Panthera tigris altaica*) on Sikhote-Alin biosphere Zapovednik, Russia. J Mammal 91:737–748

Haag T, Santos AS, Sana DA, Morato RG, Cullen L, Crawshaw PG et al (2010) The effect of habitat fragmentation on the genetic structure of a top predator: loss of diversity and high differentiation among remnant populations of Atlantic Forest jaguars (*Panthera onca*). Mol Ecol 19:4906–4921

Halstead TD, Gruver KH, Phillips RL, Johnson RE (1995) Using telemetry equipment for monitoring traps and snares. In: Masters RE, Huggins JG (eds) Twelfth great plains wildl. damage control workshop proc. Noble Foundation, Ardmore, pp 121–123

Hansen MC, Potapov PV, Moore R, Hancher M, Turubanova SA, Tyukavina A et al (2013) High-resolution global maps of 21st-century forest cover change. Science (New York, NY) 342:850–853
Harmsen BJ, Foster RJ, Silver SC, Ostro LET, Doncaster CP (2009) Spatial and temporal interactions of sympatric jaguars (*Panthera onca*) and pumas (*Puma concolor*) in a Neotropical forest. J Mammal 90:612–620
Harmsen BJ, Foster RJ, Gutierrez SM, Marin SY, Doncaster CP (2010) Scrape-marking behavior of jaguars (*Panthera onca*) and pumas (*Puma concolor*). J Mammal 91:1225–1234
Herfindal I, Linnell JDC, Odden J, Nilsen EB, Andersen R (2005) Prey density, environmental productivity and home-range size in the Eurasian lynx (*Lynx lynx*). J Zool 265:63–71
Jenny D (1996) Spatial organization of leopards Panthera pardus in Tai National Park, Ivory Coast: is rainforest habitat a 'tropical haven'? J Zool 240:427–440
Jetz W, Carbone C, Fulford J, Brown JH (2004) The scaling of animal space use. Science 306:266–268
Karanth KU, Sunquist ME (2000) Behavioural correlates of predation by tiger (*Panthera tigris*), leopard (*Panthera pardus*) and dhole (*Cuon alpinus*) in Nagarahole, India. J Zool 250:255–265
Laundré JW, Hernández L, Streubel D, Altendorf K, Laundre JK, Hernandez L et al (2000) Aging mountain lions using gum-line recession. Wildl Soc Bull 28:963–966
Lehmann MB, Funston PJ, Owen CR, Slotow R (2008) Home range utilisation and territorial behaviour of lions (*Panthera leo*) on Karongwe Game Reserve, South Africa. PLoS One 3:e3998
Lele SR, Keim JL (2006) Weighted distributions and estimation of resource selection probability functions. Ecology 87:3021–3028
Lele SR, Keim JL, Solymos P [online] (2014) Resource selection (probability) functions for use-availability data. <http://cran.r-project.org/package=ResourceSelection>
Logan KA, Sweanor LL (2001) Desert Puma: evolutionary ecology and conservation of an enduring carnivore. Island Press, Washington, DC
Macdonald DW, Mosser A, Gittleman JL (2010) Felid society. In: Macdonald DW, Loveridge AJ (eds) Biology and conservation of wild felids. Oxford University Press, Oxford, pp 125–160
March I (1993) The white-lipped peccary (*Tayassu pecari*). In: Oliver WLR (ed) Pigs, peccaries, and hipos: status survey and conservation. IUCN/SSC Pig and Peccaries Specialist Group, IUCN/SSC Hippo Specialist Group. IUCN The World Conservation Union, Gland, Swizeland pp 13–22
Marker LL, Dickman AJ, Mills MGL, Jeo RM, Macdonald DW (2008) Spatial ecology of cheetahs on north-central Namibian farmlands. J Zool 274:226–238
Matola S, Cuarón AD, Rubio-Torgler H (1997) Status and action plan of Baird's Tapir (Tapirus bairdi). In: Brooks DM, Bodmer RE, Matola S (eds) Tapirs: status survey and conservation action plan. IUCN Tapir Specialist Group. IUCN, The World Conservation Union, Gland, Swizeland pp 29–45
McBride RT, Thompson JJ (2018) Space use and movement of jaguar (*Panthera onca*) in western Paraguay. Mammalia. https://doi.org/10.1515/mammalia-2017-0040
Medellín RA (1994) Mammals diversity and conservation in the Selva Lacandona, Chiapas, Mexico. Conserv Biol 8:780–799
Medellín RA, de la Torre JA, Chávez C, Zarza H, Ceballos G (2016) El Jaguar en el Siglo XXI: La Perspectica Continental. Fondo de Cultura Económica, Universidad Nacional Autónoma de México, Ciudad de México
Mendoza E, Dirzo R (1999) Deforestation in Lacandonia (southeast Mexico): evidence for the declaration of the northernmost tropical hot-spot. Biodivers Conserv 8:1621–1641
Morales JM, Moorcroft PR, Matthiopoulos J, Frair JL, Kie JG, Powell RA et al (2010) Building the bridge between animal movement and population dynamics. Philos Trans R Soc Lond Ser B Biol Sci 365:2289–2301
Morato RG, Ferraz KMPMDB, de Paula RC, De Campos CB (2014) Identification of priority conservation areas and potential corridors for jaguars in the Caatinga Biome, Brazil. PLoS One 9:e92950

Morato RG, Stabach JA, Fleming CH, Calabrese JM, De Paula RC, Ferraz KMPM et al (2016) Space use and movement of a neotropical top predator: the endangered jaguar. PLoS One 11:1–17

Naranjo EJ, Amador-Alcala SA, Falconi-Briones FA, Reyna-Hurtado R (2015) Distribución, abundancia y amenazas a las poblaciones de tapir centroamericano (*Tapirus bairdii*) y pecarí de labios blancos (*Tayassu pecari*) en México. Therya 6:227–249

Ning H, Phin WE, Chackrapani P, Solana-Mena A, Ling ATS, Nagulendran K et al (2016) MEME – moving towards a science-based conservation of Malaysian elephants. Malay Nat J 68:191–197

Northrup JM, Hooten MB, Anderson CR, Wittemyer G (2013) Practical guidance on characterizing availability in resource selection functions under a use-availability design. Ecology 94:1456–1463

Noss AJ, Gardner B, Maffei L, Cuéllar E, Montaño R, Romero-Muñoz A et al (2012) Comparison of density estimation methods for mammal populations with camera traps in the Kaa-Iya del Gran Chaco landscape. Animal Conservation:n/a-n/a

Nowell K, Jackson P (1996) Wild cats status, survey and conservation action plan. P. in. IUCN/SSC Cats Specialist Group. The World Conservation Union Gland, Swizeland

Nuñez R (2006) Área de actividad, patrones de actividad y movimiento del jaguar (*Panthera onca*) y del puma (*Puma concolor*), en la Reserva de la Biosfera "Chamela-Cuixmala", Jalisco. Universidad Nacional Autónoma de México Mexico D.F., Mexico

O'Brien KL (1998) Tropical deforestation and climate change: what does the record reveal? Prof Geogr 50:140–153

Odden M, Wegge P (2005) Spacing and activity patterns of leopards *Panthera pardus* in the Royal Bardia National Park, Nepal. Wildl Biol 11:145–152

Odden M, Wegge P, Fredriksen T (2010) Do tigers displace leopards? If so, why? Ecol Res 25:875–881

Plowman BW, Conner LM, Chamberlain MJ, Leopold BD, Burger LW (1998) Annual dynamics of bobcat (*Lynx rufus*) home range and core use areas in Mississippi. Am Midl Nat 156:386–393

Powell RA, Mitchell MS (2012) What is a home range? J Mammal 93:948–958

Quigley H, Foster R, Petracca L, Payan E, Salom R, Harmsen B (2017) *Panthera onca*. P. in The IUCN Red List of Threatened Species 2017: e.T15953A50658693

R Core Team (2016) R: a language and environment for statistical computing. R Foundation for Statistical Computing, Vienna

Rabinowitz AR, Nottingham BG Jr (1986) Ecology and behaviour of the jaguar (*Panthera onca*) in Belize, Central America. J Zool 210:149–159

Rabinowitz A, Zeller KA (2010) A range-wide model of landscape connectivity and conservation for the jaguar, *Panthera onca*. Biol Conserv 143:939–945 Elsevier Ltd

Reyna-Hurtado R, Radachowsky J, Mcloughlin L, Thornton D, Moreira-Ramirez JF, Garcia-Anleu R, Ponce-Santizo G, McNab R, Diaz-Santos F, Elvir F, Portillo-Reyes HO, Moreno R, Meyer N, Artavia A, Amit R, Hofman MPG, Harmsen BJ, Castañeda F, Sanchez-Pinzon K, Hernandez-Perez E, Martinez W, Mérida M, Solis N, Ramos VH, Juárez D, Tot C, Lopez E, Leonardo R, Machuca O, Calderon P, Estrada C, Morales J, Guerrero S, Morazan F, Carrillo E, Cruz JC, Montalvo V, Gonzalez-Maya JF, Gomez D, Jiménez JJ, Arroyo S, Thomson I, Polisar J (2017) White-lipped peccary in Mesoamerica: status, threats and conservation actions. Suiform Soundings 15(2):31–35

Riley SJ, DeGloria SD, Elliot R (1999) A terrain ruggedness index that quantifies topographic heterogeneity. Intermountain J Sci 5:23–27

Ripple WJ, Estes JA, Beschta RL, Wilmers CC, Ritchie EG, Hebblewhite M et al (2014) Status and ecological effects of the world's largest carnivores. Science (New York, NY) 343:1241484

Ripple WJ, Chapron G, López-Bao JV, Durant SM, Macdonald DW, Lindsey PA et al (2016) Saving the world's terrestrial megafauna. Bioscience 66:807–812

Samuel AMD, Pierce DJ, Garton EO (1985) Identifying areas of concentrate use within the home range. J Anim Ecol 54:711–719

Sandell M (1989) The mating tactics and spacing patterns of solitary carnivores. In: Gittleman JL (ed) Carnivore behaviour, ecology & evolution. Chapman and Hall, London, pp 164–182
Sanderson EW, Redford KH, Chetkiewicz CB, Medellin RA, Rabinowitz AR, Robinson JG et al (2002) Planning to save a species: the jaguar as a model. Conserv Biol 16:58–72
Schaller GB, Crawshaw PG (1980) Movement patterns of jaguar. Biotropica 12:161–168
Schaller GB, Vasconcelos JMC (1978) Jaguar predation on capybara. Z Saeugetierk 43:296–301 Keywords: 2BR/behavior/Hydrochoerus hydrochaeris/jaguar/kill rate/*Panthera onca*/predation/predator-prey/prey selection
Schmidt K (2008) Behavioural and spatial adaptation of the Eurasian lynx to a decline in prey availability. Acta Theriol 53:1–16
Scognamillo D, Maxit IE, Sunquist M, Polisar J (2003) Coexistence of jaguar (*Panthera onca*) and puma (*Puma concolor*) in a mosaic landscape in the Venezuelan llanos. J Zool 259:269–279
Seymour KL (1989) *Panthera onca*. Mamm Species 340:1–9
Sikes RS, Gannon WL, T. A. C. and U. C. of the A. S. of M. Mammalogists [online](2011) Guidelines of the American Society of Mammalogists for the use of wild mammals in research. J Mammal 92:235–253
Silveira L (2004) Ecologia comparada e conservação da onça-pintada (*Panthera onca*) e onça-parda (*Puma concolor*), no Cerrado e Pantanal
Spong G (2002) Space use in lions, *Panthera leo*, in the Selous Game Reserve: social and ecological factors. Behav Ecol Sociobiol 52:303–307
Stander PE (1997) Field age determination of leopards by tooth wear. Afr J Ecol 35:156–161
Steyn V, Funston PJ (2009) Land-use and socio-spatial organization of female leopards in a semi-arid wooded savanna, Botswana. S Afr J Wildl Res 39:126–132
Swank W, Teer J (1989) Status of the jaguar -1987. Oryx 23:14–21
Tapia O, Nuñez JM (2008) Importancia del capital ecológico de la región del Corredor Biológico Mesoamericano-México: evaluación de la biodiversidad, ciclo hidrológico y dinámica de la cobertura forestal. In: CONABIO-CBMMx (ed) . Comisión Nacional para el Conocimiento y Uso de la Biodiversidad; Corredor Biológico Mesoamericano-México, Ciudad de México, pp 37–45
Thurfjell H, Ciuti S, Boyce MS (2014) Applications of step-selection functions in ecology and conservation. Mov Ecol 2:26
Tomkiewicz SM, Fuller MR, Kie JG, Bates KK (2010) Global positioning system and associated technologies in animal behaviour and ecological research. Philos Trans R Soc Lond Ser B Biol Sci 365:2163–2176
Towns V, León R, de la Maza J, de la Torre JA (2017) Marking behaviours of jaguars in a tropical rainforest of southern Mexico. Cat News 66:33–35
Treves A, Bruskotter J (2014) Tolerance for predatory wildlife. Science (New York, NY) 344:476–477
Tucker MA, Böhning-Gaese K, Fagan WF, Fryxell JM, Van Moorter B, Alberts SC et al (2018) Moving in the Anthropocene: global reductions in terrestrial mammalian movements. Science 359:466–469
White S, Briers RA, Bouyer Y, Odden J, Linnell JDC (2015) Eurasian lynx natal den site and maternal home-range selection in multi-use landscapes of Norway. J Zool 297:87–98
Worton BJ (1989) Kernel methods for estimating the utilization distribution in home range studies. Ecology 70:164–168
Zeller KA, Nijhawan S, Salom-Pérez R, Potosme SH, Hines JE (2011) Integrating occupancy modeling and interview data for corridor identification: a case study for jaguars in Nicaragua. Biol Conserv 144:892–901 Elsevier Ltd
Ziółkowska E, Ostapowicz K, Radeloff VC, Kuemmerle T, Sergiel A, Zwijacz-Kozica T et al (2016) Assessing differences in connectivity based on habitat versus movement models for brown bears in the Carpathians. Landsc Ecol 1–20

Chapter 14
Movements and Home Range of Jaguars (*Panthera onca*) and Mountain Lions (*Puma concolor*) in a Tropical Dry Forest of Western Mexico

Rodrigo Nuñez-Perez and Brian Miller

14.1 Introduction

Top predators such as jaguars (*Panthera onca*) and mountain lions or pumas (*Puma concolor*) require large areas and enough prey base for their survival (Carbone and Gittleman 2002; Woodroffe and Ginsberg 1998). Because of this, they are susceptible to human pressure and are threatened due to anthropogenic pressures, mainly habitat destruction and persecution (Cardillo et al. 2004; Ripple et al. 2014) and are good indicators of ecosystem integrity. Actually, the puma is considered as "least concern" (Nielsen et al. 2015) and jaguar as "near threatened" (Quigley et al. 2017) by the IUCN. Habitat degradation may affect predator's behavior such as home range and daily movements (Fahring 2007) and increase the risk of negative interaction with humans (Crooks 2002; Cardillo et al. 2004; Ripple et al. 2014).

Movement patterns and home range size of felines are determined by a number of factors (Burt 1943) such as body size and sex (Carbone et al. 2007; Gittleman and Harvey 1982; Lindstedt et al. 1986), foraging behavior and prey availability (Herfindal et al. 2005), habitat suitability (Conde et al. 2010; Morato et al. 2016), climate seasonability (Logan and Sweanor 2001), sociality (Azevedo and Murray 2007; Cavalcanti and Gese 2009) and search for sexual partners (Jedrzejewski et al. 2002). Understanding how large cats move and use a landscape is important for effective conservation strategies and identify priority areas for conservation and corridors (Rodriguez-Soto et al. 2011). Interaction between similar carnivores can influence patterns of movement and space use. Spatial and temporal segregation of large predators may allow coexistence, but it may increase the magnitude of their movements and the size of their home ranges. Cats can actively avoid other

R. Nuñez-Perez (✉)
Proyecto Jaguar A.C., Morelia, Mich, Mexico

B. Miller
Wind River Ranch, Denver Zoological Foundation, Denver, CO, USA

R. Reyna-Hurtado, C. A. Chapman (eds.), *Movement Ecology of Neotropical Forest Mammals*, https://doi.org/10.1007/978-3-030-03463-4_14

individuals or concentrate their activities at specific times (Scognamillo et al. 2003; Harmsen et al. 2009; Foster et al. 2013).The jaguar and puma are the largest felids in America and are sympatric across much of their range (Iriarte et al. 1990).

Jaguars extend from northern Mexico to northern Argentina (Sanderson et al. 2002; Quigley et al. 2017). Pumas occupy a wider distribution and a wider variety of habitat types than jaguars. Pumas exist from Patagonia to Northern British Columbia (Nielsen et al. 2015). Jaguars and pumas occupy a wide variety of habitats and landscape in the neotropics (Rosas and Nuñez 2014), and as other carnivores their populations and occupied areas are decreasing (Ripple et al. 2014) mainly due to persecution, habitat destruction, and fragmentation. Home range and movement pattern information for jaguar is increasing but for puma in tropical environment is scarce and more data need be collected (Laundré and Hernández 2010). For Mexico, little is known about how jaguars and pumas use landscape specially where they are sympatric. A couple of surveys have been done in tropical subdeciduos and rain forest of southern Mexico (Chávez 2009; de la Torre et al. 2017) where there are large forested protected areas and low human presence. Western Mexico is high dense human populated and the high seasonal tropical dry forest is the most common type of tropical forest and one of the most threatened with loss rate of 2% annually (Trejo and Dirzo 2000). Western tropical dry forest were jaguars and pumas inhabits suffer a marked seasonality related to rainfall. During dry season almost all plants shed their leaves and water is almost limited to highly scattered waterholes. Vegetation cover protection and availability of resources varies deeply and animals adapt to this changes. For example, in our study site, Valenzuela and Ceballos (2000) and Mandujano et al (2004) documented seasonal changes in movements and home ranges sizes for two important jaguar and puma prey species, coati (*Nasua narica*) and deer (*Odocoileus virginianus*). Preliminary data for female jaguars and pumas, show an increment of home ranges size during wet season (Núñez et al. 2002).

With main goal to determinate how two sympatric large predators, jaguar and puma, coexist and use a seasonal tropical dry forest we radiotagged jaguars and pumas in the tropical dry forest of the Chamela-Cuixmala Biosphere Reserve (CCBR). Understanding how jaguars and pumas move and use space will provide a sound conservation and management strategy. So, considering marked seasonal changes in study site, we predicted greater home ranges and larger displacement movements of jaguars and pumas during the wet season. Besides, intra and interspecific home range overlap increase too.

14.2 Methods

14.2.1 Study Site

The 131 km^2 Chamela-Cuixmala Biosphere Reserve (19 2500 N and 104 5700 W) on the coast of Jalisco (Fig. 14.1) extends easterly from the Pacific Ocean, reaching an altitude of 420 m a.s.l (Noguera et al. 2002). Terrain is rugged and hilly with

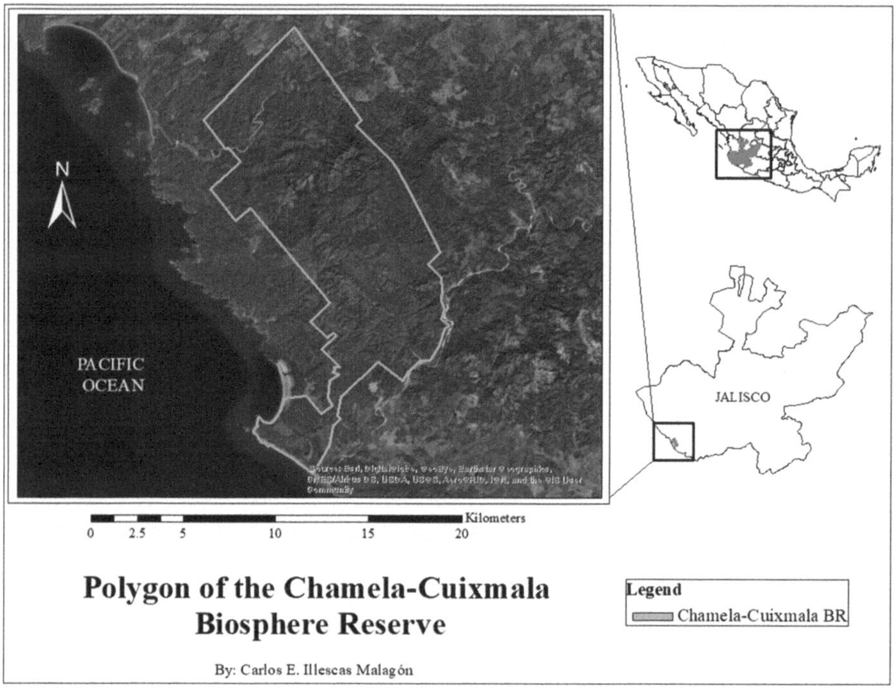

Fig 14.1 Location and polygon of the Chamela-Cuixmala Biosphere Reserve

seasonal streams, known locally as "arroyos" Streams are ephemeral, and during the dry season water is found only in scattered, isolated pools in the arroyos. The study site has a dry tropical climate (temp. 24.9°C) and marked seasonality, with 85% of the 748 mm average annual rainfall occurring from June to November and a prolonged drought from mid-February to late May (Bullock 1986). Dominant vegetation type is tropical deciduous forest, with semi-deciduous forest in the larger arroyos (Lott 1993). White-tailed deer (*Odocoileus virginianus*), collared peccary (*Pecari tajacu*), and coati (*Nasua narica*) are abundant in the area. There are no human settlements and extractive activities within the reserve, though surrounding areas are moderately used for livestock, crops, and touristic activities with a coastal highway bordering the western edge of the reserve.

14.2.2 Radiotracking

From 2000 to 2003, we captured and fitted with VHF radiocollars six adult jaguar (four females and two males) and three adult male pumas. Cats were captured by means of spring-activated "snare"-type traps (Aldrich's snares, Oregon). They were immobilized with a combination of ketamine hydrochloride with xylazine

Table 14.1 Data of jaguars captured and radiotracked and number of locations per year

				Locations per year		
Specie	Sex	Age	Weight (kg)	2000	2002	2003
Puma	Male	Adult	52	61	59	48
Puma	Male	Adult	48	53	52	–
Puma	Male	Adult	45	49	–	–
Jaguar	Female	Adult	30	45	–	–
Jaguar	Female	Adult	45	51	54	–
Jaguar	Female	Adult	30	–	49	54
Jaguar	Female	Adult	37	–	–	46
Jaguar	Male	Adult	58	–	55	51
Jaguar	Male	Adult	61	–	–	37

hydrochloride (Capture all/5) in a ratio of 1:2.5 in a dose of 4:10 mg/kg (Logan et al. 1986, 2001; Table 14.1). VHF radio collars (164 MHz) had a motion sensor (Telonics Inc., Mesa, Arizona). Nine cats were tracked, but annually, five felines were radio tracked simultaneously; and they were not the same individuals during the 3 years of study. One male puma was the only cat tracked during all 3 years, and three jaguars and a puma were monitored for 2 years. A male jaguar, two female jaguars, and a male puma were monitored for a period of 1 year only (Table 14.1).

14.2.3 Home Range

We used locations to establish movement patterns and to estimate home ranges. We used two approaches to determine locations. The first approach involved tracking with three fixed telemetry stations, while the second approach involved aerial readings in a Cessna model Hawk. Each fixed station consists of two 11-element dual-beam Yagi antennas with a null/peak system (Wildlife Materials, Carbondale, Illinois). We calculated the location of the jaguars and pumas by triangulation of the azimuths (angular directions) from the fixed antennas using the TRITEL software (Biggins et al. 1992). TRITEL generates locations in UTM coordinates. The antennas have a triangulation error of 1.6°, which indicates an error of 16 m at 1 km distance. According to the error calculated for the fixed antennas, it was estimated that the locations presented an average error of 0.27 ± 0.24 km^2 (SD). We discarded those readings that could be rebounding or erroneous, and they were not used in further analysis. To calculate area estimates, we used independent locations obtained by radiotelemetry with the Tracker 1.1 program (Camponotus 1994). An independent location is when there is a difference of 12 h or more between them, because in that period, felines can completely cross their home range (Logan et al. 2001). We estimated area used in three different ways. First, we performed home range estimations by means of 95% minimum convex polygon (MCP). The MCP is considered to be the most comparable home range estimator between different studies. We also obtained the core area, calculated as the MCP at 50%. We used a total of 764 locations to estimate the home range of 6 jaguars and 3 pumas (Table 14.1). In 2001, data were taken during the rainy season only.

The movement patterns were described in two ways: (a) The distance between two independent locations on two or more consecutive days an individual was tracked. This type of data gave us information about the linear distance traveled daily. (b) Based on the data obtained during 24 h, the distance traveled in one night was calculated in 30-min intervals, and the linear distance between the initial and ending locations was measured. This type of information allowed us to observe two types of basic behavior, (a) search for food and (b) change in area used. To determine if there were differences in the movement patterns of the jaguars and pumas, the final linear distances traveled of the two species and sexes are compared by means of a one-way ANOVA (Zar 1984).

To determine the movement patterns, in 2 years we completed 27 24-h tracking sessions, 20 sessions in the dry season and 7 in the rainy season. One thousand three hundred twenty eight readings were obtained from seven different cats. Seven hundred and fifty six locations served to determine the movement pattern of two female jaguars, one male jaguar, and three male pumas. We obtained 269 consecutive locations to determine the daily movements of 4 jaguars and 3 pumas during the dry season and 133 during the rainy season. The data of the 24-h cycles were calculated based on the information obtained in the dry season, since in the rainy season the information was limited and fragmented. Data for the rainy season are presented but not compared with dry season data. Data is presented as averages and standard deviations.

14.3 Results

14.3.1 Home Range

Over the 3 years, male jaguars maintained an average annual home range of 100.3 ± 15.0 km^2, male pumas averaged 83 ± 10.3 km^2, and female jaguars averaged 42.5 ± 16 km^2 (Table 14.2). The home range of male jaguars was significantly higher

Table 14.2 Size (km^2) of home range of jaguars and pumas in dry and wet season and annually

	2000				2002				2003		
Ind.	Dry	Wet	Anual	Ind.	Dry	Wet	Anual	Ind.	Dry	Wet	Anual
PM350	45.5	79.5	85.3	PM350	49.5	89.2	97.3	PM350	60	71.9	81.8
PM319	29.2	58.9	67.1	PM319	35.1	72.8	79.7	JM030	53.3	71.8	83
JH610	19.6	24.8	28	JH610	16.8	55.8	57	JH300	17.9	33.5	34.1
PM890	31.8	83.4	91	JM590	58.4	108.5	108	JM590	65.24	97.8	110
JH050	34.9	57.4	66	JH355	23.2	24	26.8	JH355	30.47	32.7	43.1

(68%) than that of female jaguars ($Q = 8.394$, $P < 0.001$), and also that of male pumas was higher (49%) than that of female jaguars ($Q = 7.32$, $P < 0.001$). The home range of the jaguar and the puma presented two distinct phases, one in the dry season and one during the rainy season. The average home range of female jaguars in the dry season was 23.8 ± 7.3 km^2 and 38 ± 14.9 km^2 during the rainy season. The home range of male jaguars was 55.8 ± 3.6 km^2 in the dry season and 92.7 ± 18.8 km^2 in the rainy season. For male pumas we calculated a home range of 41.8 ± 11.8 km^2 in dry and 75.9 ± 10.5 km^2 in rainy season (Table 14.2). During the dry season, the average home range of male jaguars and pumas was significantly higher than that of the females ($Q = 7.5$, $P < 0.001$ and $Q = 4.7$, $P = 0.01$), but between the male jaguars and male pumas, there was no significant difference ($Q = 3.6$, $P = 0.057$) (Table 14.2). During the rainy season, we saw the same trend. Male jaguars and male pumas had a larger home range than female jaguars ($Q = 7.75$, $P < 0.001$ and $Q = 6.58$, $P = 0.002$). Home range polygons are shown in Figs. 14.2, 14.3, and 14.4.

Jaguars and pumas kept their home ranges stable in the dry season; however, in the rainy season, they moved temporarily to areas outside the boundaries of the CCBR. Male pumas increased their home range by 44 ± 15%, jaguar females 31 ± 25%, and male jaguars 34.9 ± 5.9%. The average core area size of the male pumas during the dry season was 13.4 ± 5.9 km^2 and 21.4 ± 7.7 km^2 in rainy season. Female jaguars had a core activity of 6.6 ± 3.1 km^2 in the dry season and 8 ± 4.1 km^2 in rainy season. Male jaguars had an average activity core area of 16.2 ± 5.6 km^2 in dry and 20.9 ± 3.5 km^2 in the rainy season.

During the 3 years of study, the home range of jaguars and pumas overlapped between sexes and between species. Female jaguars showed a 2.6% ± 0.8 level of home range overlap with other jaguars and pumas in the dry season and a level of home range overlap of 3.8% ± 0.4 in the rainy season. Male jaguars overlapped their home ranges at a level of 3.6% ± 0.7 with other jaguars and pumas in the dry season and at 3.9% ± 0.3 in the rainy season. The average area of overlap between two individuals in the dry season was 13.0 ± 10.6 km^2 and 21.1 ± 17.4 km^2 in rainy season (Table 14.3). The percentage of the shared home range of one individual with another was 32.6 ± 24.4% in dry season and 33.0 ± 24.6% in the rainy season.

Considering all the individuals, the average overlap in the dry season between jaguars and pumas was 15.1 ± 10 km^2. In the rainy season, the average overlap between pumas was 25 ± 20.4 km^2. During the rainy season, the arithmetical center of the locations (UTM coordinates) varied when compared to the dry season. On average, the arithmetic centers of all cats moved 2103 ± 1024 m between the dry season and the rainy season. The male pumas moved their arithmetic center 2762 ± 1152 m, the male jaguars 2154 ± 798 m and the female jaguars 1418 ± 521 m. The arithmetic center of the jaguar and puma home ranges varied their location among years. Considering only the dry season, male jaguars and pumas moved their arithmetic center among years an average of 2389 ± 1160 m, and female jaguars moved 1570 ± 362 m (Table 14.4). During the dry season, the distances between simultaneous locations between dyads of jaguars were 4509 ± 2423 m; between pumas it was 4799 ± 2575 m; and between jaguar and puma, it was 4960 ± 2623 m.

Fig. 14.2 Home range (PCM 95%) of jaguars and pumas during 2000 dry (**a**) and wet (**b**) season

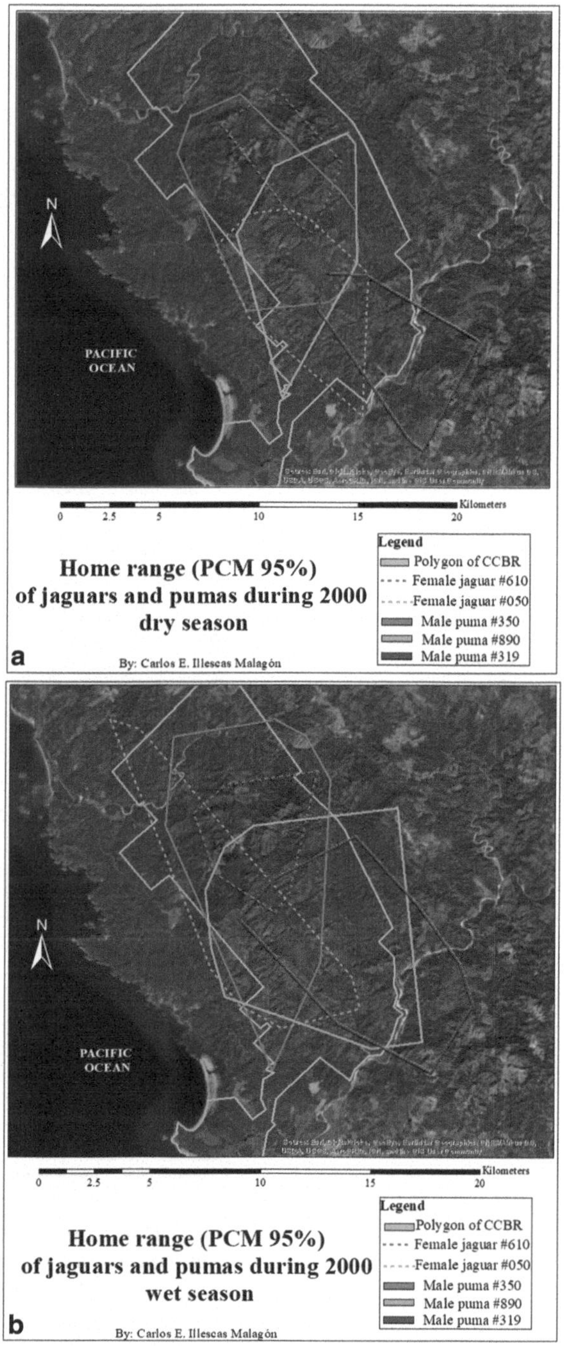

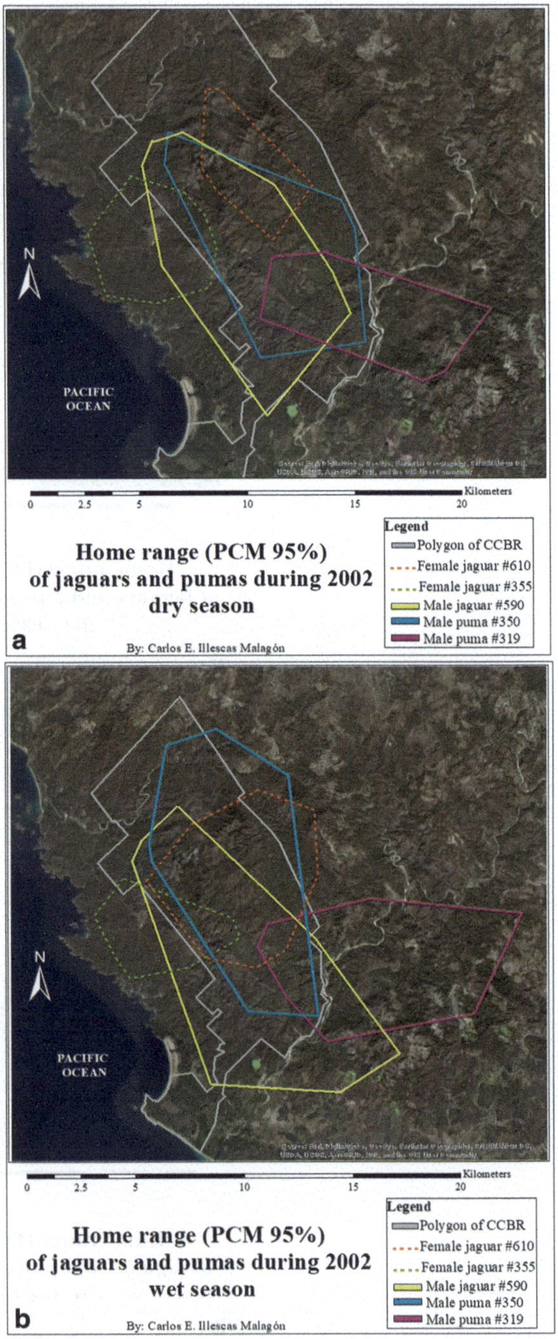

Fig. 14.3 Home range (PCM 95%) of jaguars and pumas during 2002 dry (**a**) and wet (**b**) season

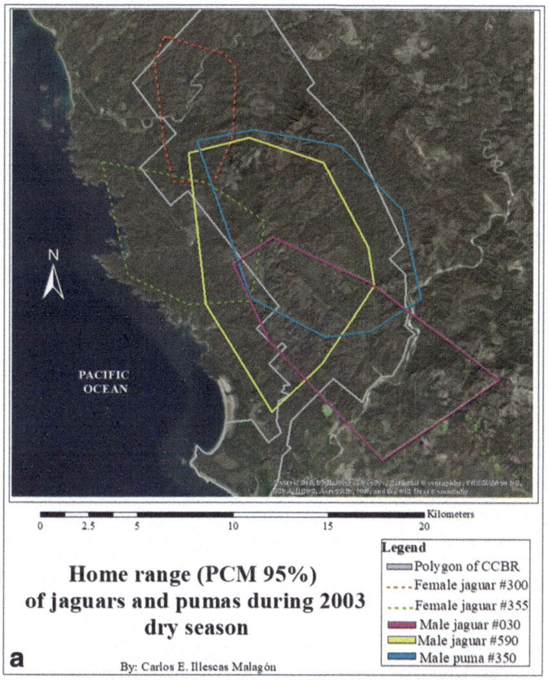

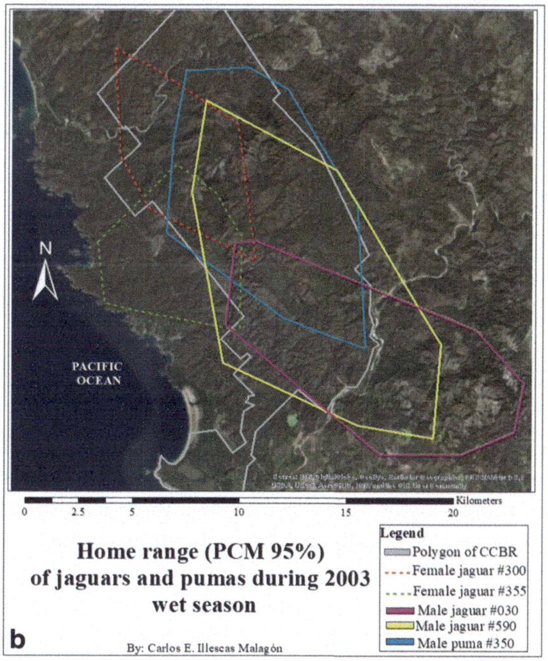

Fig. 14.4 Home range (PCM 95%) of jaguars and pumas during 2003 dry (**a**) and wet (**b**) season

Table 14.3 Home ranges overlap area (average) between jaguars, between pumas, and between pumas and jaguars during dry season and wet season

Overlap area (average)				
	Dry		Wet	
	Overlap km^2 (±SD)	% of home range (±SD)	Overlap km^2 (±SD)	% of home range (±SD)
JJ	12.2 ± 12.6	33.2 ± 33.9	13.7 ± 15	31.1 ± 20
PP	10.2 ± 11.1	26.6 ± 27.7	23.9 ± 20.4	32.4 ± 25.9
JP	15.1 ± 10	36.5 ± 19.3	22.9 ± 18.4	35.3 ± 12.9

Table 14.4 Distance (meters) between the arithmetic center in the rainy season, compared to the dry season in the same year for jaguar and puma individuals

Cat	2000	2002	2003
PM – 350	1880	4230	1944
PM – 319	3808	3100	–
PM – 890	1456	–	–
JH – 050	2050	–	–
JH – 610	758	1425	–
JH – 355	–	1940	1886
JM – 590	–	1490	1211
JH – 300	–	–	1052
JM – 030	–	–	1553

In the rainy season, the simultaneous distances between jaguars were 6527 ± 1967 m; between pumas it was 7234 ± 3460 m; and between jaguars and pumas, it was 7721 ± 1788 m.

14.3.2 Movement Patterns

Regarding movement patterns, male pumas moved over the course of one night on average 2137 ± 1402 m in the dry season and 2898 ± 1688 m in the rainy season (Table 14.5). The male jaguars traveled on average 2673 ± 1606 m daily in the dry season and 2766 ± 1973 m in the rainy season. The female jaguars moved in a period of 24 h an average of 2063 ± 1402 m in the dry season and 1959 ± 1895 m in the rainy season. In the dry season, there were no significant differences in the distance traveled among male jaguars, male pumas, and female jaguars ($H = 3.994$, g.l. = 2, $P = 0.136$). In the rainy season, there were significant differences in the distance traveled by jaguars and pumas ($H = 11.481$, g.l. = 2, $P = 0.003$), while the distance traveled by female jaguars and male jaguars was significantly different ($P < 0.05$). No significant differences were found in the distance traveled in the dry season and in the rainy season ($T = 12764.5$, $P = 0.362$). In the dry season, the female jaguar # 355 had the shortest distance recorded between two consecutive locations (344 m),

Table 14.5 Short movements. Distance (meters) traveled by jaguars and pumas in one night and final distance

	N	Traveled distance	End distance	Rate
Dry				
Jaguar H	21	4987 ± 1629	781 ± 654	6.3:1
Jaguar M	9	6112 ± 1409	1549 ± 541	4.1
Puma M	16	6451 ± 2012	1414 ± 603	4.5:1
Wet				
Jaguar H	9	6256 ± 1112	980 ± 591	6:1
Jaguar M	5	7167 ± 1206	1357 ± 375	5:1
Puma M	7	7782 ± 1474	1634 ± 545	5:1

Table 14.6 Long movements. Distance (meters) traveled by jaguars and pumas in one night and final distance

Secas	N	Traveled distance	End distance	Rate
JH	8	5401 ± 1325	2578 ± 773	2.1:1
JM	6	6309 ± 967	3901 ± 666	1.6:1
PM	11	6530 ± 1833	4269 ± 473	1.5:1

which occurred in the rainy season. The greatest distance between two consecutive points was the male jaguar # 590, who moved 6470 m, which occurred in the dry season. In the rainy season, the male puma # 319 made the longest travel (7867 m) between consecutive locations.

In a 24-h cycle, we identified two basic patterns of movement: (a) short movements and (b) long movements. The types of movements differed basically in the distance traveled with respect to the final linear distance between the initial point and the final point. In the rainy season, we also observed a movement of jaguar which may reflect a process of dispersion or search for a new territory. *Short movements* consist of movements where the linear distance from start to finish is short. Because of meandering, the actual distance traveled can be much greater than the linear distance between start and finish. As an example, on one night of the dry season, jaguar # 050 traveled 2100 m, but its final location was only 350 m from its first location. The ratio of actual distance traveled to linear distance was 6:1. The female jaguar # 610 traveled a total of 1850 m and she returned to the starting location. In the case of male pumas and male jaguars, similar patterns were observed. In one night, the male puma # 350 recorded a total of 4250 mi and finished at a final distance of 1100 m from the starting location, a ratio of 4 to 1. The male jaguar # 590, traveled 2700 m to a final distance of 550 m from its starting location, a ratio of 5 to 1. The average final distance in female jaguars was 781 ± 654 m, in male jaguars it was 1549 ± 540 m, and in male pumas it was 1414 ± 603 m. Males registered greater distances than the females. There were significant differences in the movements made by female jaguars, male jaguars, and male pumas ($F = 3.51$, g.l. = 2, $P = 0.039$) (Table 14.6). During the rainy season, there were no significant differences in the distance traveled by the two species ($F = 2941$, $P = 0.078$).

Long movements are unidirectional and constant, and the final distance traveled is a considerable distance from the starting location. One night, male puma # 319 moved 5600 m, and its final location was 4230 m from the starting location. Male jaguar # 590 traveled 5300 m and finished 3450 m from its starting location. Female jaguar # 050 traveled 3850 m and finished at 2200 m from its starting location. The linear distance for this type of movement was 2578 ± 773 m in female jaguars, 3901 ± 666 m in male jaguars, and 4269 ± 473 m in male pumas. When comparing the groups, we found a significant difference ($F = 14.304$, g.l. = 2, $P < 0.001$). Difference occurred among the female jaguars and the males of both jaguars and pumas ($Q = 7.1$, $P < 0.001$ and $Q = 5.06$, $P = 0.006$ correspondingly). Comparing the distance traveled by all of the different felines, we saw no significant difference ($F = 1.34$, $P = 0.285$).

14.3.3 Exploration Movements

After staying a period of time in the RBCC, the male puma # 319 moved 6 km north-northeast from its home range boundary in the rainy season of 2000. This individual then moved 13 km east-southeast putting him 11 km far from the home area boundary. A month later the puma returned to the CCBR. Jaguar female # 050 also moved 15 km northwest-north of its activity core in the rainy season of 2000. This individual remained a period of 1 month and then returned to the CCBR. After moving side to side within CCBR, she again moved 17 km east-southeast, remained more than a month in that area, and then returned to the CCBR.

14.4 Discussion

In tropical dry forest of CCBR jaguars and pumas both had similar movements and home range sizes. The home range of male jaguars and pumas was larger than that calculated for female jaguars and pumas, as has been observed in most wild cats. Males have large home ranges that cover some female home ranges (Seidensticker et al. 1973; Sunquist 1981; Rabinowitz and Nottingham 1986; Bailey 1993). The average annual home range of jaguar females observed in CCBR is greater than that reported for rain forest of Belize (10–11 km^2) (Rabinowitz and Nottingham 1986); and the wetlands of Pantanal, Brazil (47–83 and 50 km^2, respectively) (Crawshaw and Quigley 1991; Soisalo and Cavalcanti 2006); and smaller than in the Llanos de Venezuela (80 and 70 km^2) (Scognamillo et al. 2003), subdeciduos forest of Calakmul (133–244 km2) (Chávez 2009) and Lacandon Rain Forest, Mexico (181 km^2) (De la Torre et al. 2017). Male jaguars home range are larger than those estimated or rain forest of Belize (28–40 km2) (Rabinowitz and Nottingham 1986) and similar to Pantanal (93–130 km2) (Crawshaw and Quigley 1991) and Llanos of Venezuela (100 km2) (Scognamillo et al. 2003) and smaller than male jaguars of

Calakmul (191–725 km2)(Chávez 2009) and Lacandon Forest (290 km2) (de la Torre et al. 2017). Female pumas in the CCBR (Núñez et al. 2002) home range that varied from 25 to 60 km^2, which is consistent with that reported for South America (23–47 km^2) (Scognamillo et al. 2003) and subtropical areas of Florida (39 km^2) (Comiskey et al. 2002) but larger than reported for de la Torre et al. (2017) (21 km^2) in Lacandon Forest. Female pumas of the temperate and semiarid environments had considerably larger ranges (60–685 km^2) (Seidensticker et al. 1973; Anderson et al. 1992; Franklin et al. 1999; Pittman et al. 1999; Logan and Sweanor 2000) than tropical dry forest of CCBR.

Male pumas had an average home range similar to that observed in the tropical and subtropical areas such as Calakmul (78 km^2) (Chávez 2009), the plains of Venezuela (17–100 km^2) (Scognamillo et al. 2003), Lacandon Forest (82.2 km^2) (de la Torre et al. 2017), and the Florida everglades (60 km^2) (Comiskey et al. 2002). Like female pumas, home range sizes of male pumas in the CCBR were lower than the estimated average for the temperate areas of South America and North America (from 100 to 800 km^2) (Franklin et al. 1999; Seidensticker et al. 1973; Anderson et al. 1992; Logan and Sweanor 2000).

Although jaguar and puma males are larger than the females, this does not explain the variation in home range size. Hopkins (1989) found no correlation between the size of the individuals and puma home range. This difference in home range parameters has been related to the polygynous behavior of males (Seindensticker et al. 1973; Kitchener 1991). Males maintain a larger home range to increase their access to females, thus increasing the chance of breeding and protecting those females from breeding with other males (Sunquist 1981; Kitchener 1991; Logan and Sweanor 2000). Females, on the other hand, look for areas with dense prey and safe places to rear young (Kitchener 1991). In addition to dispersion of prey, the physiographic nature of the landscape also affects size and form of home ranges (Gittleman and Harvey 1982; Sandell 1989; Grigione et al. 2002).

The reproductive status of females is also a factor that influences the size of the home range. Female jaguars and pumas with kittens initially use much less territory, but they expand their movements as their kittens age. For the first 6 weeks after birth in the dry season, females reduced their home range from about 25 to 10–15 km^2, and they return to the same site nightly. Thus, the reason prey density is important to females. After 6 weeks, the female gradually starts moving the kitten to different sites. When the kitten is close to dispersal, the female may be using twice the habitat that she would occupy without a kitten. This type of movement pattern has also been observed in other felines (Logan and Sweanor 2001; Grigione et al. 2002).

All the monitored cats maintained most of their home range within the boundaries of the CCBR, although males of both species occasionally left the Reserve when patrolling their ranges. The home range estimated on the basis of the minimum convex polygon underestimates the real size of the home range, since it does not consider the irregularity of the landscape (Powell 2000). The CCBR is a hilly area, and three of the monitored females (# 050, 355, 300) used mainly slope areas, while the largest female # 610 used mainly summit areas. Males of both species, they used both hills and peaks. The home range of felines is not static, and the ranges of the felines vary from 30% to 45% smaller in the dry season than they are in the wet season. The term of rainy season extends past the time of precipitation to include the time that the forest and the understory maintain its foliage and water is still available in most of the natural water pools.

The seasonal change of the home range of the carnivores has been explained mainly as a response to the change in the distribution and abundance of the prey (Sunquist 1981; Sandell 1989; Bailey 1993; Grigione et al. 2002), as well as climatic factors such as rain (Quigley 1987) or snowfall (Grigione et al. 2002). Results obtained on relative abundances for white-tailed deer in the two seasons support this fact. In the dry season, preys are concentrated near streams and pools due to the scarcity of water (Mandujano et al. 2002; Valenzuela and Ceballos 2000). This means felines can invest less time and energy in searching for prey by staying near the isolate sources of water during the dry season (Wolff 2001). Even though carnivores can survive dry conditions (Schmidt-Nielson 1997), they stay near water to efficiently hunt prey. Furthermore, the cover offered by plants is minimal during the dry season because in the dry tropical forest plants lose their leaves. Thus jaguars and pumas, like other felines, rarely explore new areas without the safety of plant cover (Sunquist 1981; Quigley 1987; Beier 1995).

The abundance of water during the wet season creates a wider dispersion of prey, which is reflected in the increased size of feline home ranges for hunting. During the rainy season, water is not limited and cats can move larger distances; besides dense vegetation offers protection cover and favors greater movements and incursions in other areas (Sunquist 1981; Quigley 1987; Beier 1995). The variation between the home range arithmetic center of the dry season and the rainy season was lower in females than the males, as well as between the arithmetic centers of the dry season from 1 year to the next. This suggests that female jaguars are more faithful to home range than males and underscores the value of high-quality habitat for females.

Animal movements show important information about a species and its natural history; it's an important matter of animal behavior (Nielsen 1983). One type of movement is characterized by maintaining a short distance between the starting location and the ending location. This could suggest a search of food. The short movements varied between female and male jaguars with females traveling less than males. The second type of movement was characterized by having a long distance between the starting location and the ending location, and it could be explained as a movement of change to a new hunting area. A third type of erratic and sometimes unidirectional movement could be indicative of dispersal and/or change of home range. This type of movement has been observed previously in the study area, which was a prelude to dispersion and/or home range change (Núñez et al. 2002). These movements have also been identified in other feline studies (Beier 1995; Logan and Sweanor 2001).

Generally these "exploratory outings" happened 1 or 2 times and were during the months of highest rainfall (Aug–Sept). During these trips, jaguar # 030 and puma # 319 visited areas with moderate disturbance and were always located in areas with vegetation high cover. Beier (1995), Logan and Sweanor (2001), and Quigley and Crawshaw (2002) found a similar pattern in the dispersal of pumas and jaguars.

Unlike these authors where the dispersing individuals were juveniles or subadults, male jaguars and pumas of this study were adults with a defined home range.

When comparing the locations of the jaguars and pumas, both species showed greater association to streams than to the deciduous forest. While jaguars associated most with the main streams, pumas also used the small tributaries that ventured into the hills. Perhaps using tributaries entering the forest could explain black iguana (*Ctenosaura pectinata*) in the puma diet (Núñez et al. 2002).

Streams have several advantages over the deciduous forest. The streams offer suitable cover for good hunting opportunity as the rocks and nooks make stalking easier. Approximately 90% of the deer found eaten by felines were located on banks of streams. During the dry season, water only exists in a few isolated pools so deer and coatis concentrate their activities near these places (Mandujano et al. 2002). Streams also offer a clean traveling path, and big cats wild cats like to use those paths to move when they are not hunting (Schaller 1967; Taber et al. 1997). During the rainy season, jaguars and pumas continued using the streams too.

Results obtained in this study show a broad spatial overlap. The most influential factor in the coexistence of species is the diversity and abundance of available prey (Seidensticker et al. 1973; Rosenzweig 1966; Donadio and Buskirk 2006). In the CCBR, the abundance of large prey such as the white-tailed deer and the pecari is high, and in addition medium-size preys such as the coati and the armadillos are in adequate numbers and allow a high food niche overlap between both species in CCBR (Núñez et al. 2000). Thus, the coexistence of the jaguars and puma is viable. In particular the abundance of the white-tailed deer provides sufficient biomass to meet the needs of the two predator species. The availability of other medium-size species also contributes to the ecological coexistence; the puma consumes more iguanas than the jaguar (Núñez et al. 2000). The mutual avoidance by distance also contributes to the coexistence and to reduce antagonisms between the species. However, more detailed analyses, using quickly evolving and advanced technology and analytical tools from movement ecology, are urgently needed on the movements of jaguar and pumas across landscape, particularly across differing degrees of habitat disturbance and fragmentation. Thus, detailed studies on movement patterns may also greatly contribute to the understanding of the social and spatial organization of the wildcats.

14.5 Conclusions

The jaguar and puma home range size is similar; and the home range of males was greater than that of females. The overlap in home range between jaguar pumas was extensive and there was no spatial segregation at this level. The overlap between jaguars and pumas was reduced and spatial segregation was observed at the core area level. The home range of jaguars and pumas was greater during the rainy season than the dry season. Jaguars and pumas use mainly the streams to carry out their activities. In the rainy season, the activity of the jaguars and pumas increased. There was no significant variation in the distance traveled by male jaguars and pumas. The distance traveled by males was greater than that of the females. There was no significant change in the distance traveled during the dry and rainy season.

14.5.1 Implications for Conservation

The CCBR is an important refuge for wild cats and their habitat on the Jalisco coast and could be considered as a source population site. Without the proper protection in the area adjacent to CCBR, the possibility of losing the jaguars and pumas in the medium term is a reality. The size of the home range of jaguars and pumas shows that the CCBR is too small to protect a large population of jaguars and pumas. The Reserve could accommodate the home ranges of one to two males or three to four females without overlapping. The small size of the CCBR exposes felines to a greater encounter with humans. In most of the individual's home ranges, some portion remained outside the limits of the reserve. According to Woodroffe and Ginsberg (1998), carnivores that inhabit small reserves are more likely to perish due to the edge effect and greater contact with humans. The CCBR is a refuge for a limited number of cats, and it is necessary and urgent to increase the size of the reserve to accommodate a larger population. Populations with few individuals and/or isolated from other populations are more susceptible to negative environmental effects such as stochastic and genetic effects. Some survey suggest that under low pressure, 50 individuals are necessary to maintain a viable population for medium-long term (100 years) (Sollmann et al. 2008). To maintain a viable population, CCBR must increase in size or/and maintain corridors to keep connectivity with other populations. Because of territoriality, unequal sex ratios, and unequal contribution to the next generation, effective populations sizes are smaller than actual population numbers.

Habitat fragmentation and persecution are the greatest threats to the conservation of large carnivores (Woodroffe and Ginsberg 1998). Habitat fragmentation is occurring at an accelerated rate the coast of Jalisco. Twenty percent of the original forest along the coast of Jalisco was transformed between 1973 and 1989 and continues at an annual rate of 1.9% (Miranda 1998). Approximately 30% of the original forest of the coast of Jalisco remains preserved (Arriaga et al. 2000). Given this scenario, establishing a large protected area is impossible, but a network of smaller reserves connected through corridors would help maintain a long-term cat population and increase the likelihood of genetic exchange (Beier 1995; Eizirik et al. 2002). Data on movements are useful for corridor design. In Jalisco, there is still the possibility of maintaining habitat corridors with the Sierra de Manantlán Biosphere Reserve and the south of Nayarit with the Sierra de Vallejo Reserve and thus guarantee the conservation of a long-term population. Streams and their associated vegetation in the CCBR are where the felines concentrate their activities and move, so it is important to protect this type of vegetation to favor the dispersal of juveniles looking for new places to settle.

Acknowledgments We would like to thank to Goldsmith Family for their interest and logistic and economic support for this project. Thanks to Efren Campos, Alvaro Miranda and Marciano Valtierra of Cuixmala Ecological Foundation for their support. Thanks to Denver Zoo Foundation, and Chamela Biological Station (National University of Mexico) by funding and logistic support provided to perform this study. SEMARNAT (environmental agency) granted permits for capturing the jaguars and pumas. U.S. Fish and Wildlife Service assisted with the telemetry antennas. Thanks to Dr. Fred Lindsey of Wyoming University for his advising and support for this project. Special thanks to Saul Vazquez and Drs. Katherine Renton, Alejandro Salinas, and David Valenzuela for their support during radiotracking activities, field work and friendship. Thanks to Dr. Kathryn Stoner by tutoring my master degree at National University of Mexico. Biol Ivonne Juarez helped during desk work and biol. Carlos Illescas drew maps.

References

Anderson AE, Bowden DC, Kattner DM (1992) The puma on Uncompahgre Plateau, Colorado, Technical Publication No. 40. Colorado Division of Wildlife, Denver 116 pp

Arriaga L, Espinoza JM, Aguilar C, Martínez E, Gómez L, Loa E (eds) (2000) Regiones terrestres prioritarias de México. Comisión Nacional para el Conocimiento y Uso de la Biodiversidad. México. Blackwell Science, Cambridge, MA 470 pp

Azevedo FCC, Murray DL (2007) Evaluation of potential factors predisposing livestock to predation by jaguars. J Wildl Manag 71:2379–2386

Bailey TN (1993) The African leopard: ecology and behaviour of a solitary felid. Columbia University Press, New York

Beier P (1995) Dispersal of juvenile COGUAR in fragmented habitat. J Wildl Manag 59:228–337

Biggins ED, Herebury RL, Godbey JL (1992) Instructions for radio-tracking and using program Tritel. US Fish and Wlidlife Service, Fort Collins

Bullock SH (1986) The climate of Chamela Jalisco and trended in the south coastal region of Mexico. Arch Meteorol, Geophys Bioclimatol Ser B 36:297–316

Burt WH (1943) Territoriality and home ranges concepts as applied to mammals. J Mammal 24:236–352

Camponotus AB (1994) Tracker ver. 1.1 wildlife tracking and analysis software. User manual. Primera edición, Suiza

Carbone C, Gittleman JL (2002) A common rule for the scaling of carnivore density. Science 295:2273–2276

Carbone C, Teacher A, Rowcliffe J (2007) The costs of carnivory. PLoS Biol 5:1–6

Cardillo M, Purvis A, Sechrest W, Gittleman JL, Bielby J, Mace GM (2004) Human population density and extinction risk in the world's carnivores. PLoS Biol 2:E197

Cavalcanti SM, Gese EM (2009) Spatial ecology and social interactions of jaguars (*Panthera onca*) in the southern Pantanal, Brazil. J Mammal 90:935–945

Chávez C (2009) Ecología poblacional y conservación de Jaguar (*Panthera onca*) y del puma (Puma concolor), en la region de Calakmul, y sus implicaciones para su conservacion en la Peninsula de Yucatan, Mexico. Ph.D. Thesis. University of Granada, Spain. 66 pp

Chetkiewicz C-LB, Boyce MS (2009) Use of resource selection functions to identify conservation corridors. J Appl Ecol 46:1036–1047

Comiskey EJ, Bass OL Jr, Gross LJ, McBride RT, Salinas R (2002) Panthers and forests in South Florida: an ecological perspective. Conserv Ecol 6(1):18

Conde DA, Colchero F, Zarza H, Christensen NL, Sexton JO, Manterola C et al (2010) Sex matters: modeling male and female habitat differences for jaguar conservation. Biol Conserv 143:1980–1988

Crawshaw PG, Quigley HB (1991) Jaguar spacing, activity and habitat use in a seasonally flooded environment in Brazil. J Zool Lond 223:357–370

Crooks K (2002) Relative sensitivities of mammalian carnivores to habitat fragmentation. Conserv Biol 16(2):488–502

De la Torre A, Nuñez J, Medellin R (2017) Spatial requirements of jaguars and pumas in Southern Mexico. Mamm Biol 84:52–60

Donadio E, Buskirk SW (2006) Diet, morphology, and interspecific killing in Carnivora. Am Nat 167(2006):524–536

Eizirik E, Indrusiak CB, Johnson WE (2002) Jaguar population viability analyses: evaluation of parameters and a case study of a remnant population in a South American subtropical rain forest. In: Medellin RA, Chetkiewicz C, Rabinowitz A, Redford KH, Robinson JG, Sanderson E, Taber A (eds) Jaguars in the new millennium. A status assessment, priority detection, and recommendations for the conservation of jaguars in the Americas. Universidad Nacional Autonoma de Mexico/Wildlife Conservation Society, Mexico, pp 505–522

Foster VC, Sarmento P, Sollmann R, Torres N, Jacomo A, Negroes N, Fonseca C, Silveira L (2013) Jaguar and puma activity patterns and predator-prey interactions in four Brazilian biomes. Biotrop 0(0):1–7

Fahring L (2007) Non-optimal animal movement in human-altered landscape. Fun Ecol 21:1003–101

Franklin WL, Johnson WE, Sarno RJ, Iriarte JA (1999) The ecology of the Patagonia puma, *Felis concolor* Patagonia, in southern Chile. Biol Conserv 90:33–40

Gittleman JL, Harvey PH (1982) Carnivore home range size, metabolic needs and ecology. Behav Ecol Sociobiol 10:57–63

Grigione MM, Beier P, Hopkins RA, Neal D, Padly WD, Schonewald CM, Johnson ML (2002) Ecological and allometric determinants of home-range size for mountain lions (*Puma concolor*). Anim Conserv 5:317–324

Harmsen BJ, Foster RJ, Silver SC, Ostro LET, Doncaster CP (2009) Spatial and temporal interactions of sympatric jaguars (*Panthera onca*) and pumas (*Puma concolor*) in a Neotropical forest. J Mammal 90:612–620

Hopkins RA (1989) Ecology of the puma in the Diablo Range, California. Dissertation. University of California at Berkeley, Berkeley, CA, USA

Herfindal I, Linnell J, Odden J, Nilsen E, Andersen R (2005) Prey density, environmental productivity and home-range size in the Eurasian lynx (Lynx lynx). J of Zool Lond 265:63–71

Iriarte AJ, Franklin WL, Jonhson WE, Redford KH (1990) Biogeographic variation of food habits and body size of the America puma. Oecol 85:185–190

Jedrzejewski W, Schmidt K, Okarma H, Kowalczyk R (2002) Movement pattern and home range use by the Eurasian lynx in Białowieża Primeval Forest (Poland). Ann Zool Fenn 39:29–41

Kitchener A (1991) Natural history of wild cats. Cornell University Press, Ithaca 280 pp

Logan KA, Thorne ET, Irwin LL, Skinner R (1986) Immobilizing wild mountain lions (*Felis concolor*) with ketamine hydrochloride and xylazine hydrochloride. J Wildl Dis 22:97–103

Logan KA, Sweanor L (2000) Puma. In: Demarias A, Krausman PR (eds) Ecology and management of large mammals in North. Prentice-Hall, Upper Saddle River, pp 347–377

Logan K, Sweanor L (2001) Desert puma. Evolutionary Ecology and Conservation of an Enduring Carnivore. Island Press, Washington.

Lott E (1993) Annotated checklist of the vascular flora of the Chamela Bay region, Jalisco, Mexico. Occasional Papers of California Academy of Sciences, USA. No. 148. 60 pp

Laundré J, Hernández L (2010) What we know about pumas in Latin America. In: Hornocker M, Negri S (eds) Cougar, ecology and conservation, The University of Chicago Press. Chicago, USA, pp 76–90

Lindstedt SL, Miller BJ, Buskirk SW (1986) Home range, time, and body size in mammals. Ecol 67(2):413–418

Mandujano S, Gallina S, Arceo G, Sánchez-Rojas G, Silva-Villalobos G (2002) Venado Cola Blanca (*Odocoileus virginianus sinaloae* J. A. Allen). In: Historia Natural de Chamela. Noguera F, García A, Quesada M (eds). Instituto de Biología UNAM, Mexico. pp 415–422

Mandujano S, Gallina S, Arceo G, Peréz J (2004) Variación estacional del uso y preferencia de los tipos de vegetación por el venado cola blanca en un bosque tropical de Jalisco. A Zool Mex 20:45–67
Miranda AG (1998) Deforestación y fragmentación del hábitat: consecuencias ecológicas sobre la fauna de mamíferos de la selva tropical estacional. Report CONABIO. 46 pp
Morato RG, Stabach JA, Fleming CH, Calabrese JM, de Paula RC, Ferraz KMPM, Kantek DLZ, Miyazaki SS, Pereira TDC, Araujo GR, Paviolo A, De Angelo C, Di Bitetti MS, Cruz m P, Lima F, Cullen L, Sana DA, Ramalho EE, Carvalho MM, Soares FHS, Zimbres B, Silva MX, Moraes MDF, Vogliotti A, May JA, Haberfeld M Jr, Rampim L, Sartorello L, Ribeiro MC, Leimgruber P (2016) Space use and movement of a neotropical top predator: the endangered jaguar. PLoS One 11(12):1–17
Nielsen ET (1983) Relation of behavioural activity rhythms to the changes of day and night: a revision of views. Behav 89:147–173
Nielsen C, Thompson D, Kelly M, Lopez-Gonzalez CA (2015) *Puma concolor* (errata version published in 2016). The IUCN Red List of Threatened Species 2015: Downloaded on 28 June 2018
Núñez PR, Miller B, Lindzey F (2000) Food habits of jaguars and pumas in Jalisco, México. J Zool Lond 252:373–379
Núñez R, Miller B, Lindzey F (2002) Ecología del jaguar en la Reserva de la Biosfera Chamela-Cuixmala, Jalisco, México. In: Medellín RA, Equihua C, Chetkiewicz CLB, Crawshaw PG Jr, Rabinowitz A, Redford KH, Robinson JG, Sanderson EW, Taber AB (eds) El Jaguar en el nuevo milenio. Fondo de Cultura Económica, Universidad Nacional Autónoma y Ediciones Científicas Universitarias, México, pp 107–126
Noguera F, Vega J, García A, Quesada M (2002) Historia Natural de Chamela Instituto de Biologia, UNAM, México D.F. México, pp 568
Pittman M, Guzman G, McKinney BP (1999) Ecology of the mountain lion on Big Bend Ranch in the Trans-Pecos Texas. Final Report. Texas Parks and Wildlife, Texas, USA
Powell RA (2000) Animal home ranges and territories and home range estimators. In: Boitani L, Fuller TK (eds) Research techniques in animal ecology: controversies and consequence. Columbia University Press, USA. pp 65–110
Quigley HB (1987) Ecology and conservation of the jaguar in the Pantanal region, Mato Grosso do Sul, Brazil. PhD dissertation, University of Idaho, Moscow, ID, USA
Quigley HB, Crawshaw PGJ (2002) Reproducción, crecimiento y dispersión del jaguar (*Panthera onca*), en la región del Pantanal Brazil. In: Medelín RA, Equihua C, Chetkiewicz CLB, Crawshaw PG Jr, Rabinowitz A, Redford KH, Robinson JG, Sanderson EW, Taber AB (eds) El Jaguar en el nuevo milenio. Fondo de Cultura Económica, Universidad Nacional Autónoma de México, Wildlife Conservation Society, City of México, pp 289–302
Quigley H, Foster R, Petracca L, Payan E, Salom R, Harmsen B (2017) *Panthera onca*. The IUCN Red List of Threatened Species 2017: e.T15953A50658693. doi:https://doi.org/10.2305/IUCN.UK.2017-3.RLTS.T15953A50658693.en. Downloaded on 28 June 2018
Rabinowitz A, Nottingham BG (1986) Ecology and behaviour of the jaguar (*Panthera onca*) in Belize, Central America. J Zool Lond 210:149–159
Ripple W, Estes JA, Beschta RL, Wilmers C, Ritchie E, Hebblewhite M, Berger J, Elmhagen B, Letnic M, Nelson M, Schmitz O, Smith D, Wallach A, Wirsing AJ (2014) Status and ecological effects of the world's largest carnivores. Science (New York, NY) 343:1241484. https://doi.org/10.1126/science.1241484
Rodriguez-Soto C, Monroy-Vilchis O, Maiorano L, Boitani L, Faller JC, Briones MÁ, Núñez R, Rosas-Rosas O, Ceballos G, Falcucci A (2011) Predicting potential distribution of the jaguar in Mexico: identification of priority areas for conservation. Divers Distrib 17:350–361
Rosenzweig M I (1966) Community structure in sympatric carnivore. J Mammal 47(4):602–612
Rosas O, Nuñez R (2014) El jaguar y el puma. In: Wildlife of México. Valdez R, Ortega A (eds) Colegio de Posgraduados, SLP, Mexico, pp 85–96

Sandell M (1989) The mating tactics and spacing behavior of solitary carnivores. In: Gittleman JL (ed) Carnivore behavior, ecology and evolution. Cornell University Press, New York, pp 164–182

Sanderson EW, Chetkiewics B, Medellin R, Rabinowitz, Redford, Robinson J, Taber AT (2002) Un análisis del estado de conservación y distribución de los jaguares a través de su área de distribución. In: Medellín RA, Chetkiewicz C, Rabinowitz A, Redford KH, Robinson JG, Sanderson E, Taber A (eds) El jaguar en el nuevo milenio. Una evaluación de su estado, detección de prioridades y recomendaciones para la conservación de los jaguares en América. Universidad Nacional Autónoma de México, México, pp 555–584

Schaller GB (1967) The deer and the tiger. University of Chicago Press, Chicago, USA

Schmidt-Nielsen K (1997) Animal physiology, adaptation, and environment, 5th edn. Cambridge University Press, Cambridge, UK

Scognamillo DG, Maxit IE, Sunquist M, Polisar J (2003) Coexistence of jaguar (*Panthera onca*) and puma (*Puma concolor*) in a mosaic landscape in the Venezuelan llanos. J Zool Soc Lond 259:269–279

Seidensticker J, Hornocker MG, Wiles WV, Messick JP (1973) Mountain lion social organization in the Idaho Primitive Area. Wildl Monogr 35:1–60

Soisalo MK, Cavalcanti SMC (2006) Estimating the density of a jaguar population in the Brazilian Pantanal using camera-traps and capture–recapture sampling in combination with GPS radio-telemetry. Biol Conserv 129:487–496

Sunquist M (1981) The social organization of tiger (*Panthera tigris*) in Royal Chit wan National Park, Nepal, Smithsonian Contribution to Zoology No. 336. Smithsonian Institution Press, Washington, DC

Sollmann R, Torres N, Silveira L (2008) Jaguar conservation in Brazil: the role of protected areas. Cat News 4:15–20

Taber BA, Novaro A, Neris N (1997) The food habits of sympatric jaguar and puma in the Paraguayan Chaco. Biotrop 29(2):204–213

Trejo I, Dirzo R (2000) Deforestation of seasonally dry tropical forest. Biol Cons 94(2):133–142

Valenzuela D, Ceballos G (2000) Habitat selection, home range and activity of the white-nosed coati (Nasua narica) in a Mexican tropical dry forest. J Mamm 81(3):810–819

Wolff F (2001) Vertebrate ecology in Caatinga: A. Distribution of wildlife relation to water. B. Diet of pumas (Puma concolor) and relative abundance of felids. St. Louis. Tese (Mestrado) – University of Missouri, USA. 65 pp

Woodroffe R, Ginsberg JR (1998) Edge effects and the extinction of populations inside protected areas. Science 280:2126–2128

Zar JH (1984) Biostatistical analysis. Prentice Hall, USA. 619 pp

Chapter 15
Next Moves: The Future of Neotropical Mammal Movement Ecology

Rafael Reyna-Hurtado and Colin A. Chapman

This book compiles a remarkable array of studies dealing with Neotropical mammal movement patterns and therefore presents a unique opportunity to analyze the state of the art of movement ecology of some of the rarest and secretive species that are top predators, important prey to those predators, and/or critical to maintaining the ecological services of the forest ecosystems they inhabit. In this last chapter, we attempt to summarize lessons learned from all chapters and advance the field with respect to our understanding of the causes and consequences of animal movements in tropical forests.

Tropical ecosystems and its wildlife are in peril globally due to conversion for agriculture and cattle ranching. These forests are being degraded and destroyed at a very high rate. Between 2000 and 2012, tropical forest loss increased by 2101 km^2 per year (Hansen et al. 2013). For example, 20% of Amazon forest has been destroyed in the last 40 years, and in the last 20 years, Indonesia has lost a third of its forest primarily due to the expansion of palm plantations (Hansen et al. 2013; Global Forest Watch; https://www.globalforestwatch.org). In many areas, forests have been left standing, but hunting has eradicated the wildlife. This phenomenon was first described in Neotropical forest almost 25 years ago by Redford (1992), and

R. Reyna-Hurtado (✉)
El Colegio de la Frontera Sur (ECOSUR), Department of Biodiversity Conservation, Lerma, Campeche, Mexico

The Wildlife Conservation Society (WCS), Bronx, NY, USA
e-mail: rreyna@ecosur.mx

C. A. Chapman
Department of Anthropology, McGill University, Montreal, QC H3A 2A7, Canada

School of Life Sciences, University of KwaZulu-Natal, Scottsville, Pietermaritzburg, South Africa

Shaanxi Key Laboratory for Animal Conservation, Northwest University, Xi'an, China.
e-mail: colin.chapman@mcgill.ca

R. Reyna-Hurtado, C. A. Chapman (eds.), *Movement Ecology of Neotropical Forest Mammals*, https://doi.org/10.1007/978-3-030-03463-4_15

he called these areas the "empty forests." There is not doubt that tropical forest and its wildlife are fragile and will disappear if we do not protect them, yet these forests harbor remarkable levels of biodiversity, are resilient ecosystems that can recover if we improve protective measures and management, and are simply amazing.

In this book many of the authors were motivated by the curiosity of understanding how and why an animal moves in highly biodiverse Neotropical forests that at the same time are disappearing at a high rate or are facing tremendous hunting pressure. Understanding how animals adapt their movement to survive in such environments and what kind of relationships or synchronies they have built with other species is fascinating and has driven the authors of the chapters in our book to spend years researching movement in the field, many times under challenging or frustrating conditions.

This is illustrated in Chap. 2 as Jordan and his collaborators were luckily, or unluckily, presented with the opportunity to compare movement of the largest of all Neotropical mammals, the Baird's tapir, before and after a hurricane hit this tropical forest in Nicaragua near the border with Costa Rica. Amazingly, tapirs seem to do well after the disturbance and were able to feed on the super rich variety of herbaceous vegetation that quickly grew after many trees were toppled over during the storm. However, the species' movements were more restricted after the hurricane, probably due to areas becoming impossible to reach because of the obstacles created by the fallen trees and branches on the forest floor. Comparing normal patterns (before hurricane) versus altered movement patterns (after hurricane) provides insights into how animals cope with a disturbance event.

Another excellent example of animal resilience is the chapter by Serio-Silva and collaborators (Chap. 9, this volume) who describe the terrestrial locomotion for black howlers in southern Mexico in an area where forest fragments are increasingly being destroyed and the distances among feeding locations are increasing. However, this flexibility and apparently adaptive behavior comes at the cost of increased risk of predation by domestic dogs that are common in the area. In the absence of conservation interventions, howler monkey in the area will soon be isolated within large forest fragments or groups of smaller, adjacent fragments that are near each other, corralled by feral dogs. These small genetically isolated groups will not be viable and the species will become locally extinct.

Four chapters deal with a species that present some of the most intriguing movement behaviors of the Neotropics, the white-lipped peccary. This species perform some of the longest and most diverse movement patterns ever described for a mammal. Researchers in all Neotropical forest are well aware that white-lipped peccaries movement patterns are not replicated annually and remain as unpredictable up to date (Reyna-Hurtado et al. 2012). Another characteristic that makes this species fascinating is that it is highly social and moves in very large groups that sometime include hundreds of individuals (Reyna-Hurtado et al. 2016). The opportunity to include four chapters about white-lipped peccary movement ecology was unique and very valuable. White-lipped peccaries movements are affected by hunting pressure as demonstrated by Moreira et al. (Chap. 3, this volume) where a group living in a hunted area always keep moving, even in the dry season when other groups living in protected areas just congregate around the few remaining water sources. The fear of hunters likely drives this unique movement pattern. In Darien National Park, Panama, white-lipped peccaries can afford to remain in relatively small areas of old growth tropical forest that provide sufficient resources (Meyer et al., Chap. 6, this volume), while in French Guiana white-lipped peccaries congregate around old growth forest with occasional trips to the edge of savanna areas (Richard-Hansen et al., Chap. 5, this volume). In the Pantanal of Brazil, white-lipped peccary groups presented similar behaviors and increased their home range in areas that have less forest than in forested areas (Jorge et al., Chap. 4, this volume). While these chapters offer great examples of this species' ability to modify its behavior in response to man-made disturbances and in a matrix of different habitat types, they also highlight that additional research is needed to determine under what conditions this species cannot be resilient in terms of its movement patterns and what conditions may lead to this species becoming locally extirpated given that it is one of the Neotropical mammals most sensitive to threats.

Grotta-Neto and collaborators (Chap. 7, this volume) present an interesting review about what is known on the movement of a poorly known group of mammals, the tropical forest deer. This group is composed of approximately ten species, mostly of the genus *Mazama*. They are shy species which are difficult to capture to fit them with GPS devices. However, studies of *Mazama gouazoubira* in the Pantanal of Brazil have shown that they can modify their ranging patterns to transverse non-preferred habitats and perform straighter movements when needed. Along the same lines, the white-tailed deer (*Odocoileus virginianus*) inhabiting the swampy areas of Campeche, Mexico (Contreras-Moreno et al., Chap. 8, this volume), was studied using radiotelemetry techniques that revealed deer move more in the dry season than in the flooded season.

Conditions in tropical forests make it extremely difficult for researchers to conduct behavioral studies on the species that inhabit them. This is partly the reason why movement and social behavior of woolly monkeys have been a mystery for so long. Nevertheless, Ellis and DiFiore (Chap. 10, this volume) managed to maintain simultaneously visual contact with four identifiable groups, obtaining detailed behavioral information. Their research indicated that groups could tolerate and share large areas, while keeping spatial cohesion within the group. They also found that increases in movements occurred mostly in the mating season and changes in food availability did not affect movement to a great extent. This study was unique as this took place in a protected Ecuadorian forest with very limited disturbance. Woolly monkeys are unfortunately highly prized for their meat by subsistence hunters (Alvard et al. 1997).

With the same species (woolly monkeys) but in a rugged terrain of Colombia, Garcia et al. (Chap. 11, this volume) tested the performance of GPS collars and found that the deployment of GPS collars, although expensive, was worth the expense due to the large geographic scale that could be sampled and to the quality of the data that could be obtained. Using movement ecology as a theoretical framework, Posadas et al. (Chap. 12, this volume) followed groups of capuchin monkeys and found that ranging patterns are shaped by food availability, but in many instances food availability was difficult to estimate, especially with respect to invertebrates.

To close this book, we were fortunate to include two studies dealing with the largest predator of the Neotropics: the jaguar. The development of new movement models and analytical tools have allowed researchers to make precise inferences regarding the space use and movements of this secretive species. De la Torre and Rivero (Chap. 13, this volume) used autocorrelated kernel methods to determine the home range of five jaguars in the Lacandon Forest of southern Mexico and the important areas for each individual and used this information to make conservation recommendations. Movement of jaguars and mountain lions was studied with great detail in Chamela-Cuixmala a protected dry tropical forest in Western Mexico by Nuñez and Miller (Chap. 14, this volume) who found both large cats like to travel along streams and are moving in relatively large areas that make them vulnerable to hunters if they leave the protection of the reserve.

Humans have always been fascinated about animal movement. Since Aristotle (Nathan et al. 2008) we have tried to understand how and why animals move. Species that live in tropical environments are fascinating, and new technology allows us to have a glimpse in movement behavior of secretive species that one rarely sees in the wild (Chapman and Reyna-Hurtado, Chap. 1, this volume). GPS-equipped collars are extremely useful allowing researcher to follow animal paths when observers cannot. With the development of new technology, there became the need to develop better ways to analyze movement patterns, and the chapters of this book are great examples of some of the most innovative statistical techniques and latest models available. Many chapters have advanced beyond home range estimation to elucidate fine temporal movement patterns such as daily distances traveled or core areas or to estimate areas that an animal visits repeatedly. Speed and distance traveled daily are also estimated when consecutive fixes could be collected.

This book has presented a selection of studies from researchers that have overcome logistical challenges and have displayed an amazing level of patience to achieve their goal of acquiring movement data of tropical mammals. The different chapters are also nice examples of thoughtful research design and smart use of statistical methods needed to evaluate animal movement patterns.

The planet now has fewer and fewer places where animals can move without being influenced by significant human perturbations. Large Neotropical protected areas that are not significantly affected by human actions are very rare. This presents us with two opportunities/necessities in the context of research on Neotropical mammal movement ecology. First, we need to get more information about wildlife movement in well-conserved areas that resemble original conditions where species evolved. Alongside this, we need to monitor animal movement patterns in perturbed areas. This is essential to understand the impact of human activities such as logging, hunting, tourism, and other forest-related activities. Changes in behavior can have a large effect on population dynamics and ecosystem function (e.g., elks in Yellowstone changed behavior when wolves were introduced; Laundre et al. 2001). So, studying changes in behavior in contexts with and without significant human influence will allow us to identify when an animal is under stress due to human activities and what the implications of this might be.

Advances in technology, in statistical science, and in spatial modeling, along with the patience, courage, and passion of Neotropical mammal researchers, are changing what we know about Neotropical mammals and their movement patterns. We hope that this book will inspire young researchers to advance what we have learned to date about the movement of Neotropical mammals and to use this information for the conservation of nature, so that future generations have the chance to be amazed when, for example, a group of more than a hundred white-lipped peccaries cross paths with them, moving in the way that they have done throughout their evolutionary history. The movement could be motivated by the need of water and food, or maybe they are migrating to a new area or escaping from a jaguar – it will be the next generation's task to determine why and where are those animals moving.

References

Alvard MS, Robinson JG, Redford KH, Kaplan H (1997) The sustainability of subsistence hunting in the Neotropics. Conserv Biol 11(4):977–982

Hansen MC et al (2013) High-resolution global maps of 21st-century forest cover change. Science 342:850–853

Laundré JW, Hernández L, Altendorf KB (2001) Wolves, elk, and bison: reestablishing the "landscape of fear" in Yellowstone National Park, USA. Can J Zool 79(8):1401–1409

Nathan R, Getz WM, Revilla E, Holyoak M, Kadmon R, Saltz D, Smouse PE (2008) A movement ecology paradigm for unifying organismal movement research. Proc Natl Acad Sci 105(49):19052–19059

Redford KH (1992) The empty forest. Bioscience 42(6):412–422. https://www.globalforestwatch.org/dashboards/country/IDN

Reyna-Hurtado R, Chapman CA, Calme S, Pedersen E (2012) Searching in heterogeneous environments: foraging strategies in the white-lipped peccary (*Tayassu pecari*). J Mammal 93:124–133

Reyna-Hurtado R, Beck H, Altrichter M, Chapman CA, Bonnell TR, Keuroghlian A, Desbiez AL, Moreira-Ramírez JF, O'Farrill G, Fragoso J, Naranjo EJ (2016) What ecological and anthropogenic factors affect group size in white-lipped peccaries (*Tayassu pecari*)? Biotropica 48:246–254. https://doi.org/10.1111/btp.12269

Index

R. Reyna-Hurtado, C. A. Chapman (eds.), *Movement Ecology of Neotropical Forest Mammals*, https://doi.org/10.1007/978-3-030-03463-4

The manufacturer's authorised representative in the EU is Springer Nature Customer Service Centre GmbH, Europaplatz 3, 69115 Heidelberg, Germany. If you have any concerns regarding our products, please contact ProductSafety@springernature.com

Printed and bound by CPI Group (UK) Ltd, Croydon, CR0 4YY

30/09/2026

02225440-0007